Faced with the growth of industrial forestry in natural and plantation forests in the southern hemisphere and the restructuring of
forestry in the northern hemisphere, the forest industry is undergoing tremendous change. *Logging the Globe* investigates the transfor-
mations and their ecological, social, and economic impact.

Patricia Marchak examines issues particular to the northern and
southern regions and the global effects of these trends, using British Columbia, Brazil, Chile, Indonesia, and Thailand as full case
studies and Malaysia, Myanmar, and other Southeast Asian regions
as shorter studies. She also devotes a chapter to Japanese forestry
and the paper industry in Japan. Marchak analyses the restructuring of the global division of labour; the effect of Japanese demand
for pulp; changes in employment, production, land policies, and
markets in northern countries; deforestation; plantation forestry;
and the influence of European, North American, and Japanese
companies on tropical forests and peoples. She considers whether
industrial forestry is sustainable and suggests ways in which global
demand for forest products can be met in more efficient and more
sustainable ways.

Logging the Globe provides a global picture of a critically important
environmental and social issue. It will be of great interest to profes-
sionals in the industry, policy-makers and environmental activists,
and those with an interest in environmental and social issues.

M. PATRICIA MARCHAK is dean of arts and professor of sociology,
University of British Columbia.

Logging the Globe

M. PATRICIA MARCHAK

McGill-Queen's University Press
Montreal & Kingston • London • Buffalo

© McGill-Queen's University Press 1995
ISBN 0-7735-1345-0 (cloth)
ISBN 0-7735-1346-9 (paper)

Legal deposit fourth quarter 1995
Bibliothèque nationale du Québec

Printed in Canada on acid-free paper

McGill-Queen's University Press is grateful to the Canada
Council for support of its publishing program.

Canadian Cataloguing in Publication Data

Marchak, M. Patricia, 1936–
    Logging the globe

    Includes bibliographical references and index.
    ISBN 0-7735-1345-0 (bound) –
    ISBN 0-7735-1346-9 (pbk.)

    1. Logging – Economic aspects 2. Logging –
Environmental aspects. 3. Forests and forestry –
Economic aspects. 4. Forests and forestry –
Environmental aspects. I. Title.

SD131.M37 1995   338.1'7498   C95-900581-1

Typeset in New Baskerville 10/12
by Caractéra production graphique, Quebec City

Contents

Maps

Tables

Figures

Measurements and Abbrevations

Ton/Tonne: The measure for tons throughout the text is metric; that is, one ton equals 2,204.6 pounds or 1,000 kilograms. In the pulp business, metric tons are sometimes described as "tonnes" to distinguish them from the United States measure of "tons," which equals 2,000 pounds or 907.9 kilograms. In so far as it is possible to identify measures used in various international contexts, the text uses conversions from the American measure.

Hectares/Acres: All measures given in acres in the original have been converted to hectares: 1 hectare (ha) to 2.471 acres.

CM: cubic metres.

MCM: million cubic metres.

Acknowledgments

Conference organizers are vital participants in the gestation process for books of this kind. They provide the opportunities for people of like interests to meet, test ideas, and receive some critical feedback before going into print. I want to thank the organizers of many conferences geographically as far apart as Quesnel, British Columbia, and San José, Costa Rica, between about the mid-1980s and the mid-1990s. I also want to express my thanks to various individuals who have invited me to give guest lectures or speak to public audiences, whether as the W.S. Main lecturer at Berkeley or as something rather less prestigious – though providing me with good feedback – in any number of forestry towns.

The general argument on the globalization of the industry was published under the title "For Whom the Tree Falls" in *B.C. Studies* (1991). The chapter on traditional forestry communities in the Kyoto Prefecture, Japan, is a continuation of an article I published in the *Journal of Business Administration* (1991). I wish to thank my Japanese colleagues, guides, and interpreters for their help in conducting this fieldwork in 1989, in particular, professors Manabu Morita, Ryoichi Handa, Yoshiya Iwai, and Hiroshi Matsuda.

The chapter on Indonesia draws heavily on fieldwork carried out there in 1990, during which I was given access to the mills and plantations owned by Djamzu Papan. I am most grateful to him and to my guides. I particularly wish to thank Dr M.H. Matondang for his great kindness in arranging my visit to Sumatra, and Asriel Jasria for the tour of plantations there.

I am much in debt to Sr Marsellos Pillair, vice-president of the Associacão Nacional dos Fabricantes de Papel e Celulose. He kindly arranged for me to visit mill sites and plantations throughout central and southern Brazil. I am also indebted to numerous generous contacts in each company, who provided me with information, and to members of Greenpeace and other groups critical of forestry practices in Brazil, for their contribution to my understanding.

In Chile I was helped by academics, foresters, and individuals in various offices. I would like to offer very special thanks to both CODEFF and the Chilean forestry service for providing me with data on request and to CMPC for its generosity in providing a special tour of its plant at Laja.

My visits to other countries span half a decade, and the individuals who provided information at various times are too many to identify here individually. As is always true of large projects, a good many people have been involved in one way or another.

Throughout forested countries today, there are social scientists or foresters with a social-science bent who study the expansion of the forest industry and deforestation/afforestation more generally. These individuals have become known to one another and have created an informal network, exchanging information and reading drafts of proposed works. Among the many I have been fortunate to meet or with whom I have corresponded are several I wish especially to thank for their generosity in sharing information and ideas: John Dargavel of Australia, Ian Penna, formerly of Australia and now in Japan, Michael Roche of New Zealand, Jussi Ramoulin of Finland, and Antonio Lara of Chile. There are also journalists who have invested time and talent in learning about the forest industry, and of these I especially want to acknowledge the interpretations of Thailand's forest policy published by Ann Danalya Usher and the work of Ben Parfitt in British Columbia.

I am also indebted to my able research assistant, Jacqueline Tracey, who gathered data on the BC forest industry and did a literature search on forestry in other regions during the summer of 1989. These data have been woven into this book, as into earlier articles. She subsequently ran off to Australia to take a PHD with John Dargavel. Eric Leinberger is the map maker, and I am in his debt for the talented renditions of places discussed in this book. Eileen Oertwig, my secretary at UBC, has patiently covered for me when the book took precedence over dean's work, and I do appreciate her help.

Since completing the manuscript, I have met several newcomers to the field: students who are preparing fascinating theses or beginning post-PHD research on tropical and temperate forest regions. Their work cheers me greatly, because the sociology and political economy of forestry have been sparsely populated fields of inquiry in the past. I am persuaded that the most critical issues are social, political, and economic and that when we finally address and solve these, the ecological issues will no longer be great problems. If we fail to solve them, the ecological problems will cease to be solvable. So I welcome these newcomers.

Funding for the research on global forestry came from the Social Science Research Council of Canada. I want to express to them my most sincere thanks.

Finally, my continuing appreciation to Bill Marchak for his patience and understanding with a wife who persists in writing when the sun shines and the little forest of Galiano awaits new seedlings.

Logging the Globe

1 Globalization and Restructuring of the Forest Industry

Swiddeners, farmers, ranchers, miners, builders, warriors, and generals; the impossibly poor, the implausibly rich – these are deforesters of record. Among them have been and still are national governments exercised about their territorial sovereignty or seeking ways of avoiding redistribution of agricultural land by clearing forests to give subsistence plots to the poor. Both literally and mythically, they are explorers in search of their El Dorados, conquistadores for whom natives, as well as trees, are impediments along the way. They seek fuel; they seek food; they seek solace, medicine, or power. For many the forest is an alien and frightening place. For those to whom the forest was home, conservation and lengthy rotations of subsistence crops are no longer possible. Others have claimed the land, and forest dwellers are now labelled encroachers.

The forest industry is a relatively recent arrival to the deforestation process. Until logging techniques were mechanized and there were mass-production technologies for sawing wood, the forest industry was not capable of mass destruction. Until roads were built, access to forests was restricted. Until transportation methods accommodated long-distance carrying of lumber and then of pulp, producers and remote markets were not linked on a regular basis. Until new techniques for pulping came on stream, paper was produced from non-wood fibres and, much later, from a relatively small range of coniferous woods. Tropical and subtropical forests, most of the northern boreal forest, and many hardwood forests around the world were not worth logging for industrial purposes, and if they were destroyed, it was generally because they were in somebody's way rather than because they were valuable.

In the late nineteenth and the first half of the twentieth century, northerners established a full-scale, mass-production forest industry based on temperate-zone coniferous (softwood) forests. The industry produced lumber and wood-based pulp, newsprint, paper, and various speciality paper products in Russia, throughout Scandinavia, in some countries of western and eastern Europe (depending on whether any forests were still standing by then), and finally across North America. The producers were northerners, and so were most of the buyers. The forest industry became the mainstay of national economies in Sweden, Finland, and Canada, and equally significant to local economies in various regions of the United States and western Europe. In the period from about the 1950s to the 1990s, northern forest products were sold in increasing quantity on a wider world market, meeting the growing demand for wood products, especially in the developing countries of Asia and Latin America.

Japan, with forests as its only natural land resource, also established forest industries in the late nineteenth century. It is in the northern hemisphere, and its forests are in the temperate zone, but its wood products were contained in the domestic market before the Second World War; it was not one of the "traditional" northern participants. Its natural forests were largely destroyed during and immediately following the war, but national reforestation programs were instituted, and by the 1980s new coniferous, as well as hardwood, stands were being thinned and harvested for local construction markets. These stands were insufficient to meet the needs of Japan's growing paper industry. Imported wood, wood chips, and finally pulp were essential to its success.

JAPAN'S ROLE IN GLOBALIZATION

Perhaps it is to Japan that we should attribute the current phase of globalization; certainly, it has been Japanese companies, urgently scouring the world for wood, who have pioneered global sourcing practices. But American and European companies are also reaching across oceans for new resources, and Japan's Asian neighbours are not far behind. Besides, globalization is not really a new development, even for forestry. Timber has always been sought by sailors to construct vessels. The British navy earned its imperial honours because it had good wooden warships. European trading companies plundered their colonies of decorative wood. And after all, northerners have sold wood on global markets for over a century.

Even so, the current globalization has a certain difference: it is no longer a matter of plundering exotic forests for occasional timbers. It is now a full-scale industrial operation. In tropical forests, it involves clearcutting. On cleared land it encompasses planting seeds through to manufacturing pulp, even paper, in locations where the species being planted does not grow naturally. The global plantation-based forest industry produces manufactured wood products for world markets, and it does so in southern locations – Brazil, Chile, Indonesia, Thailand, South Africa, and other southern hemisphere countries – as well as in the Iberian Peninsula and the southern United States.

Japanese companies are active investors in this new global industry. As early as the 1960s, Japanese paper manufacturers began to change the face of world forestry by the sheer scale of their search for new raw materials beyond the country's borders. The population was becoming affluent, and its demand for paper and other wood products was growing. By the 1970s, Japan had the world's second largest paper industry (the United States had the largest), and its produce was sold almost entirely in its domestic market. As well, the Japanese people needed construction timber for housing, and for this too they looked for raw materials beyond the country's borders.

Initially, Japanese companies sought logs. They were procured in neighbouring Southeast Asian countries, Russia, the United States, Canada, and New Zealand. Some logging continues in Malaysia, Papua New Guinea, and Myanmar, but other neighbouring countries finally instituted log bans. As the log trade diminished, wood chips were procured in North America and Chile. Then the wood-chip market waned, and the companies began to buy market pulp and to invest in pulpmills offshore. Among the places receiving this investment are the boreal forest lands of northern Canada and Russia, where aspen trees, earlier spurned as potential fibre sources and even regarded as "junk trees," are now marketable. The next stage, just beginning, is for the companies to manufacture paper outside Japan for the home market.

RESTRUCTURING THE TRADITIONAL NORTHERN INDUSTRY

Although such large companies as Weyerhaeuser sell logs from the American Northwest to Japan, traditional northern temperate-zone softwood sources are in decline. The decline results from resource depletion in some regions and the effects of acid rain in others. Beyond these factors are rising costs of logging ever more remote regions, increasing demands for improved forestry practices or the cessation of logging, and the anticipated impact of cheaper products from the southern hemisphere. In the Pacific region, old-growth forests have been overexploited; elsewhere, temperate forests are suffering the combined impacts of urban sprawl, acid rain, and other forms of pollution. Even where planting has been undertaken, in the United States and the Scandinavian countries, second growths are often disappointing, and they take a long time to reach maturity if they are conifers. An industry dependent on a raw material that does not reach commercial size within the lifetime of one investor tends to lose its attraction after the original harvest, the more so when reforestation and silviculture are expensive and labour-intensive activities. Before the 1970s the eucalyptus plantations were unproven alternatives; by the 1990s they, together with other hardwoods and southern pines, are obviously the future fibre sources for pulp: investment will inevitably move in that direction.

Restructuring the northern industry means reducing its reliance on the production of lower-value products, especially pulp and newsprint, which can now be manufactured elsewhere. Scandinavian producers, anticipating the economic union of Europe, began phasing out their least productive mills in the 1970s and investing in higher-value manufacturing in mills located elsewhere on the continent. Other European, American, and Canadian companies gradually followed this lead. Restructuring also means changing the labour force and dependent communities. Employment levels began dropping in the late 1960s as mills were automated. Although production continued upward, employment levels decreased steadily thereafter. Now, claiming the need to regain competitive strength on global markets, companies attempt to create more flexible labour supplies. Unions have lost members and strength, and are now losing capacity to define and delimit jobs.

THE NEW PLANTATIONS

Brazil may well be the "forest giant" of the twenty-first century. It has the land base, the labour force, eager governments, and local investors. A director of the major Spanish company CEASA* claims that the entire pulp requirements of the world could be met by fast-growing pulpwood species on just 3 per cent of Brazil's land area. Given all

---

* CEASA is the acronym for Celulosa de Asturias SA, a Spanish company owned by Wiggins Teape. The speaker is Robert A. Wilson, commercial and forestry director of CEASA, at a market pulp conference in Vancouver, BC.

the advantages of southern plantation pulps, he anticipates a buying spree by southern companies as they take over northerners, inherit their customers, and use what is left of their forests as backup for papers that need long (coniferous) fibres. Brazil is the most promising of the new plantation economies, but there are others with stakes, including Indonesia and Chile; and there are the more-established plantations of Spain and Portugal. None of the others has as much available land as Brazil, but all could be major players in forestry of the next century.

Deciduous trees (hardwoods) grow more rapidly than softwoods, and they shoot up at incredible rates in southern climates. They are generally easier to plant and sustain. To utilize them as pulpwood sources, the industry developed appropriate technologies, and by the 1970s these were on stream in many geographical regions. As well, biotechnology research was undertaken to experiment with diverse species and develop hybrids especially useful for pulping or for construction timber. The eucalyptus tree in several of its five hundred or so varieties was named the winner in the pulp process: it grows to commercial size in five to seven years in warm climates, it can be coppiced to produce three crops from one root, and it is successfully grown in plantations and requires relatively little by way of infrastructure or management once planted. Plantations produce even-spaced, even-aged, homogeneous crops; their output is standardized, predictable. The trees can be cloned and their growth cycles perfectly programmed. Infrastructure and harvesting costs are low.

In some warm-climate regions, other hardwoods such as acacia, gmelina, and albizzia have similar growth rates and attractive commercial properties. In more mountainous regions or temperate climates of the southern hemisphere and in more southern parts of North America and Europe, exotic pine species became the new softwoods of the industry, grown just as eucalyptus is grown, in huge plantations, and reaching commercial size within thirty years. The southern United States, the Iberian Peninsula, the Caribbean, Chile, New Zealand, and Australia are all prime plantation regions for pines.

In addition to low wood costs, labour costs also favour southern-based plantation industries, with the result that factory prices for their pulps are substantially lower than for northern-based pulps. Softwood fibre costs in 1992 were nearly three times higher in Sweden and Finland, and nearly twice as high in British Columbia, as in Chile. The cost advantage for hardwoods was even more marked. Labour, energy, chemicals, and freight costs varied by region, and fluctuating exchange rates are disadvantageous to the southern countries, but the overall advantages in the early 1990s were clearly in the southern climates (Wright, 1992).

Plantation eucalyptus or acacia pulps are cheaper to produce, and since more pulp can be extracted from a lesser land area, they are also claimed to be more ecologically efficient users of land. In any event, plantation forestry has distinct advantages over logging in tropical forests. From a strictly industrial point of view, it offers standard-sized products, whereas tropical forests are notoriously hard to log selectively, produce many species that are useless for either construction or pulpwood, and can rarely be replanted. From a more ecological perspective, United Nations agencies, aid and commercial banks, and international companies hail plantations as the antidote to deforestation. Where plantations are embedded on land that was earlier degraded or is not otherwise capable of sustaining a forest, a persuasive argument is that plantations preclude the necessity of cutting natural stands and reduce the incentives for doing so.

BOREAL FOREST

Land, labour, fibre source – all of these are presently cheaper in Brazil and Indonesia than in Sweden and Canada. Economic theory tells us, however, that as plantations are established in the hot countries, the price of land and labour will increase. Trees will be planted as long as more profitable uses of higher-priced land are not devised, but the great cost advantages now enjoyed in the south will diminish over time. Northern countries have some land that, based on only the economic models of industrialism, are not useful for other purposes. In particular, they have the boreal forest lands. Some forest economists argue that low land costs for these areas will make them increasingly competitive with southern regions, despite the length of time needed for growing new crops. The problem with this argument is that time is a crucial variable in the entire industry. Within the half-century or more that it will take to grow a new hardwood crop in the boreal latitudes, southern producers will have become dominant in the industry. By then, indeed, substitutes for wood-based pulp and construction may well preclude the utility of any further investment in forestry. The fact that some northern land is of no value for other crops may mean that it will be either left alone or reforested, but it does not mean that the northern industry will regain strength in the lifetime of the planters. What the boreal forest now provides is a natural fibre that has cost nothing to produce. Although the low land value is a reason for its availability, the low cost of the fibre in the 1990s is the major attraction.

THE ISSUES

Globalization, restructuring, deforestation, and expansion of plantations are the basic processes described throughout this book. The
objective is to examine the processes, the context, and the outcomes
for both traditional northern forest regions and new southern regions.
An artificial forest can be grown almost anywhere today: technical
issues are not at the heart of this industry. The core issues are ecological, social, cultural, political, and economic. What kind of world is
being carved out with the destruction of natural forests? with the
creation of artificial ones? with the depletion of fuelwood and the
accretion of pulpwood? Where do people fit into the picture? what do
governments do? how do the interests of the rich and the poor
intersect and clash in forests? What is the nature of property rights in
forests? Are there multiplier effects from forest-industry development?
who are the beneficiaries, who the victims of growth? These questions
underly this study, and it becomes clear that there is no single answer
to them. Globalization and restructuring produce both benefits and
losses; in some regions the benefits are shared by local communities
and workers; in others, only the costs are shared.

Ecological Concerns

Though plantations are becoming a new source of wood, both temperate and tropical forests continue to be logged. Not much remains
of the great coastal rainforests of the North American Pacific Northwest, and that is a hotly contested issue in this day of struggles over
forests. The original temperate forests elsewhere in North America
and Europe are gone; second growths are sometimes successful as
crops, but rarely a match for the original as habitat for multiple species
of flora and fauna. Logging in the boreal region is opening up a new
frontier for the industry, but it is a frontier with scarcely any established scientific knowledge about the land's capacity to sustain this
activity. Virtually nothing remains of the Atlantic araucaria forest in
Brazil. In the tropical zones of Asia, Africa, and Latin America, forests
are declining at rapid pace.
    Forests are essential for the earth's oxygen supply, and they serve
multiple other functions that no other life-form can supply. They are
sometimes renewable, but where soils are fragile or fail to carry the
necessary nutrients for regrowth, they are not renewable. Further,
when the canopy of the old-growth forest is gone, micro-temperature
changes can trigger widepread ecological system changes, so that the
land becomes a desert. The struggle over forests is not a simple fight

over economic versus aesthetic or spiritual values. The more fundamental question is, is logging of natural stands ever sustainable? Many
critics would argue that forestry as practised in North America over
the past century is not sustainable and that logging in tropical forests
is even more dangerous. Beyond the current estimates of tree stocks
(as a quantitative measure of sustainability), some scientists are beginning to express doubts about the possibilities of achieving reliable
standards by which to establish sustainable harvest levels.

Three ecologists recently addressed the issue and observed: "The
great difficulty in achieving consensus concerning past events and *a
fortiori* in prediction of future events is that controlled and replicated
experiments are impossible to perform in large-scale systems" (Ludwig, Hilborn, and Walters, 1993:17). If it is impossible to predict
outcomes for large systems, then it is equally impossible to know in
advance what will result from logging in natural forests. In the view
of these ecologists, we should look to history rather than science for
evidence, and history informs us that deforestation produces very
serious ecological and social problems.

Logging in old-growth forests may mean destroying a non-renewable
resource – at the least, non-renewable in the lifetimes of present
inhabitants of the earth. But plantations are not exempt from ecological concerns either. They are not forests; they are agricultural crops,
seeded entirely for commercial purposes. They are not habitats for
wild flora and fauna, and they often displace such habitats. They
displace other vegetation, often including trees but also many plants
that are or may be vital to total ecosystems. Plantations are monocultures, and the lack of biodiversity is of concern. They typically have
sparse canopies and so do not protect the land; they cause air temperatures to rise, and they deplete, rather than increase, the water-
table. They are generally exotic to regions. While the initial planting
may be free of natural pests and diseases, that situation will not last,
and plantation regions may not be in the position to combat scourges
yet to arrive.

The question is not whether plantations are good or bad in the
abstract, but rather whether the defects are so great that plantations
should be avoided; alternatively, whether the advantages – especially if
plantations reduce the propensity to deforest all remaining natural
stands – are sufficient to justify more research and investment. As long
as the world demands wood products, and especially wood-based
papers, trees will be logged. Plantations are preferable to natural
forests for such logging, provided they are established on land margin-
alized or denuded by earlier agents: ideally, land for which no better
use can be detected and which would benefit from rehabilitation.

Property Rights and Market Controls

Forests were common property under the joint stewardship of inhabitants at early stages of most societies, and they were viable as long as inhabitants were not numerous and had shared interests in sustaining the land. In remote forested regions such systems may still be found. Privatization of land superseded common property arrangements wherever the land itself became a commodity, where trees became valuable beyond subsistence needs, or where the grazing of sheep or other animals promised rich profits to whoever could control the land. The classic examples are the enclosures of Europe, which some writers erroneously label "the tragedy of the commons" (Hardin, 1968). The tragedy, in fact, occurred after the commons were privatized and precisely because, as private land, they were overexploited by owners who marketed their produce (Ciriacy-Wantrup and Bishop, 1975; Fortmann and Bruce, 1988; May, 1986, 1992; Marchak, 1991). The human tragedy is experienced today by indigenous peoples in many tropical forest areas, where ancient lands managed by them as common property are suddenly declared to be state or private land.

Private ownership in forests takes two very different forms: private woodlots owned as family farms and corporate estates. The first is the characteristic ownership pattern in Scandinavia and Japan. In both cases, farmers are subject to national regulations to ensure adequate management and harvesting. Corporations exist in both territories and increasingly own some land in timber estates, but historically they have purchased their wood supplies on open, competitive log markets. In both regions, total forested land has increased over the past half-century, and reforestation rates are high.The corporate model is characteristic of North America, though in the United States, corporations are more often owners of land, while in Canada they own long-term harvesting rights to crown (state) land. The record of corporations in conservation and reforestation is mixed at best; there is no clear argument to be made on the basis of historical evidence about the superiority of either outright corporate ownership or the combination of state ownership and corporate control of land. Judging from the historical record, however, one can observe that large, integrated corporations operate in their own interests, not in the interests of communities or regions where they are logging. Further, their interests are defined within the marketplace and may change as markets change.

Markets and private property rights are inextricably linked together. Private owners require security of tenure to plan investments and produce market goods. Neoclassical economic theory argues that such individuals will, collectively, produce, sell, and buy enough to maintain a steadily growing economy in which everyone benefits, even though each person acts only in his or her own interests. Markets are means through which they can exchange goods and services without creating a huge bureaucratic government; therefore markets are the quintessential democratic medium. With reference to forests, neoclassical economic theory argues that wood will be used if there are markets for it; it will merely be destroyed if there are no markets, and especially if there are markets or other uses for land currently forested. The more valuable the wood becomes, the more probable that governments, companies, and individual owners will utilize and protect it. This argument is at the core of the Tropical Forestry Action Plan (discussed in chapter 8) and is familiar as well to northern, traditional forestry regions.

The neoclassical argument is flawed in numerous respects. The market provides economic incentives to log forests and disincentives to protect or conserve them. It is true that if land has economic value and trees have none, forests may be logged incidentally; but it is also true that if trees have economic value, forests will be logged intentionally, and either way the forest is destroyed if market forces are the guide. Market controls have different impacts where the marketable good consists of trees from natural forests than where it is widgets or wine, the classical examples in economic theory. Neither trees nor indigenous cultures are replaceable in the fashion of widgets. Spiritual and aesthetic values are not marketable but are destroyed if trees are cut. Further, individual trees are embedded in forest ecosystems, and cutting trees inevitably impacts on the whole system and all its inhabitants. Where an entire forest is destroyed, there are environmental and social impacts far beyond the property originally covered by that forest. Economic theory has used the term "externalities" for this wide range of incidental effects and costs, but when the externalities are of such magnitude, the theory is manifestly inadequate.

Markets are sometimes posed as the opposite of government controls. In fact, governments and markets are necessary twins in modern industrial systems; governments protect private property rights and ensure the observance of ground rules for market participants. They everywhere have tended to act largely in the interests of the most powerful groups in their territories; sometimes, indeed, in the interests of external powers. Except where small woodlot farming was well established before the expansion of the forest industry, governments have tended to allocate resource rights to large corporations with manufacturing capacities. These corporations have subsequently held enormous power to influence government action because their bargaining position is so strong. If they exited a region or curtailed production, whole areas dependent on sales of the resource and its products would be harmed, unemployment would rise, and government revenues would plummet. Once captured in this way, governments are rarely able to extricate themselves from the long-term obligations they have incurred through the granting of land or harvesting rights. They provide further subsidies, reduce tax obligations, and create new incentives to keep the industry operating in their territories. The system becomes self-perpetuating. We see this process in British Columbia and the Pacific northwestern United States over the past century; it is just beginning in many tropical countries.

Industrial Restructuring and Its Unintended Consequences

The postwar boom in industrial countries was built on the "Fordist model" of economic growth. The term is a shorthand used by social scientists to include assembly-line mass-production systems, unionized labour in large company structures, high wages that sustain mass-consumption levels, and social welfare systems that provide safety nets and Keynesian economic policies. This model was characteristic of forestry regions just as much as of manufacturing cities between the 1950s and late 1970s.

Globalization of all industrial sectors, beginning with labour-intensive manufacturing companies that sought cheap labour supplies in underdeveloped countries, has undermined the Fordist system. Companies confronting the impacts of new world competition, lower labour costs elsewhere, new fibre sources in unexpected countries, and volatile investment patterns and markets look for solutions. First, they may seek cheaper raw materials. Scandinavian companies have succeeded in gaining these in other countries, and other-origin companies may follow that example. Where they still have access to resources, however, manufacturers are likely to hold on until these are depleted. If they choose to stay in their country of origin, their next strategy is to seek cheaper labour supplies. Restructuring of labour began with automation of mills in the late 1960s, but the tempo of downsizing increased rapidly after 1980. The second phase of restructuring is the creation of greater flexibility in the use of labour supplies. Often that is possible only if unions are broken, and much of the restructuring of the 1990s has further undermined already weakened unions.

At issue for northern forestry regions are both material outcomes and matters of principles and rights. Both investors and workers benefited from the long postwar boom. Investors can use their profits to gain a foothold in the new industry elsewhere. Workers tend to be geographically immobile and in any event do not have the capacity to take advantage of moves to southern climates. Investors are not generally embedded in rural communities; workers are. In the market economy there are few obligations for investors when an industry from which they have profited is in decline. The costs of unemployment and a much eroded welfare system are borne largely by workers. The bargaining positions are thus uneven: if workers reject the "flexing" of their work, they may lose everything.

For the region as a whole, there are various and often unanticipated consequences of restructuring. One effect is the shrinking of the middle class, a phenomenon appearing throughout the industrial countries. Another effect is that groups whose voices were silenced during boom times gain the capacity to speak on behalf of forests and aboriginal and other communities. Governments gain some freedom from organized capital and labour to reconsider tenure systems, allocation rules, and alternative uses of land. While societies would not voluntarily choose to experience such sudden economic decline, one can at least see the possibility that in the midst of much pain a renewal of sorts might take form.

Economic Development and Its Social Costs

Northern industrial countries are at the end of a growth phase. Now it is the turn of developing countries, but the social costs of establishing huge plantations are much more prominant than any similar costs were for northern countries at the same early stage of development. Where there are still indigenous peoples, both forestry in natural stands and plantations deprive them of land. Typically, they are dispersed or pushed to the margins of the forest. Inevitably, they return, now labelled encroachers, because they need fuel and subsistence plots in order to survive. Deforestation becomes self-perpetuating as long as impoverished people have no economic alternatives.

The dilemma of the poor was eloquently phrased by an Indonesian visitor to Japan. He was campaigning against a Marubeni plan to log native mangrove forests in Irian Jaya for a wood-chip export venture and was visiting the Japanese environmental group JATAN. He said, "Japanese [businessmen] want to develop land owned by indigenous people and make some sort of a 'new world.' The problem is that we indigenous people cannot survive in such a 'new world.'" JATAN reported this in connection with its field research on a Marubeni wood-chip venture in southern Chile called "Terranova" (1993:10). New worlds, like the "new world economy," are not designed to benefit everyone.

Advocates of wood-chip or pulp plantations claim that the wealth created through these plantations will eventually improve the standard of living for all. Multiplier effects for regional development, technology transfers, and access to foreign markets are all in the package of promised returns. However, investors generally ask for development funds and tax incentives to begin plantations and mill construction. In a study for the Inter-American Development Bank, Stephen McGaughey and Hans Gregersen argued that the lengthy gestation periods for plantation growth, uncertainty of profits, high-risk mill investments, and other risks should be offset by public subsidies (1983:24). In addition, they claimed, there are public benefits "that do not accrue to the private investor but benefit the region or society at large." Governments, said they, should "provide direct and indirect incentives to the private sector" to obtain these anticipated benefits. In fact, subsidies for start-up costs, taxation incentives, land grants, and relaxation of labour laws and other legislation have all been provided by one government or another.

Critics argue that the risks are overrated and the benefits to private investors undervalued in such arguments. Plantation species, for example, do not have especially lengthy maturation periods; the risks of investing in them are not unusually high. Multiplier effects should not be taken for granted because capital invested in a resource industry has no compelling relationship to capital invested in supplier industries. Where external capital is concerned, it is more commonly the case that the machinery suppliers see potential new markets for their products while the pulp buyers see potential new sources from which they can manufacture other end products. Indigenous capital might be invested in linked machinery industries over time, but machinery industries have long learning curves if investors begin from scratch, and technology transfers are not easily obtained. Manufacturers of mill equipment operate substantial international businesses, and virtually every wood-based pulpmill in the world purchases basic mill machinery from a handful of companies in Europe, the United States, and Japan. Overall then, the more likely outcome of plantation forestry is that industrial countries will benefit from the import of relatively cheap raw materials (wood-chips from Chile to Japan, for example), relatively cheap semi-processed materials (such as pulp from Brazil to the European or Japanese partner papermaking company), and the new business for machinery and engineering or other professional expertise (Finnish and Canadian consulting-company fees, for example).

Imported mill machinery is an extention of a business established in industrial countries. Many service industries are similar. Consultants, planners, financiers, mill-construction companies, engineers, and technical advisers of other kinds are often linked via the World Bank and other private banks to the construction of new mills and the planting of new regions. They undertake virtually no risk, and their payment is guaranteed by international banking rules. Sandwell or H.A. Simons of Canada and Jaakko Pöyry of Finland are regularly on hand, brought in by companies, governments, the World Bank, or development banks to conduct feasibility and technical studies. Supplies are frequently purchased from the industrial countries, either because only they manufacture them or because the buyers are obliged to use their loan funds in this way.

These linkages are not unique to forestry; this industry follows others in maintaining an international division of labour whereby northern industrial regions provide technical expertise and machinery, while southern regions provide land and labour. The international division in this particular industry used to be among northerners, with the United States and western European countries producing most of the higher-value products and the more northern European countries and Canada producing the lower-value raw materials.

Assuming plantations are preferable to forestry in natural locations, one has to ask whether, even in plantations, economic returns are sufficient to justify the precluded alternative uses of land. The public purse is tapped when new projects are started; it is persistently involved in expansion phases; then, when a downturn occurs, governments are expected – indeed often obliged, because by then regional economies are dependent on the industry – to come to the rescue. The infrastructure of roads and townsites is a major public expense, as is the public bureaucracy required to maintain industrial forests and manage a system of harvesting rights. Even where forest land is privately owned, it has almost everywhere been granted free or sold at low cost, and governments are invariably involved in the financing of projects. There are returns in the form of employment, yet even that becomes dubious when technologies displace labour. There are export earnings, but these are not automatically – and in many cases, not at all – distributed to the benefit of the plantation region or country.

Before the end of the 1970s, some of the strongest advocates of plantation forestry were rethinking their position. Jack Westoby, who was one of these in the 1960s when he was director of the Department of Forestry Programme Coordination and Operations in the Food and Agriculture Organization of the United Nations (FAO), noted later that things had not worked out as anticipated: "nearly all the developments which were taking place were enclave developments; multiplier effects were absent; welfare was not being spread; the rural poor were getting poorer, and their numbers were increasing." "The poor are driven into the forests, and the forests are disappearing, because land, wealth, other resources, and power are concentrated in the hands of an elite determined to hold on to their privileges at any cost." Westoby chastised ecologists who see only trees and prattle on about "spaceship earth." Nine-tenths of the ship, he pointed out, "is given over to sumptuously furnished first-class cabins occupied by a handful of people, while everybody else is crammed into the engine room and the bilges" (1983:2–4).

A different viewpoint emerges in Brazil, where plantations and pulpmills constitute a thriving economy, and several branch-plant and licensed manufacturing companies are producing pulpmill machinery and adapting logging machinery obtained elsewhere to local conditions. Brazil is not representative of countries involved in the plantation economy, but that may be because it is in all respects further ahead in its development of a pulp plantation economy. If it can succeed in changing the power relations between south and north, then other countries might follow. That optimistic version notwithstanding, Westoby is correct in noting that the vast majority of projects have provided few beneficial spin-off effects and done much damage to both land and people.

As much of what we read here implies, private ownership, corporate empires, captive governments, clearcutting, mass production, mass consumption, and global market systems are linked together in time: they belong to the last half of the twentieth century in traditional northern forest regions. What is now happening with globalization is that the same forms are being established for the twenty-first century in new plantation lands, but with the one difference that forestry is now an agricultural enterprise.

A Note on Population Pressures

Population density that exceeds the carrying capacity of land is an obvious threat to forests, as well as to all else on earth. However, the history of deforestation in almost every instance discloses a curious pattern: population growth has tended to follow, not precede, deforestation. The regions most threatened today are not densely populated. The rate of population growth no doubt needs to be curbed, but we should be sceptical of arguments that assume deforestation or other forms of ecosystem destruction are due entirely to population growth.

Alternatives

Plantations may be less damaging to the earth than logging in natural forests. However, they do incur ecological and social costs, and developing regions, rather than investors, are taking the risks of their establishment. In natural forests, societies could improve forestry practices and conserve areas through substantial increases in resource rents (known as stumpage) and the cessation of practices that encourage poor wood utilization. In countries where logging concessions are presently granted on the basis of corporate pressures, personal alliances, and bribery, improvements in logging practices could be achieved simply by introducing open, competitive bidding. Small woodlot farming has its own defects, but where combined with national forestry regulations and genuinely competitive log markets, it appears to be a superior method of managing forest land.

Although the World Bank and the FAO have achieved less noble stature than was hoped of them and must accept responsibility for many of the disastrous forestry development schemes of the 1960s to 1980s, critical attention to their activities has resulted in some beneficial modifications. In any event, international regulations and pressures are potentially powerful deterents to poor forestry practices, and international agencies and banks will be required to bring these to bear. International agreements and regulations regarding logging and trade are beginning to take form. There are proposals for the establishment of forest reserves under United Nations control or other international jurisdiction. Enforcement of unambiguous and universal regulations respecting logging practices, silviculture, and reforestation might take the form of market boycotts. Restrictions on log imports, log-export bans, and producers' associations and cartels are all potential means of changing the global trade in logs and the more damaging logging practices.

International controls are proposed on the assumption that wood products will continue to be tradable items in the foreseeable future. But there are already substitutes for wood and paper. Wood-free papers have been produced in poor countries for a long time: their market share could well increase elsewhere. The eucalyptus has become the darling of the industry within a scarce twenty years; other vegetable fibres may have even greater potential for producing paper cheaply.

Non-industrial deforestation continues because people need fuel and food-producing land in order to survive. Conservation of forests would require that non-wood fuel sources and alternative food sources be made available to masses of poor people. Cessation of construction, farming, ranching, mining, damming of rivers, and many other activities that incidentally destroy forests would be essential; genuine protection of forest reserves, likewise. The creation of alternative fuel and food sources has to be the first priority, because without these, no protective measures can be sustained.

ORGANIZATION OF THIS BOOK

Northern dilemmas and issues are tackled in part I, which consists of four chapters. Chapter 2 is concerned with the nature of the northern forest, acid rain, technological developments in the forest industry and its impact on employment and communities, and other changes in the environment and socio-economic context of the northern forest industry. Chapter 3 focuses on restructuring already under way by the 1990s. This provides a region-by-region analysis of changes in the northern hemisphere; data on levels of concentration and information on the leading northern companies; data on imports, exports, production, consumption, and prices for wood products; and finally a case study of a border conflict between the United States and Canada over lumber sales.

Chapter 4 provides a case study of British Columbia, a leading producer of temperate coniferous wood products in Canada. This discussion illustrates the defects of development on the "captured," or "client-state," model and the direction companies are taking to restructure. It also encounters in the public debates under way in BC some of the alternatives to reliance on corporate forestry. The fifth and final chapter of part I is about the forest industry in Japan and especially the global expansion of Japanese papermaking companies. It also provides a discussion of the impact of free trade in logs on rural mountain villages in Japan.

The whole of part II is concerned with tropical, subtropical, and southern temperate regions, almost all of which are touched by or totally integrated into an export trade centred in Japan. Chapter 6 is about the nature of tropical forests in general and forest dwellers. Its focus is on indigenous and land-poor people, especially where they have been or are still engaged in shifting agriculture. It also contains a discussion of pre-plantation deforestation in the Amazon region. Chapter 7 deals with tropical pre-plantation industrial forestry and provides short case studies of many countries where logging occurs. These include Myanmar, where teak logging amounts to land rape, Malaysia, where the intention of the government is to deforest tropical areas in order to plant fast-growing species, and other Asian and Latin American regions. We turn to the Tropical Forestry Action Plan and plantation forestry in chapter 8. This chapter includes discussion of the arguments for and against eucalyptus or other plantations and data on their locations, extent, and viability to date. Short case studies are provided of New Zealand and Australia, neither of them tropical countries but both having plantations of long standing.

The next four chapters are lengthier case studies of countries with either industrial logging in subtropical or tropical forests or with plantation forests: Thailand, Indonesia, Brazil, and Chile. Thailand was chosen as an example of a developing country in which leaders, backed by international agencies and funding, have tried to establish a plantation and pulpmill economy, even though the country has few conditions likely to provide for success in that attempt. Its villagers strongly oppose further inroads on what is left of communal land. Its high deforestation rate and both legal and illegal logging practices have caused extensive soil erosion and flooding, and the proposed plantations have already resulted in numerous scandals. Like Thailand, Indonesia is trying to move into the plantation-pulp export business, and like Thailand, it has overcut its tropical forests, created soil erosion and flooding problems, evicted poor people from forest areas, and allocated logging rights on considerably less than a competitive and open basis. Unlike Thailand, it has a substantial land base. It is now the world's top plywood producer, and it might well become a major pulp producer, albeit at high cost to its people.

In both Brazil and Chile, forestry enjoys commercial success, and though controversies are many, the industry is growing. Brazil, in particular, is expanding its plantation areas and pulp-production capacity at a very rapid pace. It is gradually winning new market shares, not only in Japan and Europe, but also in the United States. It is a well-organized, state-of-the-art industry, and its land base is mainly in areas that were logged prior to the development of plantation forestry. Chile's industry uses more of the natural temperate forest, and its plantations tend to replace recently clearcut stands.

The case studies were chosen so that a full range from underdeveloped to rapidly developing, from small to large, from crude to sophisticated, could be examined. The focus in each study derives from the particular nature of the country. However, in all the cases presented, in more or less the same order from one chapter to the next, information is given about the natural forest species and their extent, the historical context of forestry, the indigenous peoples and the landless poor, plantation forestry, and pulp and paper production. If there are organized protest groups or social movements, these are discussed, together with critiques of forest practices. The final two chapters provide more detail on companies, because the companies in Brazil and Chile have already become very important participants in world forestry, either as domestic companies or as joint ventures with established foreign companies and investment groups. Employment and environmental policies of such companies as Aracruz and Klabin will provide the benchmark for other companies in the next two or three decades; their model is already being taken as *the* model in Indonesia, Thailand, and Malaysia.

Differences between case studies are not all intentional. There is the inevitable problem of differences in accuracy, other quality, quantity, and specificity of the available data sources. Not all countries have sophisticated data-gathering capacity, and even where it exists, there is considerable debate about such basic matters as rates of deforestation.

There are no case studies for countries of Africa. South Africa expanded its plantation area and presence in industrial forestry over the 1980s, but it is still not a major participant in the global industry. Other areas of Africa produce very little and export, if at all, only decorative woods. The rainforests of West Africa fed a thriving timber trade in decorative woods in the 1950s, but overexploitation of both wood and profits left the region bereft of wealth and with a severely damaged forest.

The concluding chapter outlines a number of proposals that might reduce the rate of deforestation. It also summarizes some of the issues raised in the rest of the book relating particularly to property rights, the role of governments, economic development and multiplier effects, and the relationship between local communities and global economies.

DATA SOURCES AND INTENDED READERSHIP

Sources of data include both primary fieldwork and secondary data or published literature. My fieldwork on forestry goes back to the mid-1970s in British Columbia and to the mid-1980s in New Zealand and Australia. It includes research in the late 1980s and the 1990s in Japan, Thailand, Indonesia, Brazil, and Chile.

The book is intended for a general reading public, as well as for academics in the social sciences and forestry. Specialized knowledge of forests and forestry is not required. A critical sense will be required, however, because, as this introductory chapter should have warned readers, there is very little consensus on either how and whether industrial forestry should occur or on the merits and demerits of globalizing this resource industry.

PART ONE

Northern Forests
Northern Industry

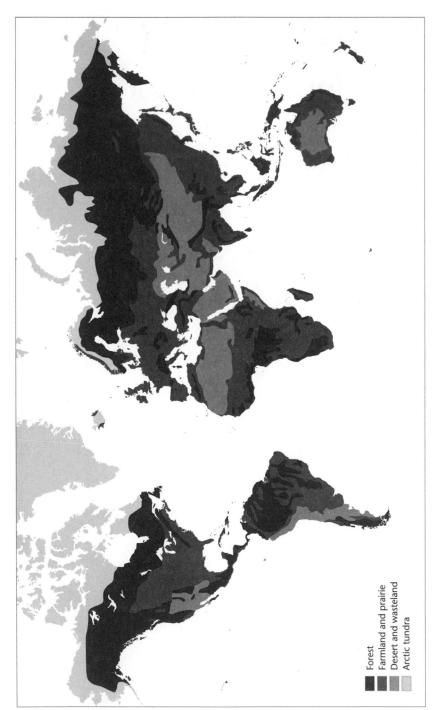

Forest
Farmland and prairie
Desert and wasteland
Arctic tundra

Map 1   Principal vegetation types of the world

2 Northern Forestry
in Changing Conditions

Northern coniferous forests still grow three-quarters of all industrial wood.* Northern countries still produce most of the world's lumber, pulp, and paper. But the north's potential for continued growth is limited in some regions by a deficiency of wood and more generally by the lengthy period of growth for its trees. As well, its own technologies and the external environment are fundamentally changing the northern forest industry. It has automated its production systems and reduced employment in rural areas. Its logging practices and pulpmill emissions have engendered intense negative public opinion. Technologies of other industries and socio-demographic changes affect its markets. This chapter focuses on the changes; the next looks at the industry's responses.

FOREST TYPES AND ECOLOGY

First, a description of northern forests. Those of most interest to industry over the past century are the temperate-zone, coastal rainforests and interior forests at latitudes between about 40 and 60 degrees north. The dominant species in these latitudes are spruce (*Picea*), fir (*Abies*), hemlock (*Tsuga*), and cedar (*Thuja*), each with subspecies

---

* "Northern" here refers to temperate, boreal, and subtropical zones of the northern hemisphere, but "traditional northern" forestry refers only to operations in the temperate zone and primarily to those in coniferous forest regions.

Figure 2.1
Global timber inventory by region: conifer growing stock (per cent of total)

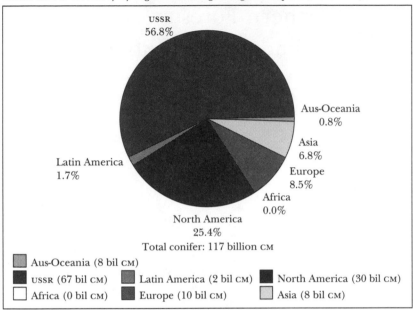

Source: Backman and Waggener, 1991: fig. 1.

adapted to various climatic, altitude, and soil conditions. Each region
has characteristic species either not found or less frequently found
elsewhere. Douglas fir (*Pseudotsuga menziesii*) is an example; it has been
a particularly valuable source of wood in the coastal temperate rainforests of the Pacific region of North America. Larch (*Larix*) and many
species of pine (*Pinus*) grow in drier coniferous regions. Hardwoods
such as beech, aspen or cottonwood, birch, and oak typically make up
a smaller proportion of the mature forest, but may predominate either
in the early stages of forest growth or after the decay or destruction of
old-growth trees. Shrubs, ferns, herbs, fungi, and fallen trees are essential to the total ecosystem. So are insects, birds, and many forest animals.

At the northern margins of the temperate zone is the boreal, or
taiga, forest. It covers thousands of miles and half of the total forested
area of Canada, as well as stretching right across northern Europe and
the former USSR in latitudes between 60 and 70 degrees. Typically, it
includes trembling aspen (*Populus tremuloides*), black poplar (*Populus
trichocarpa*), tamarack (*Larix laricina*), spruce (*Picea glauca*), and jack
pine (*Pinus banksiana*) along its many lakes and rivers. The land in
the boreal zone is similar to that of the tropics: extremely fragile and
largely unknown in a scientific sense. While some European and

Figure 2.2

Global timber inventory by region: deciduous growing stock (per cent of total)

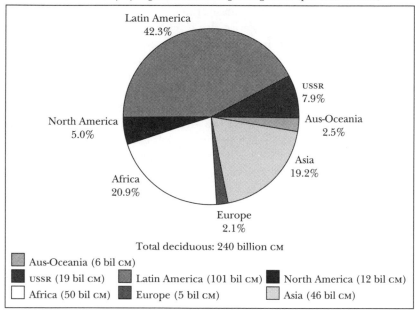

Latin America
42.3%

USSR
7.9%

Aus-Oceania
2.5%

North America
5.0%

Asia
19.2%

Africa
20.9%

Europe
2.1%

Total deciduous: 240 billion CM

Aus-Oceania (6 bil CM)

USSR (19 bil CM)   Latin America (101 bil CM)   North America (12 bil CM)

Africa (50 bil CM)   Europe (5 bil CM)   Asia (46 bil CM)

Source: Backman and Waggener, 1991: fig. 1.

Russian boreal zones were logged, the North American forest was left intact before the 1980s because it was uneconomical to log by comparison with more accessible coastal and more southern interior regions.

At the southern margins of the northern temperate zone are the semi-arid temperate to subtropical regions of the southern United States, the Iberian Peninsula in Europe, and parts of the Mediterranean region. The southeastern states were a major location for the U.S. forest industry in the late nineteenth century, then went into decline, while the Pacific coastal region became dominant. Now a new plantation-based forest industry is flourishing in the same southeastern states. The chief of the U.S. Forest Service noted that its timber is the "most important agricultural crop in the South" (U.S., Dept of Agriculture, 1988: foreword). Its species are *loblolly* pine and various hardwoods. In Spain and Portugal, pine and various species of eucalyptus are grown in plantations. The marginal regions to the far north and the south of the temperate zone (with the exception of the southern United States) were not engaged in the "traditional" northern forestry industry; all are becoming more prominent as traditional regions slip into decline.

Some General Characteristics of Forests

Forests are important for the maintenance of the oxygen-carbon dioxide balance of the earth because the trees absorb carbon through the process of photosynthesis and give off oxygen into the air. Though there is some debate about how vital they are for the emission of oxygen, most writers presently accept the view of forests as the "lungs of the earth" (for the contrary view, see Kimmins, 1992:194). There is less debate about the beneficial role of forests for the regulation of water cycles. Trees pull water from the soil through their roots and eventually return it to the atmosphere as water vapour, thus preventing the drying of the soil and the cycles of floods and droughts that already afflict many deforested regions.

Wildfires have a role in the development of forests. Even where they cover a large area, they rarely kill or cause fall-down of all trees, and the remaining biomass is essential to regrowth. They open the canopy in old-growth forests, allowing sunlight to reach new trees and shrubs. Windthrow has similar effects. Some ecologists claim that clearcutting is similar in its impact to these natural events, but that view is contested on the grounds that clearcutting does not leave any trees intact or allow for a multilayered canopy to develop. In addition, clearcutting tends to destroy soil, in part because of road building and the use of heavy machinery, but also because of slash burning and microclimate changes.

The ecosystem of all forests is a complex web of interconnected living and dead organisms. In the temperate regions, the web tends to have high variation over space yet low density of variation in species within the same area, a condition called *beta* diversity by ecologists; *alpha* diversity consists of high-density variation within the same area and little change over space, a condition typical of tropical forests. Variation from region to region of a temperate forest may be a consequence of diverse historical conditions, such as fire, insect infestations, or other natural changes, together with soil and altitude conditions. Species diversity changes over time as the structure of the forest changes. The advantage of diversity, apart from the values associated with aesthetic taste or a generalized belief in diversity as a good thing in itself, is that it provides the forest with resilience to diseases and pests.

Old-Growth Coniferous Forests and Spiritual Values

Old-growth coniferous forests are prized because they contain high species biodiversity, the trees are large and varied, the forest canopy is multilayered, there are many decaying trees and other biomass on the forest floor, and there are numerous other species for whom the forest is a natural habitat. A definition would normally include an age condition, but the precise age would differ with species. A 700-year-old western red cedar is old; so is a 500-year-old Douglas fir, though both species grow to still more venerable ages. Forests of less than such advanced ages may also merit consideration as old growths because the species in their soil and altitude are not likely to reach several centuries or because they remain small, even if elderly. Indeed, in general perception a stand of 200-year-old cedar or Douglas fir constitutes sufficient age to merit the term "old growth." More critical to a definition would be the structure of the forest, usually evidenced through a multilayered canopy.

A researcher and forest ecologist describes a 250-year Douglas fir forest in western Washington in these terms: "The most striking features of these forests are the size of the trees, and the fact that they support a relatively unique group of birds and animals – some of which cannot survive outside these areas. Wildlife often found in such 'old growth' areas include goshawks, spotted owls, hairy woodpeckers, pacific giant salamanders, pine marten and fishers. Bald eagles, hawks, squirrels, deer and other wildlife can also be found." (Meyer, 1990:3). Such forests are extraordinarily beautiful, and since each is a unique development over centuries, they cannot be duplicated by human managers. For conservationists, this in itself is a reason for preserving them. There are many other reasons, including particularly the sustenance of genetic pools that cannot be recreated in managed and plantation forests, and the sustenance of wildlife habitat and migration routes for many species. There may also be economic reasons, such as the development of tourism. Spiritual sustenance is a major reason for those who love forests.

Forests may nourish the soul, please the eyes and nose, and provide a habitat for wildlife; they may be vital to the maintenance of an oxygen-carbon dioxide balance of the atmosphere and to the regulation of water cycles. Yet despite their noted virtues, forests have been assaulted everywhere throughout history. In fiction, in mythology, and in history, they have been depicted as dark abodes for evil spirits. "We are not out of the woods" is a metaphor for frightening situations.

There are human groups, however, who have lived in forests over many centuries and who do not share that fear. Indigenous peoples the world over tend rather to speak of forests with awe and respect as spiritual homes. This is true of the indigenous peoples of the northwest coast of North America. Chief Simon Lucas of the Hesquiat Band on Vancouver Island describes the spiritual value of forests in these terms: "Most importantly, old growth forests are areas which are full of life; in which 'life' in all its wonderful diversity is immanent ... Within forests, we are completely surrounded by life; within forests, we can renew our spiritual bonds with all living things. Can we, or any people, survive without a strong spiritual basis to our lives?" (BC, Ministry of Forests, 1990).

In 1994 the Haisla Tribal Council turned down much needed employment for its band members as loggers in the Kitlope Valley of northern British Columbia. West Fraser, the company holding timber licences and offering employment, withdrew voluntarily shortly afterwards, allowing the government to name the entire 317,000-hectare valley as protected territory under co-management authority between government and Haisla. The explanation for the Haisla's refusal to take jobs or cooperate in the exploitation of this last unlogged rainforest watershed in BC is that they consider the valley their spiritual home. They will log selectively and on a sustainable basis, but will not log on the industrial model typical of northern forestry. Hank Ketchum, president of West Fraser, may have been persuaded of the Haisla claims; other members of the non-indigenous population may not be as supportive. A less enthusiastic approach to forests is that of some geologists or others who take very long views of natural history and who point out that what is now called "old growth" is still a fairly recent development. Left to itself, they say, a forest will decline, and other forms will eventually take its place. Forests that are not destroyed by such natural forces as fire, earthquakes, or diseases and are not harmed by human activity may survive and replace themselves over a very long time after reaching an initial "climax" stage of growth, but they do not exist in perpetuity. The argument may be true, but it is similar to an argument that might be made about all human endeavour. After all, the human species may be as temporary on the earth as any other.

These two views of forests are central to debates now raging in most northern forested regions. Although sometimes characterized as a fight between urban ecologists and loggers, expecially when groups are caught in the media spotlight contesting control of particular territories, the struggles are both deeper and broader. Neither ecologists nor loggers are homogeneous groups; neither owns the territories under dispute; rarely are members of these groups as simple-minded in their support of their position as media imagery suggests. Beneath land-use conflicts are vital issues of sustainability for human populations, as well as for forests and wildlife, and changing conditions in the marketplace and the cultures of northern societies.

AIR POLLUTION AND ACID RAIN

Logging is the obvious target of attack for those who wish to preserve or conserve the earth, but at least as serious a threat to forests is air pollution from numerous other sources. According to the *Ecologist*, some 600,000 tons of sulphur per year land on Sweden, not more than a sixth of which could possibly be due to Swedish sources. The Black Forest in West Germany was the first to manifest signs of a fatal disease, then the Bavarian forests in the south and southeast; rapidly and relentlessly, the sickness spread to the north and to the south. Within the first six months of the 1980s, according to this source, half of the total forested area of West Germany became infected, and in quick succession, a third of Switzerland's forest, half of Luxembourg's, a third of Czechoslovakia's, and at least a quarter of France's (Bunyard, 1986b:4–6). Ecologists debate the specific causes: coal- or oil-fired generation of power, city sewage seeping into the waters, intense animal husbandry, motor vehicle emissions, aluminum, steel, and other heavy industries all contribute to the disease.

Further studies have supported the general argument propounded in the *Ecologist*. While cautious in their estimates, authors of one major study contended that "forest decline is a serious problem in much of Europe. Many countries are experiencing declines where the volumes in damaged stands are more than 15 to 20 percent of the growing stock of exploitable closed forests, and where these volumes exceed annual fellings by many times (often 5 to 10 times)" (Nilsson and Duinker, 1987). Another analyst emphasized the complexity of interactions between air pollutants, natural contexts, and internal forest dynamics as causes of air pollution (Prinz, 1987). Further studies have shown that there are different patterns and intensity of damage for various species and at various elevations; blanket generalizations will not do (e.g., Clarkson and Schmandt, 1993). Czechoslovakia and East Germany have suffered particularly heavy damage. Even with downward adjustments in the estimates of damage, there is general acknowledgment that acid rain, alone or in combination with a number of other factors affecting soil nutrients and forest context, has become a major problem.

TIMBER RESERVES

Accompanying tables and figures provide current estimates of forest reserves, but several caveats are in order. Estimates of past reserves and deforestation tend to be unreliable because measuring instruments

Table 2.1
Temperate regions: estimated subdivision of forest into exploitable and unexploitable (million hectares)

Region	Exploitable[1]	Unexploitable
Europe	133.0 (89.1%)	16.3
Former USSR	414.0 (54.8%)	340.9
Canada	112.1 (45.3%)	135.1
USA	195.6 (93.3%)	14.0
Australia	17.0 (42.7%)	22.8

Source: FAO, "The Forest Resources of the Temperate Zones. Main Findings of the UN-ECE/FAO 1990 Forest Resources Assessment," ECE/TIM/60 (n.d.), summary table 3, p. 8.

[1] Exploitable as defined by FAO includes forest or other wooded land in regular and sustainable use as a source of wood; available and accessible. However, there is no universal agreement on the definition. Previous FAO surveys have used such terms as "productive forest" or "operable forest" to refer to the same thing.

were crude before the advent of satellite photography. Further, species whose commercial value has changed over time affect estimates of growing stock. Thus measures of incremental growth are dubious. What is or is not exploitable is subject to definition, and as the table indicates, the definitions are not universally the same. The ideal measurement would combine qualitative characteristics with volume data, but current measuring tools are not up to that task. For an indication of the qualitative dimension, we need to discuss particular regions (as in the following chapter), but readers should be warned that entirely comparable data for various regions are simply not available.

Though land may be classified as "exploitable," industry cannot count on all the wood. Some timber is theoretically exploitable but inaccessible, too distant from markets, or not within reasonable distance from good transportation routes; some is decadent or immature and not economic to log. What remains can be logged, but harvesting on a sustainable basis requires long-term management. Finally, we will have to measure, not according to a single year's standing timber, but according to the amount of timber that could be harvested on a sustainable basis over a full century, since that is at least how long it takes to grow replacement trees. When we take these measures, none of them extreme or outrageous, into account, we discover that the sustainably exploitable timber in northern forests is much less than the usual simple estimates of how much is still standing.

The northern industry cannot meet the world's market demand on its present reserves; indeed, European producers cannot meet their home market demands. Though Scandinavian countries have reforested for nearly a century and other western Europeans since the end

Figure 2.3
Temperate regions: distribution of forest and other wooded land

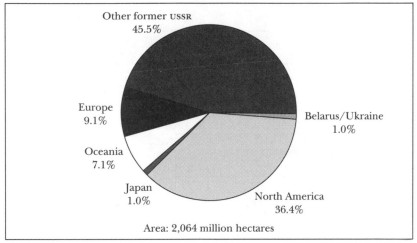

Area: 2,064 million hectares

Source: FAO, "The Forest Resources of the Temperate Zones: Main Findings of the UN-ECE/FAO 1990 Forest Resources Assessment," ECE/TIM60 (n.d.), summary fig. 2.

of the Second World War, Europe consumes more than it produces of timber (called roundwood once it is logged), lumber, and pulp, and the limits of timber production have been reached. In 1991 Europe as a whole produced about 10,000 cubic metres (CM) less wood, another 10,000 CM less lumber, and about 6 million tons* less pulp than it consumed. Only the Nordic countries, supplementing internal wood and pulp sources with imports, produced more paper and paperboard than they consumed, giving Europe as a whole a favourable balance (data given in the tables in chapter 3).

In some parts of North America, timber reserves still look large on paper, but they are ever more distant from mills and markets, of low quality, and increasingly expensive to harvest. In other parts of the continent, harvesting rates have exceeded regeneration capacities. In the former USSR and eastern Europe, remaining timber reserves are distant from markets, infrastructure is poor, and the quantities of high-quality timber are subject to debate.

* All tonnage measures are metric. One metric ton (tonne) equals 2,204.6 pounds or 1,000 kilograms. Short (U.S.) tons, which measure 2,000 pounds or 907.9 kilograms, are converted where identified. Readers should note, however, that original sources do not always stipulate the measures used. As a measure of capacity, 40 cubic feet of timber may be considered an equivalent to a metric ton.

Figure 2.4
Net annual increment per hectare on exploitable forest in selected countries
(cubic metres overbark per hectare)

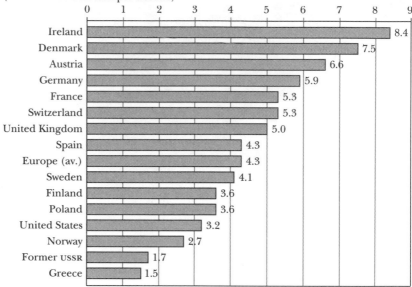

Source: FAO, "The Forest Resources of the Temperate Zones: Main Findings of the UN-ECE/FAO 1990 Forest Resources Assessment," ECE/TIM60 (n.d.), summary fig. 12.

Even if the forests were as limitless as many people once believed, the basic condition is that coniferous trees in northern latitudes take at least a century to reach maturity. Nature actually takes closer to four hundred years. If the managers of denuded regions were to begin a program of massive replanting in the year 2000, even using plantation pine or poplar species, they could not produce a harvestable crop on a competitive basis with eucalyptus varieties. They would have to undertake expensive and ecologically problematic intensive management and pray for a miraculous decline in industrial pollutants. They would have to pay for fire damage and pest and disease control. The trees need seed nurseries, silvicultural treatment, thinning, pesticides, and nutrients, and when finally grown, they must be accessed over mountains by means of expensive logging roads. And even intensive management is not guaranteed to be workable. In short, the replanting and management of a northern forest, especially a northern mixed-softwood forest, is a costly undertaking. It ties up capital for a very long time – generally longer than the lifespan of those who do the planting – a condition not likely to attract enthusiastic investors if they have alternatives. Deficiency of trees and slow growth cycles are reasons for the search beyond northern temperate regions for new fibre sources; they are among the reasons for turning the forest industry into an agricultural enterprise in southern regions.

TECHNOLOGICAL CHANGE, EMPLOYMENT, AND COMMUNITIES

While in long historical perspective, industrial forestry is not the main culprit in the destruction of forests, the cultural appreciation of forests expressed by Simon Lucas is inconsistent with much of the history of logging. This is especially so in North America, where, until very recently, the industry depended almost entirely on nature to provide the trees. In its crudest form, the industry is a mining operation, extracting a resource and moving on to other sites. Sometimes natural regeneration takes place, sometimes it fails because the soil is so damaged, the microclimate so altered, or the streams so silted that succession species cannot take root.

Logging and Sawmill Practices

Initially reliant on animal as well as human power, axes, and relatively simple saws, logging technologies soon included tractors, chainsaws, multi-function logging machines, feller-bunchers, chippers, and specially adapted barges and trucks, the net results of which are capacities to fell forests even on steep northern terrains at rapid pace and with decreasing labour input. Sawmills, likewise, have moved through phases from relatively simple handsaws carried into the woods, through mechanical saws combined with labour in factory-like assembly-line production systems, to highly automated systems in which human input consists of monitoring computerized sorting and sawing systems.

In northern countries, where concentration of ownership and integration of facilities have both moved in step with technological change, conifers are generally logged and shipped to huge mills. Logs suitable for construction lumber may be processed through sawmills, with chips from the sawmills moving on to pulpmills. (In British Columbia, where this separation is mandated by legislation, there are persistent questions about sawlogs that somehow end up in pulpmills; elsewhere the separation is not mandated because the difference in tree qualities may be less great.) Even in the most modern northern sawmills, much of the wood is not utilized. One researcher, David Tillman, estimates that on average a sawmill recovers about 40 per cent of the small log (1985:3). The remainder is sent on to pulpmills as chips. In eucalyptus pulpwood operations, whole logs are processed in pulpmills because the wood is unsuitable for lumber. This is also the case for aspen logs in the boreal regions.

Clearcutting on a massive scale began in the 1940s in North America and was directly related to the emergence of huge integrated companies with private land grants or harvesting rights to enormous areas. Continuous cutting in contiguous regions one year after another, a form common in the Pacific coastal states during the 1950s and 1960s, is one form. Small and dispersed clearcuts have also been tried, but these created interspersed blocks of cut-over land and old-growth forest, and they were almost as harmful to wildlife and overall habitat as continuous clearcuts.

Over time, the industry has become more sophisticated, either preparing the soil for natural regeneration or replanting trees before moving on. Crude and managed forms coexist today, but there is far more public concern over the crude extraction process and more scepticism about management techniques than in the past. Now that much of the earth's forest cover is gone, its disappearance is causing concerns about the atmosphere and human survival, and the forest industry finds itself at the centre of the conflict over logging and preservation.

Pulping Technologies

The same coniferous forests that supplied ideal building materials in northern Europe and North America also provided pulpwood, and northerners developed pulping technologies for the dominant species in their territories. Pulp is the basic ingredient for papers and paperboards, and most (not all) pulps are derived from wood. Technological development in pulping has made it both possible and economical to process hardwoods which were not suitable sources of fibre two decades earlier. By the 1990s, eucalyptus and other hardwood pulps were in high demand, and state-of-the-art pulpmills were being constructed at rapid pace throughout the tropical and subtropical regions of the world. As world demand grows for newsprint, telephone directories, tissues, paper napkins and diapers, and writing and book papers, the demand for suitable market pulp grows with it.

Pulp may be created from wood fibres by mechanical or chemical methods, each having specific advantages and disadvantages for different kinds of paper. Mechanical methods, which predated chemical processes, reduce wood to fibre by grinding or beating. Before the 1960s, the chief use of mechanical pulp was for newsprint, but since the development of a method known as thermomechanical pulping

(TMP), mechanical pulps have gained a wider range of applications. The TMP method was developed in the 1970s, though known earlier and experimented with in Sweden during the late 1960s. Technical knowledge improved in the following years, and by 1980 there were 170 plants worldwide using TMP processes. Mechanical pulps as a type have less strength than chemical pulps, but new production techniques have increased pulp strengths and the difference between types of pulp is now less marked. In general, mechanical methods are more friendly to the environment, and mill construction is less costly.

Chemical processes are still in greater use than mechanical ones, and their effluent is of greater concern. Various chemical combinations have been tried. They all remove lignin and hemicelluloses from wood, freeing the fibres and allowing them to be reunited in the form of a mushy pulp. Sulphite pulping methods were superseded by sulphate (kraft) processes during the 1950s. These increased efficiency and reduced chemical wastage. Sulphate pulp is used as a strengthener in mechanically pulped newsprint, but kraft sulphate mills have high initial capital costs and are profitable only at annual production levels of about 300,000 tons (Hartler, 1987; Kellison and Zobel, 1987). A number of production changes were introduced in the late 1960s that increased the energy and materials-utilization efficiencies in pulp mills.

Combinations of mechanical and chemical pulping (chemi-thermo-mechanical methods, or CTMP) are now in place. The TMP and CTMP mills can operate at appreciably lower capital costs and mill size, and have lower specific wood consumption levels than bleached kraft sulphate mills. They can be economically located where wood supplies and domestic demand are relatively small, but they require high energy inputs and are not viable where energy is expensive or unavailable.

In terms of geographical restructuring, the message is fairly straightforward: various methods of pulping are now on stream, and companies or countries have access to processes that will suit their conditions, their wood sources, and their energy supplies. The early methods were developed specifically for northern conifers. As these became scarce, but also as the potential of fast-growing hardwoods became apparent, new methods were researched and finally put into practice. Pulp now can be made from many wood fibres in vastly different conditions.

It should be noted here that pulp can also be made, and traditionally was made, from non-wood sources. Other vegetable and even non-vegetable sources are viable in some economic and geographic conditions, and may become more important as wood sources decline or become ever more controversial. At the present time, about 7 per cent of the world's pulp is derived from non-wood sources.

Employment: Swedish and Canadian Trends

Employment in all sectors of the northern forest industry has declined since 1960. The basic cause is technology that reduces or eliminates labour input. Feller-bunchers in northern woods, high-powered automatically controlled saws in sawmills, and computer-controlled operations in pulpmills have displaced workers.

Lars Lonnstedt (1984) examined Swedish forestry employment over the twentieth century, concluding that even with sustained yield practices, employment had declined steadily since 1940 in logging and since the late 1960s in processing. Swedish manufacturing was increasing, but the jobs were not in forestry regions. Even with the most benign assumptions about trends and supplies, Lonnstedt predicted that employment in Swedish forestry would decline by some 3,000 jobs per annum. He analysed data produced by the Swedish National Insurance Board and the relevant labour union for the period between 1910 and 1980, and he demonstrated that the peak period in logging employment occurred between 1930 and 1940. The reduction in logging has amounted to about 2 per cent yearly, in spite of wood removals increasing by about 1 per cent a year until the early 1970s. Within the forest industry as a whole, the peak in employment was reached in 1960, despite an average increase in production volume of about 4 per cent annually until 1975.

Official statistics published by the Swedish government in 1992 indicate that the average number of employees in wood-processing establishments declined through 1950 to 1989, with the largest drop occurring in the 1980s. At the same time, the number of establishments declined markedly. Those producing pulp and paper also declined, but the number of employees remained fairly stable throughout the same period. Remaining mills were much larger, and each employed more workers.

The same general trends are apparent in other countries of Europe and in North America. Where large companies gained control of the emerging industry in North America, wage-labour was employed in woodlands as well as mills. The early history is replete with unhealthy living conditions, high accident rates, and low wages for bunkhouse men; conflict and militant rebellion were among the responses, together with unionism that spanned the scale from radical to business. Business unionism was the more expensive, though it bought labour peace. It rested less on rhetoric and more on sober calculations

Table 2.2
Sweden: number of establishments, average employment per establishment, and average total employed in wood-processing and pulp and paper industries, 1950–1990

	Wood processing		
Year	Number of establishments	Average employment per establishment	Average total employed per year
1950	3,501	20	70,020
1970	2,492	32	79,744
1980	1,734	41	71,094
1988	1,303	43	56,029
1989	1,292	44	56,848
1990	1,143	48	N/A
	Pulp and paper		
Year	Number of establishments	Average employment per establishment	Average total employed per year
1950	353	148	52,244
1970	275	215	59,125
1980	225	271	60,975
1988	217	247	53,599
1989	220	257	53,900
1990	258	210	N/A

Source: Selected data from Sveriges Officiella Statistik, Skogsstyrelsen Jonkoping, 1993, *Skogsstatistisk Arsbok, 1993*, Tabell 12.6: Antal Arbetsstallen inom Skogsindustrin och Arbetsstallenas Genomsnittstorlek under Perioden 1950–1990.

of shareholders' earnings and employee bargaining power. Inevitably, companies sought ways of reducing the number of workers. Logging, sawmill, and pulpmill machines were constantly upgraded and transformed to use less human labour. In Canada as in Sweden, the cycles of the wood-processing and pulp and paper industries differ somewhat, with the numbers employed in wood processing (lumber) declining earlier. Logging can be more accurately measured because the workers are typically hired on a wage-labour basis, whereas in Sweden and in most of Europe they are or were woodlot owners and casual rural workers.

More significant than absolute employment data may be simultaneous measures of employment or man-hours and production volume. As the accompanying table and figures indicate, employment in logging in Canada has steadily declined since the mid-1960s, though production increased until 1988. Employment was more variable in sawmills, but did not grow very much over the late 1970s to mid-1980s and has since levelled off. Production has grown substantially

Table 2.3
Canada: production and employment data in logging, sawing and planing mills, and pulp and paper industries in selected years, 1970–1990

Year	Roundwood		Saw & planing		Pulp & paper		
	Production 1,000 m³	No. employees	Production 1,000 m³	No. employees	Pulp 1,000 t	Paper[1] 1,000 t	No. employees
1970	117,324	52,230	26,588	48,776	16,609	11,252	80,371
1975	111,497	53,622	27,305	49,156	15,156	10,066	84,046
1980	150,649	54,370	44,995	66,278	20,054	13,388	86,872
1985	161,985	45,962	52,109	58,961	20,481	14,425	77,562
1990	166,966	43,786	54,909	53,294	22,836	16,465	77,768

Source: Forestry Canada, Selected Forestry Statistics Canada, 1991, table IV–9 for roundwood production data 1970–89, and 1990 roundwood production data from Statistics Canada cat. no. 25–202; table IV–13 for employment data 1970–89, and 1990 employment data from Statistics Canada cat. no. 25–201; Table II–7 for lumber production from 1970–90; table IV–18 for employment data from 1970–89, and 1990 employment data from Statistics Canada cat. no. 35–250; table II–13 for Wood Pulp Production 1970–90; table II–16 for paper and paperboard production 1970–90; table IV–26 for employment data from 1970–89, and 1990 employment data from Statistics Canada cat. no. 31–203. [1]Paper includes all paper and paperboard production (as defined in Selected Forestry Statistics Canada, 1991).

in the same period, taking place in fewer, but much larger establishments. Employment in pulp and paper mills grew in the 1970s, when many new mills were constructed, and has since levelled off, while production has continued to increase.

In both countries, the peak employment periods for the industry were in the late 1960s and early 1970s. The oil crisis of 1974–75 cast a shadow on this, as on other energy-demanding industries, but production moved upward again after the industry had implemented various energy-conserving strategies, including more efficient use of wastewood as fuel. The "second spurt" wavered in the early 1980s, grew again, and then levelled off, and growth has declined since the mid-1980s. The long-term trends inform us that early employment growth depended on a labour-intensive technology in an expanding industry. The industry is no longer labour-intensive, and it is no longer expanding in many traditional regions. Employment trends are not simple functions of sustainable yields, though sustainable yields would be preferable to the "cut and run" syndrome. Employment trends are also not a function of strong market demand, though more people are employed when it exists. Even where the industry is producing more than it did in the 1960s and 1970s, employment has declined. The basic cause of decline is the development of technologies that displace labour in the production process.

Not shown in data on trends are changes in the composition of jobs. As mills became automated, the jobs occupied by unskilled labour were most frequently phased out. Workers had to be able to read computer printouts and learn how to operate new kinds of machinery, so higher educational levels were demanded. In pulpmills, except for tradesworkers, an increasing proportion of jobs are available only to university-trained engineers. In the woods, foresters have supplanted loggers in some jobs. The same impacts have occurred in salaried jobs in management and clerical occupations: incumbents need to be computer-literate and generally require advanced education. Among the social impacts of such changes is the loss of access for first-generation immigrants who do not speak the dominant language and have low educational levels, together with others who have not completed high school or gained trades skills. In many rural regions these people were once major sources of labour in the forest industry.

The Impact on Communities

Community stability is often cited as a goal of forest policy, whether in North America, Japan, or developing countries. Few in industry or government choose to consider seriously the contradictions inherent in a global marketplace between such a goal and the overriding goals of export-oriented business. If productivity is the goal and it can be achieved through labour-saving technologies, then no matter what business leaders say, communities consisting of forest workers will decline.

Sustained-yield policies have occasionally been extolled as the means to achieve community stability. The linkage was a bedrock assumption of two royal commissions in British Columbia in the 1940s and 1950s and was still current in the 1980s (e.g., Reed, 1987). Yet R.N. Byron, in a 1976 PHD thesis, provided data to show that market conditions and the degree of diversification in towns were crucial variables for community stability; sustained-yield forestry was not germane. He did not examine changes in technology, and some of the most significant of those changes were only then beginning to have an impact. Swedish analysts J. Randers and Lars Lonnstedt (1979) not only questioned the linkage with sustained yield, but they suggested that stability of forestry and stability of regions might be contradictory goals. They pointed out that when the industry can no longer grow by extracting huge quantities from a new forest region, the only way of increasing productivity is to reduce employment.

Most North American forestry communities were created within the last half-century. A few are outgrowths of earlier logging or agricultural settlements, but many were created specifically as bedroom communities for large integrated sawmill and pulpmill operations. Since small woodlots were never established, these towns from their inception depended on corporate employers. Often such employers have investments elsewhere, and local mills are merely the sources of wood for more advanced manufacturing operations in larger market regions. In these bedroom towns, much that happens is beyond local control. Fundamental economic decisions about them are generally made in boardrooms far away (Marchak, 1983; Robbins, 1988; Beckley, 1994).

Indeed, the preservation of community is itself a controversial issue in North America. While Japanese forest owners take for granted that the sustenance of their ancient villages is of paramount importance (see chapter 5), some residents and many employers in the rural industrial towns of Canada and the United States assume that geographical and social mobility is normal or desirable. One employer in the BC interior town of Quesnel responded to my questions about unemployment issues there: "Who cares about this place? It's not worth saving, and the kids will have a better life in the big world beyond it" (my field notes, 1988). This attitude was pervasive in the 1960s and 1970s, when jobs were plentiful and forestry workers were highly transient. In the late 1980s and early 1990s, it is a much more problematic stance: there are few jobs in "the big world beyond it," especially for young people with limited skills and education or for immigrant workers with limited command of the dominant language. Before the more general downturn in the economies of industrial countries, declining employment in the forest industry was a less pressing concern for governments than it became in the 1990s. Restructuring of the industry at this stage necessarily involves retraining and relocation of workers, though attempts to do so in Sweden and BC are not yet examples of great success.

At a more general level, we might speculate on the relationship between declining employment in this industry and rising hostility toward logging and pulpmill operations. Introduction of new technologies may temporarily improve the financial balance sheet for the owners of industries, but the benefits are not shared with displaced workers. Societies absorb such damage from a single industry if other employment opportunities are expanding, but northern economies have not been capable of absorbing displaced workers in forestry or other industries in the closing two decades of the twentieth century. Contract loggers and truckers may fight for their livelihoods against environmentalists, whom they characterize as rich kids who know nothing about the woods. Yet when the entire industry is laying off workers and neighbours are migrating to urban centres in search of work, loyalty to large and technologically sophisticated companies must surely be tested to the limits. Further, the society as a whole has a much reduced stake in the success of an industry that no longer provides employment on a scale comparable to its profits. Indeed, since some companies are investing in manufacturing plants elsewhere, they must be removing profits made in declining regions. While there is rarely a one-to-one relationship between social conditions, a reasonable hypothesis in this case would be that the social defence of the forest industry erodes in direct (if not quantitatively precise) relationship to employment levels.

Employment levels are not the only variables affecting community support and worker loyalties. Thomas Beckley (1994) conducted an intensive study of a New England mill town; he concluded that a protracted and bitter strike over "flexible" restructuring had left a very long legacy of distrust. The company won its flexibility but lost the loyalty of its workforce. The same fight is going on in British Columbia as this book is written (see chapter 4). Beckley also noted that the New England mill was once perceived as a local company, but it was taken over first by a petroleum company and then by the national forestry giant Boise Cascade, and the sense of belonging eroded for workers. The transience of management under the aegis of large national or international companies is another factor that destroys the sense of community and affective relationships between management and mill workers. Beckley's thesis is consistent with evidence in other studies (e.g., Marchak, 1983; Robbins, 1988; Hayter and Holmes, 1994) showing both the vulnerability and the transitions of resource-dependent towns.

Change in the Technological Context

Technology in forestry is changing, imposing new demands and altered conditions on the northern industry; simultaneously, the technological and socio-economic contexts for forestry are altering. Electronic data interchange (EDI) and, more generally, computer-based inventory control and financial transactions are among the contingencies likely to affect the forest industry. Computerization in its first phase increased, rather than decreased, paper demand. Computer paper, in fact, became a major product of the industry. However, the second phase featuring the EDI is expected to decrease the net paper flow of business, as computers talk ever more to one another. Newspapers, specialized journals, books, and paper-based products such as tissues and paperboard may not be immediately affected by EDI, though substitution and saturation effects may have impacts. As literacy increased in the developing countries over the second half of the twentieth century, demand for these products expanded. In the near future, it is expected to continue upward, which explains the push by every company to increase mill capacity, thereby courting a familiar and recurring feature of this industry: overproduction.

While electronic data processing has increased the demand for papers in the short run, other developments have decreased, or threaten to decrease, demand for wood and paper products. Steel, aluminum, and plastic have already displaced lumber in many aspects of housing construction. Vinyl siding on single and multi-family buildings is commonplace where wood was once essential. Many simple items, from utensils to sports equipment, are now made from synthetics or mineral-based materials. In paper, as well, there are numerous substitutes in general use – polystyrene cups and plastic packaging materials, for example. In short, the forest industry is rapidly changing, but so are other industries. Technological change is a multilayered development, and no industrial sector undertakes its transformation in isolation.

Demographic and Socio-Economic Changes

Demographic and economic changes may be expected to affect consumption of wood, as well as other products. Income distribution in most northern countries was fairly stable between 1950 and 1970, and then changed substantially between 1970 and 1990. The top 10 per cent gained greater shares of the total wealth; the lower 40 per cent were relatively poorer. There was a general downward shift in income for the middle 50 per cent; indeed, sociologists speak of the shrinking of the middle class. Such a change has implications for the forest industry because it is that population in the industrial countries that purchased lumber, newspapers, paper, and books.

In the 1950s the majority of households consisted of small nuclear families with one parent employed for income. Now there are many single-parent families, and more households consist of several unrelated single persons. Where two-parent families still maintain households, both parents tend to be in the workforce. Many households, however, have no regular income-earners, and many single persons or single parents are unemployed and without earnings. As well, the northern industrial countries have aging populations and low fertility rates. All demographic and economic conditions of these kinds will affect consumer capacities and habits.

The changes are social and cultural, as well as demographic and economic. Where northern countries were earlier homogeneous in language and culture, they are now heterogeneous, and traditional mass-circulation newspapers have lost their constituencies. Illiteracy is a continuing problem, and illiteracy in the dominant language a separate, but related issue, its prevalence rising in North America as the population becomes more diverse in cultural origin and language. This change might reduce consumption rates for mass-produced traditional wood and paper products, but it might also bring about an increase in demand for more specialized items suitable for targeted populations, known as "niche" markets. Roman Hohol (1991), consultant to H.A. Simons Ltd of Toronto, told a conference of producers of groundwood papers in 1991 that "quality will become a much more important determinant than price in determining consumer choice." For those with the wealth to choose quality, that might be the case; for those without, the outcome would be a decline in consumption.

The lumber market is more complicated. Housing construction rates vary with general economic conditions, and the market has always reflected these cyclic turns. The growth of multi-family housing complexes and the development of numerous alternative construction materials contribute to an apparent long-term decline in demand in the United States. The cost of lumber, related to diminishing supplies of high-grade coniferous logs from the northern U.S. states, is a major factor in this decline. We return to this issue in a discussion of the U.S. – Canada border war over Canadian lumber exports in the next chapter.

PULPMILL POLLUTION AND
PUBLIC OPINION

The sulphur odour emitted by pulpmills used to be laughed off by local folk in many forestry regions as "the smell of money." No more. Companies are under attack for their emissions into the air and even more for effluents that enter the streams and oceans. Air emissions are less foul-smelling today, though pulpmills are still among the producers of acid rain. Companies are pressed even more urgently today by the problem of pulpmill effluent in water.

During the 1970s and 1980s, governments imposed regulatory measures on pulpmills to protect water from certain identified pollutants. A Canadian study in 1988 indicated that these had been effective in bringing about some reductions (Sinclair, 1991). However, some of the regulations had not been complied with, and the industry's explanation was that investment requirements were excessive and might result in mill closures. With about 175 communities dependent on the industry, such statements are taken seriously by governments.

William Sinclair, after examining the investment habits of the major firms, concluded that such an explanation was inadequate. In his opinion these companies were "not economically fragile" (1991:94). Further, in his view, "the great majority of mills could adopt the best available effluent controls and technology" at an affordable cost, but the industry tended to invest in added capacity instead, when the economy was buoyant. Sinclair notes:

The costs imposed on society by inefficient pulp and paper manufacturers cannot be fully appreciated unless it is understood that the pulp and paper industry uses about 35 per cent of the nation's total annual forest production. It is the nation's largest industrial user of water and the second largest user of energy ... the more efficient a mill operation, the less water, energy, and timber resources required to sustain its operations." (1991:101–2)

Public opinion was not in favour of the pulp industry in 1992. Shellfish harvesting had been stopped along the BC coast because the fish were contaminated by dioxins. Governments had been slow either to impose or to ensure compliance with regulations on dioxins and other toxic substances from mills. Pervasive non-compliance with standards had been documented in Canada (Sinclair, 1988; Brown and Rankin, 1990; see also Greenpeace Canada, 1989; Lyons, 1989). An industry plan, crafted in 1987 to offset antagonistic public opinion in the face of U.S. government reports on toxic dioxins in fish downriver of pulpmills, was publicized after reporters obtained it through freedom of information legislation (Vancouver *Sun*, 17 Sept. 1992:B2). The public mood encouraged further regulations and steeper penalties for offending pulpmills.

But in late 1992 new research by the National Water Research Institute in Burlington, Ontario, and the Pulp and Paper Research Institute of Canada indicated that bleach-free mills were as dangerous to fish as those emitting organochlorines from bleaches. There were well-documented changes in sex organs and other features of fish swimming within waters close to the mills, but they occurred irrespective of chlorine content. Since all the attention of governments had been focused on chlorines and mills were either refitting at huge expense or being built with massive detoxicant systems to deal with chlorines, this news was worrisome. If not chlorines, what? And should the renovations and special treatment plants still be required?

Preliminary results suggested that the toxic materials were actually naturally produced lignin, which, when pulped and emitted from mills in concentrated effluent containing phenols, resin acids, and fatty acids, contaminated the water. Further studies in 1994 again indicated that chlorine and the family of organochlorines which had been viewed as the culprits were not the cause of fish deformities. Of the new problem, one Canadian environmental scientist noted: "You aren't putting into the environment anything that has not been there before, but you are doing it in a hell of a high concentration" (quoted in *Globe and Mail*, 10 Nov. 1994). By this time the use of bleaches in pulp manufacturing had declined substantially. According to one document, Canadian mill use of bleaches had declined by some 60 per cent by 1992 (Canada, 1992a:7). Some firms are boasting that they can eliminate chlorines altogether and transform wastes into energy. At least two Canadian mills claim they have reduced effluence to zero (McInnis, 1994, ftn.17). As well, secondary treatment of mill effluent before it is discharged into rivers and oceans has become a standard procedure in both northern and southern mills.

The problem of toxic effluent is not yet solved, but scientists express the opinion that it is now understood and controllable. The industry turns again to public perceptions, since whatever science says about bleaches, an alerted public is showing a preference for unbleached papers. Here, as in so much else about this industry, we see the tendency to evade the genuine issue: a willingness to assume that the problem is public opinion rather than anything central to the industry. Public opinion in this case may be wrong or at least outflanked by scientific evidence. But the public's sense of unease is justified when fish die from effluent, whether caused by one agent or another. The supposition that the problem is controllable – the word in this industry is "treatable" – is entirely consistent with industrial assumptions throughout modern history. The objective is to control nature, not to find ways of living with it.

SUMMARY: FORESTRY
IN CHANGING CONDITIONS

That portion of the northern forest critical to the industry over the past century lies between 40 and 60 degrees north latitude. At its northern margins up to about 70 degrees north latitude lies a huge taiga belt of boreal forest now being industrially exploited, and at its southern margins are fairly new plantations of pines and eucalyptus cultivated mainly for pulpwood.

Ecological diversity is the hallmark of a mature forest, though in a northern temperate forest, diversity is measured over changing altitudes and distances rather than in densities per square kilometre. Old-growth forests in the north are prized by some and dismissed by others, but whatever the attitude of humans in their vicinity, forests serve functions ultimately essential to the sustenance of life on earth.

Forestry has been an important economic activity in northern regions. However, employment has declined in both Sweden and Canada relative to the volume of production over the 1960–90 period and absolutely over the period since about 1970. The causes are technological: labour-intensive operations have been phased out, automated mills have low labour requirements, and logging, as well as mill work, has become more capital-intensive. Companies, claiming the need to be competitive on world markets, are attempting to establish a "flexible" workforce, but this may be perceived as union busting. One study of a New England town and current events in British Columbia indicate that while flexibility may be achieved, the price is paid in loss of worker loyalty.

Rural communities dependent on forestry have lost population as well as jobs, and the long-standing belief that sustained forestry would result in stable communities has been challenged. Since public subsidies for forestry have been justified in terms of employment and community stability, these trends may underlie the growing antipathy toward the forest industry.

Technological changes in other industries and socio-economic changes in northern societies also affect the forest industry and its markets. Finally, pulpmill effluence has been shown to cause severe deformities and genetic changes in fish. Dioxins were believed to be the cause, but scientific studies now indicate that the cause is naturally occurring acids and other chemicals from lignin emitted from mills in very high concentration. The recommended solution is to reduce or recycle effluent through secondary treatment before it is discharged into open water. Public opinion continues to blame bleaches, and the industry is influenced by public opinion and government regulations against the use of chlorines in the pulping process. While opinion may be incorrect regarding the specific cause, concern regarding the deleterious impact of effluent on fish is not misplaced.

3 Restructuring
the Northern Industry

In forestry as in most other industries, the development of capacities to mass produce goods led to increasing levels of concentration at each new stage of industrial development. A century ago, small, family-owned businesses and rural farms were overwhelmed by large companies that were able to form because of social and technical changes in their economic context. Then it was steam-powered cable logging and railroad hauling that made the difference to forest companies. Those with access to bank funding could adopt the new technologies and grow beyond the local market; those who could not tended to fall by the way, eventually taken over by their larger cousins. Some of the largest American firms at the end of the twentieth century emerged in its first decade as these changes gave them openings, including Scott, Mead, Crown Zellerbach (now Cavenham), Kimberly-Clark, International Paper/St. Regis, and Weyerhaeuser. European companies, no longer competing behind national borders, have a much larger domestic market than ever before. Stora Koppärberg, Svenska Cellulosa, Kymmene, Enso-Gutzeit, Metsä-Seria, and Feldmühle are now continental papermakers.

These same companies are now globe-trotters in search of cheaper raw material and new markets. They are flanked by formidable competitors with home bases in Japan, Korea, or Taiwan and emerging forest giants in Brazil, Chile, and Indonesia. They are accompanied by the service industry, machinery manufacturers, and engineering consultants seeking new markets linked to forestry.

European and North American companies differ in their approach to global changes. Historically, Scandinavian companies were resource

extractors for much of Europe, just as Canada was for the United States. They produced the raw material and pulp and shipped them to the more populous markets of France and Germany, where they were manufactured into lumber and paper. Unlike those in Canada, however, Scandinavian producers did not gain control of woodlands. Small woodlot owners still own half or more of the forests in European countries, foreign ownership never characterized the European resource industries to the degree typical of Canada, and huge integrated companies were not dominant in forestry. As well, European governments retained a more directive function *vis-à-vis* natural resources than governments in North America. These historical differences have enormous impact on how these northern regions deal with technological, employment, and market changes. Scandinavian countries began the restructuring process in the late 1960s and early 1970s, taking advantage of the economic and political changes in Europe as well as changes in technology. North American regions were slow to respond to change. They have tended to seek new frontiers rather than undertake fundamental restructuring.

The giant firms of an earlier era provided employment to growing numbers of wage-workers. Logging and milling were central activities in the rural regions of all northern countries. Now, because of automation, the globe-trotters offer much less employment in northern regions even while producing much more lumber, pulp, and paper. The decline in employment began before the development of plantation economies. The emergence of global competitors, however, has challenged northern employers to reduce labour costs even further. Flexibility is the key here: companies challenge unions to allow workers to undertake a wider range of jobs. The struggle over flexibility and employment levels is particularly heated in North America, where huge companies have control of woodlands as well as manufacturing plants, frequently have shallow roots in rural communities, and are embedded in the adversarial system of company-union bargaining. The transition in Scandinavia, and more generally in Europe, has been less bitterly contested. Governments in Sweden and Finland intervened early in the 1970s to institute worker retraining schemes and created incentives to other manufacturing industries to locate plants in forestry regions.

THE CONTEXT TO RESTRUCTURING: REGIONAL PROFILES

Regional profiles provide a sense of how the forest policies and practices of the past affect responses to present challenges.

Northern Europe

Forest renewal became a major national policy in Scandinavian countries at the turn of the century. In Sweden this policy included substantial new research undertakings, a National Forest Survey, and improved wood-processing operations. Sweden now has a second-growth forest at least twice the size as in 1900, though smaller than in pre-statistics eras and lacking many flora and fauna species of the original forests. The annual increment through the 1980s was reported to be 100 million cubic metres (MCM), while the annual cut was only 68 MCM. Official statistics reported an annual plantation of 500 million seedlings for commercial purposes (Andreason, 1978; Raumolin, 1985; Remrod, 1989). There is general acknowledgment that replanted species are inferior to the original cover and biodiversity is much reduced. Whether continued clearcutting on short rotations can be sustained in the southern regions is uncertain, and no long-term knowledge exists about the impact of clearcutting in the boreal region because it began there only in the 1950s. But even the critics agree that reforestation has been more seriously undertaken in Sweden and the other Scandinavian countries than in Canada.

A history of active intervention in forestry by government, combined with land-ownership practices that modify the power of large companies, allowed the Swedish government to intervene again in the mid-1970s, this time to deal with a recognized resource deficiency combined with mill overcapacity. Some analysts recommended even more intensive forest management, but forestry policy-makers recognized that this brings with it numerous ecological issues, including demands on water resources that more ecology-conscious citizens are unprepared to accept. Instead, the government imposed a law in 1976 that required companies to submit detailed plans, including clear indications of timber reserves to be used, for any mill expansions or new facilities. This law had the twin effects of reducing the rate of fellings and increasing the annual increment of standing timber.

Anticipating potential benefits of reduced tariff barriers for the entry of their manufactured products into the European Common Market, Swedish planners urged the forest industry to phase out less productive units, rationalize larger units to produce higher value-added products, and source more wood outside Sweden. To counteract rural depopulation as the restructuring was undertaken in the 1970s, companies worked with government officials to establish small industries and retrain workers in rural areas. The number of operating papermills was reduced from sixty-eight to fifty-six, but production between 1970 and 1985 rose by 12 per cent for pulp and 60 per cent for paper (Noble, 1986:54). During this restructuring, Sweden was able to maintain a strong presence in export markets by devaluing the krona in 1981 and 1982. It produced more newsprint and fine paper in the late 1980s than in the 1970s, but did so with less home-grown timber.

Swedish policy had plenty of domestic critics. In addition to those who condemned the reduced quality of the forest, there were others who believed that Sweden's industrial policy was contrary to its economic interests. For example, J. Hansing and S. Wibe (1993) argued that the rationing of timber in the 1970s resulted in excess supplies. Timber growth, in their view, was much higher than originally estimated and demand much lower. They decried the closure of older plants and delays or stoppages in the expansion plans for other mills; they were displeased with the import of increasing supplies of wood and argued that Sweden had lost its strength in forestry markets because of misguided laws that forestalled market forces.

Partly in response to such criticisms, but as well with reference to successful reforestation and establishment of Swedish companies elsewhere in Europe during the 1980s, a national committee was appointed to conduct a review of public forest policy. Its report was submitted in 1992, and a new forest policy came into force at the beginning of 1994. The new policy emphasizes environmental protection combined with expansion of traditional and more innovative wood industries. The objective is to harvest about 100 MCM per annum on a sustainable basis. With exceptions where unique ecosystems require special conservation measures, the policy rejects exclusive zoning for protected areas, timber utilization areas, and parkland. Sweden's new policy is to integrate commercial and non-commercial lands, maintaining a balance between environmental, recreational, and industrial values. The state regulatory regime has been reduced. State forest management has been assigned to a new public corporation, AssiDoman AB. This corporation now effectively owns 15 per cent of the productive forest land base and is the major producer of sawn timber and pulp and paper products, but 49 per cent of the share capital was sold to the public in 1994 (Haley, 1994; Skogsstyrelsen, 1994a and 1994b).

Investment in Finland's forest sector also declined after 1975, and restructuring was undertaken. Many sawmills closed, and older pulpmills were phased out. Pulp exports also declined, but new investments were made in more advanced papermills. Taking advantage of the new market conditions in Europe, Finnish papermakers exported increasing quantities of groundwood speciality papers. Forest economist Jussi Raumolin (1986b:96) notes that because Finland lacks both fossil fuels and ample hydroelectric power, the use of thermomechanical pulp has created yet another conflict between the industry – which would benefit from nuclear power capacity – and environmentalists. Despite the closure of older mills, there is still overcapacity relative to forest supplies in Finland, and while foresters still pursue the notion of more intensive management, the defects in that plan are becoming ever more controversial. At the same time as these changes occurred, Finnish engineering companies – always closely allied with the forest industry – became dominant in the production and export of machinery and equipment.

The European Community (Western Europe)

Western Europe lost much of its forest cover over several centuries, and population pressures have precluded extensive reforestation. However, after 1945 all countries introduced measures for reforestation and afforestation, both for commercial and conservation purposes, and although land area devoted to forestry has not changed substantially, the volume of wood per hectare has increased. The net annual increment for the EC countries between 1940 and 1980 was 127.6 MCM, 58 per cent of this in coniferous stock (E.G. Richards, 1987: table 4:7). By the late 1980s, Denmark, Ireland, the Netherlands, and the United Kingdom all had below 15 per cent of their land surface classified as forest and other wooded land; Belgium, France, the former Federal Republic of Germany (FRG), Italy, and Luxembourg had between 22 and 31 per cent so classified. Coniferous species predominate in Germany and Belgium; deciduous in France, Italy, and Luxembourg. These forests are not now capable of supporting large-scale wood-processing industries; instead, the companies draw on new fibre sources in southern Europe and other imported raw materials (E.G. Richards, 1987:101–3).

The former FRG undertook extensive reforestation and industrial reorganization in the immediate postwar period. A national policy included strict cutting rules, reallocation of fuelwood to industrial roundwood, and new forest laws that led to the improvement of degraded stands, intensive silvicultural management, enrichment plantings, reforestation programs, and effective control of forest damage incurred during and immediately after the war. Most forest land is owned by some 400,000 small farmers, whose economic survival has been a concern of forest policy throughout the postwar period. Since the mid-1970s, national roundwood production has been constant at about 28 to 30 MCM annually, insufficient for domestic needs. Damage caused by acid rain has affected up to half of all forests; with unification, and taking into account extensive damage in the former East Germany, the salience of the issue has increased (E.G. Richards, 1987:152–7).

After 1965, more than half of the FRG's total annual wood supply was imported, first from tropical countries and later from the Iberian Peninsula. Restructuring of the veneer and plywood industries took place as logging bans were imposed in tropical countries. The wood industry developed uses for wood residues and waste wood in the fibreboard industry as early as the 1940s and developed the particle-board industry in the 1950s. The FRG in the mid-1980s had some 3,400 sawmills, down from about 11,000 in 1950 but still far from a fully rationalized industry.

The paper industry was using waste paper as 40 per cent of its fibre source by the 1960s, and in the next decade its companies invested in North American and Scandinavian pulpmills to obtain new sources. In the mid-1980s, the FRG produced close to 10 million tons of pulp and paperboard, compared to Sweden's 8 million tons; it still had 175 mills to Sweden's 55 and Finland's 46, although by this time a number of its mills were owned by Swedish and Finnish companies. Anti-monopoly legislation inhibits further rationalization. Both the domestic and export markets are served by German producers, but while they sold over a third of their production in the mid-1980s (mostly in specialty papers), they also imported large quantities of pulp and bulk papers and were dependent on imports. Environmental concerns have imposed new restrictions on producers. Since 1980, competitors in all wood-product sectors have successfully challenged domestic companies on the German and European markets (E.G. Richards, 1987:156–64).

Other European countries, all with smaller forest holdings, demonstrate the same general patterns. The traditional division of labour within Europe has changed. Reforestation and afforestation have been undertaken in most countries, but the domestic wood supply is not capable of satisfying domestic market demand. More investment is going into higher value-added sectors, the northerners are buying into and taking over western European companies, and raw materials are being sourced in non-traditional regions.

The United Kingdom has increased its domestic wood supply through plantations on marginal agricultural land, of which 58 per cent of the 2.3 million hectares is privately owned; the remainder is owned and managed by the Forestry Commission. Even so, the UK is Europe's largest importer of forest products. In 1990 there were ninety-one paper and board mills and six pulpmills; total paper and board capacity was about 5 million tons, pulpmill capacity about 680,000 tons. Italy produces slightly more pulp, paper, and board than the UK. Like other European countries, it imports more than it exports and could not operate on its own timber reserves. Its paper industry was internationalized in the late 1980s and early 1990s, when Swedish and Spanish companies became active shareholders. France produces 73 per cent more pulp and 31 per cent more paper and board than the UK. Though imports exceed exports, the difference is not as marked as for Germany and the UK. Paper producers in France are well placed to capture the expanded European Community market, and Swedish, Finnish, UK, and American companies have all purchased shares.

The Iberian Peninsula

The most important non-traditional forestry region for Europe is its own southern region, the Iberian Peninsula. *Eucalyptus globulus* – introduced to Portugal as an ornamental tree in 1839 and then used in agriculture and the construction of railway ties and telegraph posts – was pulped there long before it became a widely marketed product elsewhere, but the Iberian Peninsula was not part of the northern trade before the 1960s. By the end of the 1980s, the region as a whole accounted for about 2 per cent of world production of pulp, nearly 7 per cent of the world-market chemical wood-pulp production and 5 per cent of world-market chemical wood-pulp exports, and some 2 per cent of world-production of paper. Spain and Portugal had become the largest world suppliers of eucalyptus pulp, at 1.7 million tons per year (Molleda, 1988). They were also becoming important consumers of paper products from European Community suppliers (Rolo, 1987; Oliveira, 1987).

Between 215,000 and 300,000 hectares of eucalyptus, representing about 11 per cent of the total forest area of Portugal, had been planted by the mid-1980s. These plantations became so controversial that eucalyptus gained the nickname "the fascist tree" amongst country dwellers. The critics claim that eucalyptus stands are causing desertification; companies counter that evidence on long-term lowering of the water-table is not established (Kardell, Steen, and Fabiao, 1986; see also chapter 8). Governments, in any case, continue to encourage investment. In total, the Iberian eucalyptus plantations cover about 850,000 hectares and produce about 7.5 MCM of wood per annum (almost all pulpwood). The top five producers control about 250,000 hectares. In Spain the state controls 150,000 hectares, and the remainder are held by private companies. The land area in plantations is growing by about 10,000 hectares per year and could eventually cover another 400,000 hectares in the peninsula (Molleda, 1988).

Portugal in the early 1990s had 16 pulpmills and 85 pulp-paper or paper and paperboard mills. Spain has 54 pulpmills producing virgin eucalyptus pulp and about 180 mills producing paper and paperboard, using this pulp together with waste paper and other fibres. Citizens in both Portugal and Spain have expressed alarm about eucalyptus wood exports to other European countries, and export controls were introduced at the end of the 1980s. Spain, and to a lesser degree Portugal, are also experiencing substantial increases in domestic-market demand for paper, and despite a growth in capacity for paper and board production, as well as market pulp, they have imported increasing quantities from manufacturing plants elsewhere in Europe. They are caught in the common dilemma of small countries with a successful raw or intermediate product to sell and high levels of foreign investment from companies with manufacturing components elsewhere.

The United States

The United States has the largest exploitable forest land area; it produces and consumes more wood than any other nation. About 280 million hectares were classified as commercial forest land and 14 million hectares as non-commercial forest in 1990. Coniferous species constitute about 57 per cent of the stock, with about 19 per cent Douglas fir and much of the remainder loblolly and short-leaf pines planted in southern states.

During the nineteenth century, American settlers devastated the forest, often with no commercial returns. The land was worth more than the trees, and old-growth forests were simply cleared and burned. Said the chief forester of the United States, "By the turn of the century the greatest, swiftest, the most efficient, and the most appalling wave of forest destruction in human history was ... swelling to its climax in the United States; and the American people were glad of it" (Pinchot, 1947:1). Michael Williams (1988:211–15), in surveying the early history of the American forest, calculates clearances in the range of 46 million hectares before 1850* and over 79 million hectares by 1909. He estimates an original volume of 5,200 billion board feet of standing timber reduced to between 2,000 and 2,800 billion board feet by the end of the nineteenth century. Biodiversity is another measure of change: Barr and Braden (1988:228) estimate that the United States once contained some 1,100 species of trees, of which 647 remained in 1970. Marion Clawson argues that the average volume of wood per acre in the early period of European colonialization was about double

* Original estimates in acres.

that of the contemporary forest (1983:197). Others contend that while this estimate may be accurate, the forests encountered by European settlers were already at their climactic stage and new growth was slow, also that reforested regions have greater capacity to ensure future supplies for industrial uses (which may also be true, but irrelevant to anyone concerned with non-commercial values in old-growth forests).

As transportation and other technologies created profitable opportunities for industrial forestry, companies formed in the northern states. The tradition of small woodlot owners never took root. Over the century of forestry operations in the United States and Canada, large companies have dominated the industry. In the United States, these companies usually owned private land and also logged by bid-contracts on national forest lands. Public forest lands, under either federal or state ownership, were intended to provide access to timber for commercial enterprises on a competitive basis. By the mid-twentieth century, large companies had integrated their operations, with American companies straddling the Canadian border. Logging, sawmilling, and pulping were typically undertaken by a single company or by subsidiary companies of a single parent. Sawmills and pulpmills were generally located at the same site, thereby facilitating the transfer of pulpwood chips.

The arguments in favour of large, integrated concerns are that they are more efficient wood collectors, are better able than small operations to fund new investments, have economies of scale and research and development capacities, have longer-term horizons and more stable labour demands, and are better positioned to establish international markets for their products. In the event of fibre shortages, they have built-in supplies. The arguments against these corporations are that they require enormous and guaranteed supplies of wood for their high-cost mills, they have such economic power that governments are unable to develop independent forest policies in changing conditions, they do not necessarily engage in research and development, they are capable of moving investments elsewhere, they modernize operations so that labour demands are constantly declining, and they dominate international markets. Arguments in favour of integrated mills won the day in North America – or better stated, arguments were created after the fact by way of justifying an organization that was designed by investors whose interests lay in large, integrated facilities. Economies of scale were important for the production of standardized long-run materials such as construction wood and pulp.

When the northern forests were gone, companies moved south; 40 per cent of southern forests disappeared between 1860 and the turn of the century. Then the industry moved west; the Pacific coastal states became the last frontier. As industrial forestry became entrenched on the coast, some reforestation took place elsewhere in the United States. Improved control of forest fires, reversion to forest of abandoned farmland, and declining demand for wood-based fuel and other wood products enabled older regions to regenerate forests.

Frederick Weyerhaeuser purchased 364,225 hectares of prime forest land in Washington State from the Northern Pacific Railway Company at the turn of the century. He subsequently put together nearly 800,000 hectares of timberland in Washington and Oregon, obtained at about $3 per hectare (M. Williams, 1989). Forestry was neither preceded nor supplanted by agricultural settlement in the Pacific region because of poor soils. Indeed, pioneer American lumbermen were disappointed when they discovered that, having cut a stand, they could not sell the land to farmers (Cox, 1983:21). Population density remained low, and forestry, together with mining and fisheries, continued as central economic activities into the late twentieth century. Even so, the last frontier was producing wood at such a rate that by 1920, production of lumber constituted some 30 per cent of the national total. Estimates of remaining reserves at that time indicated that cutting exceeded restocking in Washington, and the industry moved more of its operations to Oregon.

Technological changes made it possible to access ever more distant areas and to utilize new species from time to time, thus changing the base estimates of available timber. As the Douglas fir declined in consequence of rapid rates of harvesting, the coastal industry sought out new wood sources. Hemlock, for example, was not regarded as commercially valuable until the 1950s, lodgepole pine until the 1970s. Even as the base moved upward with the technological capacities of the industry to use substitute species, areas close to mills and towns became deforested. Since the mid-1950s the coastal region has produced about half the U.S. total of logs (M. Williams, 1989:289). This high rate of logging took place in the last sizeable old-growth forests; moreover, in forests with trees much larger and a biodiversity much richer than in most of the forests already depleted. The Pacific coast industrial expansion exceeded the sustainable carrying capacity of timber reserves by the 1960s. Logging continued nevertheless.

Of total commercial timberland in the United States, about half is in national or other public-land categories, half in private ownership. A higher proportion of timberland in the Pacific than in other regions is under federal jurisdiction. During the deep market slump of the early 1980s, timber sales by the Forest Service from federal lands indicated that over 40 per cent of sales in the four western regions, and nearly all of those in the Rocky Mountain regions, generated revenue below the cost of roads and other infrastructure required for logging (Slocum, 1986). Subsidies affect the price of logs, the price of lumber, and the incentives to develop alternative technologies. Wildlife defenders call these practices "subsidized destruction." One must ask why such practices would be deemed acceptable by the U.S. Forest Service; the only apparent answer is that its role is to sustain the industry, even if that depletes the forest and diminishes the much-vaunted role of market forces. On the other hand, sales at depressed prices allowed some industrial activity to continue: privately owned timber was being sold as roundwood to Japan.

Public concerns about declining forest reserves in the United States come up against the rights of private property owners; no government has the power to demand that forest owners protect their lands. California tried to enact regulations to restrict logging on private timber land in 1991, arguing that there was an environmental emergency. The regulations were thrown out at the County Superior Court level on the grounds that an impending environmental crisis was based on "speculation, concerns and suspicions, not evidence." It is difficult to prove that a crisis is about to occur, and the head of the Board of Forestry in California said of the ruling: "The crisis has been compounded by the political deadlock on forestry legislation. Our enemy now is not politics; it's time" (*Fresno Bee*, 20 Feb. 1992).

Opposition to clearcutting in the remaining old-growth forests on federal land in the Pacific Northwest finally concentrated on the spotted owl. Environmentalists took the federal government to court for failing to comply with the requirements of the Endangered Species Act regarding measures to protect wildlife. A study known as the Jack Ward Thomas report proposed that 2.3 million hectares of national forest land be preserved as owl habitat in Oregon, Washington, and northern California. This would have nearly halved the annual federal timber sale in the region, and the Forest Service estimated that it would cost the industry some 28,000 timber jobs over the 1990s. In 1991 the U.S. Fish and Wildlife Service said that it would limit logging on 4.7 million hectares in deference to the owl; it subsequently reduced the figure to 3.1 million hectares. Environmentalists won the ensuing court battle. New timber sales in old-growth forests were stopped pending provision of acceptable plans by the Forest Service for habitat protection. Cutting in old-growth forests on U.S. Bureau of Land Management property in Oregon was stopped in 1992.

The United States government has since appointed a team of scientists to examine the entire issue of forest ecosystem management,

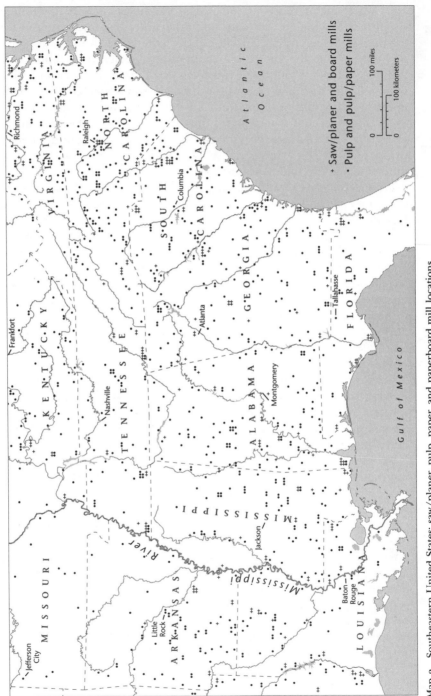

Map 2 Southeastern United States: saw/planer, pulp, paper, and paperboard mill locations

including the biological, social, economic, and political aspects (U.S., Department of Agriculture, 1983). Forty-eight alternatives were set out in the review team's published study, and ten options examined in detail. One of these was chosen by the U.S. president, and forest policy in the Pacific Northwest has subsequently changed accordingly, though not without controversy. The overall changes involve reduction of logging on public lands, substantial investment in reforestation, and forest renewal. The process and the report are remarkable departures from previous land-use conflict resolutions, and they will have long-term impacts on allocation and decision-making over resources in the United States. A court decision of 1994 upheld the plan, though further court challenges were still likely.

The plantation "third" forest" of the southeast is now the growth region. It covers nearly 74 million hectares and represents 58 per cent of the region's land area. One-third is covered by pine (primarily loblolly), 15 per cent mixed pine-hardwood stands (primarily oak), and the remainder hardwood. Several major U.S. companies have private lands in the southern states, where they rely on plantation pine fibre. Parsons and Whittmore, for example, started up a new kraft pulpmill in Alabama in 1991; it also has plantation holdings in the Iberian Peninsula.

Canada

Since Canada has 53 per cent of its total land area covered by forests and much of the remainder is barely habitable wildlands, it is not surprising that the Canadian forest industry has played a major role in the country's development. Forests are under provincial jurisdiction, and 94 per cent of forest land is still publicly owned. Of 416 million hectares of forest land, 46 million hectares (about 11 per cent) are protected as wilderness. About 112 million hectares are classified as commercial and are deemed productive and accessible.

Despite its rich resource base and low population density, Canada's record of forest management is appalling. Jussi Raumolin (1986b) undertook a comparative study of the forest sectors in Finland and eastern Canada from 1920 to 1980. He concluded that Finland, with only three main tree species as the country's chief asset, had created a carefully managed forest industry, while eastern Canada had built a relatively incoherent industry with vast lands, many resources, and diverse tree species. Finland developed sustained-yield forestry very early and always engaged in botanical research. Canada did neither. Finnish landholding was in private, but also very small, woodlots. Canada's was in public land and quickly dominated by large companies with harvesting rights. Foreign investment was welcomed in Canada, shunned in Finland.

The history of the industry in Canada is similar to that in the United States, but there was no interlude of southern activity, since there is no comparable region in Canada. From east to west, from the Atlantic provinces on through Quebec and Ontario, the industry finally arrived at the Pacific coast while that area was still a fur-trading colony under the control of the Hudson's Bay Company. It became established there in the form of small, portable sawmills before the turn of the century. Very soon these were supplanted, and large companies took over.

The deficiency of traditional timber supplies in the Atlantic region, Quebec, and Ontario was well documented in the early 1970s (F.L.C. Reed and Associates, 1974; Canada, Dept of Industry, Trade and Commerce, 1978; Science Council of Canada, 1983). The British Columbia government maintained that its renewal policies were sustaining the forest; but illusion, more than reality, was sustained because as the coastal forests were cut, industry moved into the hitherto untapped northern interior regions.

With relatively copious wood supplies, domination by large, integrated companies, many of them foreign-owned, and historical reliance on the United States market, the Canadian forest industry failed to come to grips with global changes in the 1970s, or even in the 1980s. It had done little research over the boom period and relied on the production of standard wood products. Automated technologies were put in place and larger mills constructed, but the range of products remained stable. The volume of timber felled rose over the 1950–90 period, and the production of logs, wood chips, lumber, and pulp all expanded, but employment grew slowly to the end of 1979 and then began to decline, the rate of decline increasing rapidly with the two major recessions of the early 1980s and early 1990s. Belatedly recognizing global change, the Canadian industry began to phase out older mills in the 1990s. More value-added wood products for niche markets are being developed, and newsprint and low-grade market pulpmills are on the block if there are any buyers. British Columbia provides a case history of both the dependent, peripheral state and the current restructuring (documented in chapter 4).

Restructuring is not what the move into boreal forests is about, however. On the contrary, this development re-creates peripheral resource regions for a more advanced economy. While the American industry moved south, the Canadian industry – or rather, companies from Japan, the United States, Taiwan, and Canada – moved north in the 1980s. Megaprojects were given the green light in Alberta in 1986, and public subsidies were eagerly provided to all mill-construction companies who proposed to log in the province's 220,000 square kilometres of productive commercial forest. This represents a third of the province, an area equivalent to the United Kingdom.

Two of six major pulp projects started in Alberta during the early 1990s are backed by Japanese investors (Daishowa-Marubeni International and Mitsubishi). The largest is Alberta Pacific (Al-Pac), majority owned by Mitsubishi (45 per cent), with minority shares held by Canadian shareholders (14 per cent) and three other Japanese investment groups (the remainder) (Pratt and Urquhart, 1993; McInnis, 1994; Richardson, Sherman, and Gismondi, 1993). This company received subsidies for mill construction amounting to about $275 million. Public costs for infrastructure amounted to some $75 million, and up to $125 million was established as a stand-by debenture for future expansion (McInnis, 1994). The mill has a capacity to produce 1,500 tons of bleached kraft pulp daily, or over over 500,000 tons annually, consuming 3.2 MCM of timber per year. Its aspen wood supply arrives in double-bedded trucks at a rate of one load per 90 seconds (my field notes). The wood is logged in the 73,430 square kilometres covered by Al-Pac's twenty-year renewable forestry management agreement, an area the size of Hokkaidō but with twice as much productive forest. The Daishowa-Marubeni bleached kraft pulpmill on the Peace River, completed in 1990, consumes 1.8 MCM annually. Governments paid infrastructure costs of some $65 million, and the company was granted harvesting rights to some 40,000 square kilometres.

These two mills, and another four or five up or in construction, do provide employment. But for the Daishowa mill the project engineer, H.A. Simons, notes that it is "engineered to run on minimum manpower, using a total of about 300 employees, versus 500 to 600 with comparably sized operations." The Al-Pac mill employs about 365 people on site, with another 660 employed in wood harvesting, primarily on a seasonal basis (H.A. Simons Ltd, "Peace River Pulp Mill," 1990; quoted in McInnis, 1994).

Opposition to these land grants, subsidies, mills, and especially the vast clearcutting of the boreal forest has been massive, articulate, backed by research, and persistent – also unsuccessful. The Al-Pac project was initially rejected by an independent expert panel, the Alberta-Pacific Environmental Impact Assessment Review Board, which ruled that no mill should be built until assurance was gained that the Athabasca River could absorb the effluent and that harvesting would occur in sustainable quantity. The Alberta government effectively ignored the board's ruling. The company's studies gave no comfort to the sceptics, for they had serious omissions and contradictions. Its study of the impact of effluent on fish consisted of a 96-hour immersion test and concluded that since 80 per cent of the fish survived, all was well. As one local farmer observed, "Just because they're alive doesn't mean they're fit to eat or can even reproduce again" (Nikiforuk and Struzik, 1989:67). The Alberta government's own studies of the operations of two existing pulpmills at Hinton reported a deleterious impact on the oxygen levels, and thus on aquatic life, in the Athabasca River. The Saskatchewan Research Council published a report suggesting that the existence of the boreal forest is more important for its oxygen-generating value than it would be as a source of pulp fibre. There are widespread fears that regeneration of the forest – if large-scale reforestation is even possible – would take the form of a monoculture aspen plantation. Although native and Métis peoples were not of one voice (*Globe and Mail*, 1 Dec. 1989: A8), many spoke against the developments. Treaty No. 8 Indian nations, whose land is contiguous to the developments, have expressed opposition on the grounds of health concerns. The Lubicon people, who have been trying to negotiate their land claims for many years, staged demonstrations in opposition.

To explain why a provincial government would proceed with development against such vociferous opposition, one refers to the history of resource dependence and of reliance on huge corporations, the long-standing acceptance of foreign investment, previous experience with public subsidies for private companies (Pratt and Urquhart, 1993), the lack of Canadian models for alternative kinds of government interventions, and above all, a lack of respect for the northern forest and its life-forms. The Alberta government apparently knew of no other way to cope with a changing global environment.

Including the new mills in the boreal region, Canadian capacity for pulp production is enormous. Up to twelve mills are projected for the Alberta-Manitoba region. Twenty-four are already in place in BC; as well, expanded capacity is planned at these mills, and several new ones are under construction in that province. Older mills throughout Ontario, Quebec, and the Atlantic provinces are also expanding (M'Gonigle, 1990:6). The stated rationale is an anticipated annual increase of 2 to 3 per cent in consumption of kraft pulp (Woodbridge, Reed, 1988). The objective of all this activity is to enlarge the territory for forestry (existing territory having already been depleted), restructure production units to be "more efficient," and create economies of scale still larger than those already in place; thus to become competitive again in global forest markets. The boreal forest is essential to this endeavour because it provides an expected sixty years of cutting opportunities, allowing Canada (though half of the companies, and particularly the very large enterprises, are owned elsewhere) to stave off the day when Brazil and Indonesia will dominate the pulp markets.

In the early 1990s, the federal and some provincial governments, still supporting northern expansion, publicly acknowledged timber deficiencies in the more traditional zones, rural communities in decline, and a history of lost opportunities. The process of what was termed "renewal" began with a number of glossy booklets about sustainable forests, environmental innovation, and green plans. Round tables were established. Across the country, identified "stakeholders" had their say at hastily called meetings. Companies, environmentalists, and government leaders devised a "blueprint for change" called the National Forest Strategy (1992). It expresses concern for non-timber values, aboriginal interests, small communities, and public participation in decision-making. It also emphasizes the need to change in order to sustain an economically viable industry.

The Former USSR and Eastern Europe

The former USSR has greater forest land area, more timber stock, and more apparently exploitable forest than Europe or North America. However, there is uncertainty about the state of these forests. Disregard for conservation was documented by historians and earlier writers reporting on the state of forests under the czarist regime. Between 1800 and 1914, an estimated 70 million hectares of forest were cleared in European Russia (Barr, 1988:244). Deficiencies in management, a deteriorating quality of the growing stock, the absence of conservation policies, and wasteful exploitation of accessible timber in the Soviet period have also been documented (Pryde, 1972, 1983; Sutton, 1975; Komarov, 1980; Barr, 1988; Barr and Braden, 1988). The Soviet Far East may have more abundant timber supplies than European Russia. One study of sawlog supplies in that region indicates that about a third of the commercially exploitable forest, an area of about 73 million hectares, is accessible and sufficiently stocked to be economically viable (Cardellichio, Binkley, and Zausaev, 1990; Backman and Waggener, 1991).

THE IMPACT OF CHANGES ON LEVELS
OF CONCENTRATION

Jaakko Pöyry, the owner of a global company based in Finland, who reappears throughout this saga on the forest industry in accounts of development as far removed from that country as the boreal forests of Alberta and Thailand's impoverished tropical land, notes in a 1988 publication that while ownership of the forest industry is becoming more concentrated in major producing regions, there are not yet many multinational corporations (Aurell and Pöyry, 1988). In fact, there are a number of multinationals spanning the global industry, but Pöyry has a point: in the major producing regions, there are large firms not tied in to global investment groups.

Table 3.1 lists the top forty pulp and paper companies ranked by sales in 1992 by *Pulp and Paper International*, with the number of countries where they have subsidiaries that are separately listed from the main company. It shows American and European firms still in leading positions, but it also indicates the growing strength of Japanese and other Asian competitors. Offshore investments by Japanese companies are frequently understated because they have consolidated accounts or because they operate through consortia. The largest American pulp and paper companies have extensive private timber holdings; among the giants are International Paper, Kimberly-Clark, Georgia-Pacific, Stone Container, Scott Paper, James River, Champion International, and Weyerhaeuser, with Union Camp, Mead, Westvaco, Boise Cascade, and Louisiana Pacific not far behind. Of the integrated companies, most of the same ones occupy the top ranks. Identified investment locations outside the United States run as high as twenty-five and twenty-one for International Paper and Scott Paper respectively; others ranged from two to twenty in 1992. Eight companies controlled about 40 per cent of the American paper market in the early 1990s. This is a greater degree of concentration than was the case a decade earlier, but not as much as in many manufacturing industries. These companies have international interests, and restructuring is coincident with globalization. American firms with subsidiaries in Canada include many of those named above (International Paper, Weyerhaeuser, Champion International under the name Weldwood, and many others). Firms currently under majority Canadian ownership and control include Noranda and Canadian Pacific, both in shaky financial condition during the 1980s and 1990s.

Swedish and Finnish companies began to diversify their holdings in Europe long before northern European countries joined the European Community. The corollary was increased concentration and integration within the northern industry. The major international paper companies of Sweden are Stora Koppärberg, Svenska Cellulosa, and Billerrüd-Uddeholm. Stora has established large-scale production units for printing and writing papers within the EC and has acquired eucalyptus plantations in Portugal. As well, it has expanded to North America. Like other Swedish companies, Stora created a new strategy

Table 3.1

Top forty pulp and paper companies with international holdings, ranked by sales, 1991, as listed by *Pulp and Paper International* (September 1992)

Rank by sales			No. of countries where investments located
1988	*1992*	*Company*	
2	1	International Paper (US)	25
7	2	Kimberley-Clark (US)	18
3	3	Georgia-Pacific (US)	2
6	4	Stora (Sweden)	9
15	5	Stone Container (US)	8
9	6	Scott Paper (US)	21
5	7	James River (US)	13
	8	Arjo Wiggens Appleton (UK)	10
18	9	Svenska Cellulosa (Sweden)	20
8	10	Champion International (US)	3
1	11	Weyerhaeuser (US)	3
10	12	Oji (Japan)	3
13	13	Jujo (Japan)	1
19	14	Honshu (Japan)	2
	15	SIB/MS Holdings (US)	1
20	16	MoDo (Sweden)	4
	17	United Paper Mills (Finland)	11
36	18	PWA (Germany)	7
24	19	Daishowa Paper (Japan)	2
25	20	Kymmene (Finland)	4
22	21	Union Camp (US)	5
4	22	Fletcher Challenge (NZ)	7
11	23	Mead (US)	10
34	24	Westvaco (US)	2
	25	Daio Paper (Japan)	1
26	26	Jefferson Smurfit Group (Ire)	13
12	27	Boise Cascade (US)	2
28	28	Enso-Gutzeit (Finland)	6
14	29	Noranda (Canada)	2
30	30	Buhrmann-Tetterode (Netherlands)	4
	31	Sappi (South Africa)	4
35	32	Rengo (Japan)	4
	33	Jamont (Belgium)	11
39	34	Metsä-Serla (Finland)	5
17	35	Sanyo-Kokusaku (Japan)	1
	36	Temple-Inland (US)	2
27	37	Canadian Pacific (Canada)	2
	38	Mitsubishi (Japan)	1
	39	Haindl (Germany)	3
	40	Cartier Burgo (Italy)	1

Source: Pulp and Paper International, 1992. Top 150 listings (September 1988, September 1992).

in the 1970s, selling off its holdings in steel and other industries in order to restructure its forestry operations. It modernized its paper machines, merged with or took over supplier firms, and increased capacity to produce higher-value end products for the EC market. Other large companies took similar action, while less able companies failed to attract funding for renewal.

At the end of the 1960s, Finnish pulp and paper companies had half or more shares in three British companies, one company each in Italy and Germany, and two in France. Their holdings increased in the 1970s and 1980s. By the end of the 1980s, they had acquired controlling interest in thirty-two companies in Europe; thirteen in the UK alone and others in Germany, France, Italy, and Spain (Raumolin, 1988:15–16). Finland had became the leading exporter of printing and writing paper in the world, marketing abroad nearly 30 per cent of the total by 1985. In contrast, Canada exported less than 8 per cent, slightly more than Sweden and France.

The largest company resulting from mergers is the Finnish concern Kymmene Oy, which includes the former companies Kymi-Stromberg Oy, Oy Kaukas Ab, and Oy Wilh. Schauman Ab. Kymmene now has subsidiaries in Germany, the UK, and France, and is one of the leading manufacturers of fine, printing, and writing papers in Europe. The state-owned Enso-Gutzeit, fully rationalized since 1980, has expanded into North America as well as Europe. Other Finnish companies have likewise branched out into Europe, with paper, newsprint, cardboard, linerboard, and tissue paper subsidiaries in the UK, Spain, Denmark, Greece, Germany, and France (Raumolin, 1988: table 3:25). These companies are integrated, large, and productive of high value-added and diversified products. State participation, always present in the Scandinavian industry, has increased in the form, mainly, of joint ventures with private capital; indeed, in Raumolin's view, the line between public and private is now obscure (1988:17). Finnish machinery companies have also steadily expanded. Valmet Paper Machinery has, since 1980, acquired holdings in the United States, Germany, Canada, and Italy, and it sells its wares throughout the world.

Major companies of the former FRG include Feldmühle, now allied with Stora of Sweden and half-owners of the Beghin Corbehem mill in France. Waldhof-Aschaffenberg (PWA) joins Feldmühle in the top fifty ranks for global companies as measured by sales. Both have been active in acquiring new capacities through plantations and pulpmills in Spain. Ranked well below these, but still sizeable paper companies are Haindl and Hanover Paper (*PPI*, Aug. 1988:70). As noted above, Nordic companies have been actively purchasing shares or buying out German companies; as well, International Paper (US) acquired voting shares in Zanders Feinpapiere, one of West Germany's leading papermakers.

Bowater and Reed survived the general restructuring of the industry in the UK largely because they had already moved investments outside the Islands. One of the spin-offs from that exit was the creation, in 1986, of UK Paper, subsequently taken over by the Finnish company Metsä-Seria. Wiggins Teape Appleton (owned by British American Tobacco) merged in 1990 with Arjomari-Prious of France and created one of the highest-value papermaking companies in the European Community. Other firms, both British and foreign, sold shares in the same year: divestments characterized the activity of two U.S. firms, Georgia-Pacific and James River (*PPI*, July 1991:31).

During the early 1990s, changes in ownership of French firms included Kymmene (Finland), which acquired newsprint capacities in Chapelle Darblay; NCB (Sweden), which raised its holdings in Société Française Charfa; UPM (Finland), which purchased a major large sawmill firm, Ferdinand Braun et Cie, in Alsace; SCA (Sweden), which bought out Norembal; Sonoco (U.S.), which acquired shares in Lhomme; International Paper (U.S.), which purchased shares of Smurfit and Perrier; and Amcor (Australia), which acquired shares in Thierval Sacoc (*PPI*, July 1991:36–7). Most German and French companies import raw materials, so it is not surprising that they have a presence in Spain, Portugal, and sometimes in South America or Asia, but they are not as globally oriented as Japanese firms.

Six pulpmills are located at the mouths of large rivers on the coast of Portugal, including a major operation by the Portuguese-Swedish-owned cellulose company Celbi. The Portuguese government, through Portucel, is the major investor in two mills, Caima owns two, and Arjo Wiggins Appleton, which has a 50-per-cent interest in Soporcel, owns one. Their plantations are located on land previously used for agriculture; Celbi's plantations cover some 30,000 hectares. In Spain, the government company, Ence, had two mills, and Scott Paper had one; Arjo Wiggins Appleton through its company CEASA and Sniace each had one mill in 1987. Ence subsequently sold one of its mills to Scott-Iberica, and Scott indicated its intention to build another mill. The Ence mills were producing 240,000 tons per year and 160,000 tons per year respectively by the end of the 1980s. The Finnish company Enso-Gutzeit, working with La Papelera Española, is considering another large mill in northern Spain, and Feldmühle of Germany and Tampella of Finland are also investing in Spain (Oliviera, 1987; Knight, 1988; *PPI*, July 1987:7, Jan. 1989:7, Mar. 1989:15).

Asian companies have invested in operations throughout Latin America and Asia, but many of them are organized in the forms of

Figure 3.1

Pulpmills: capacity, consumption, and production by geographical region, 1992

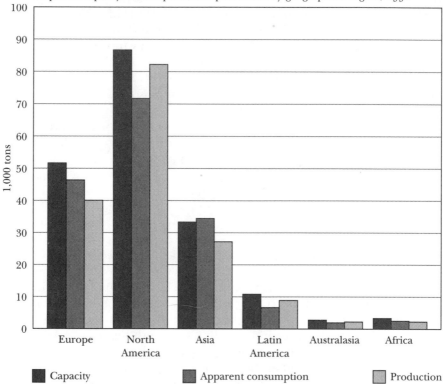

Source: Pulp and Paper International., July 1993: annual review section, 23–5.

joint ventures and consortia and may be less easily identified than the corporate forms characteristic of American companies. Those listed in *Pulp and Paper International*'s annual ranking in 1992 are Oji (Ohji), Jujo, Honshu, Daishowa, Daio, Sanyo-Kokusaku, and Mitsubishi, all of Japan. Since that time, Jujo and Sanyo-Kohusaku have merged to become Japan's largest pulp and paper company. Japan's companies are discussed in chapter 5.

Apart from the multinationals, there are new and non-traditional companies throughout the developing countries, and for the most part these, while often highly concentrated within a single country, have not yet expanded beyond domestic borders. These include Klabin and Suzano of Brazil, CMPC of Chile, and Indah Kiat of Indonesia. In other cases, the companies are majority owned by domestic investors but controlled or managed by Japanese, European, or American compa-

Table 3.2

USSR, Europe, and North America: industrial roundwood production, imports, exports, and consumption in selected countries, 1991 (1,000 m³)

	Production	Imports	Exports	Consumption
WORLD	1,599,272	127,074	123,474	1,602,872
FORMER USSR	274,300	100	14,618	259,782
EUROPE	284,018	44,700	30,527	298,190
Austria	13,989	6,441	1,291	19,139
Belgium/Lux.	4,510	4,091	2,108	6,493
Former Czech.	15,097	162	682	14,576
Finland	31,107	5,513	359	36,261
France	34,310	2,750	5,843	31,217
Germany	40,369	3,012	10,301	33,080
Hungary	3,500	1,217	1,398	3,319
Italy	4,088	6,684	52	10,720
Norway	10,077	1,592	1,044	10,625
Poland	14,334	74	845	13,563
Portugal	10,583	793	527	10,850
Romania	12,773	54	0	12,827
Spain	15,282	3,090	148	18,224
Sweden	47,300	4,977	1,173	51,104
Switzerland	3,370	756	1,245	2,881
United Kingdom	6,129	265	594	5,800
Former Yugoslavia	8,213	1,152	884	8,481
NORTH AMERICA	588,878	7,589	33,074	563,383
Canada	171,215	5,419	3,265	173,369
Mexico	7,763	151	0	7,914
United States	409,900	2,019	29,809	382,110

Source: FAO, 1993.

nies. The New Zealand firm Fletcher Challenge was a major player on the global stage for a decade or so, but since the late 1980s it has suffered fiscal problems and divested itself of many holdings.

MARKETS CIRCA 1990

Market data for the European Community countries and Asia show a common pattern: countries consume much more than they produce. Paper imported from Nordic countries and pulp from Spain supplement the European diet. Roundwood and lumber are imported as well. Japan produced about 3 million tons less pulp than it consumed in 1992, and it consumed virtually all of its own paper and paper-

Table 3.3
USSR, Europe, and North America: sawnwood production, imports, exports, and
consumption in selected countries, 1991 (1,000 m³)

	Production	Imports	Exports	Consumption
WORLD	457,477	86,172	87,529	456,120
FORMER USSR	75,500	175	4,780	70,895
EUROPE	81,709	33,537	23,878	91,368
Austria	7,234	716	3,991	3,959
Former Czech.	3,621	31	910	2,742
Finland	5,983	63	4,288	1,758
France	10,988	2,287	1,179	12,096
Germany	15,158	5,045	1,278	18,925
Norway	2,313	422	644	2,091
Poland	3,205	131	288	3,049
Romania	2,456	0	269	2,188
Spain	2,898	1,833	130	4,601
Sweden	11,536	170	6,941	4,765
United Kingdom	2,189	7,280	66	9,403
Former Yugosla.	3,204	75	793	2,486
NORTH AMERICA	158,629	30,247	46,505	142,371
Canada	52,040	1,486	36,980	16,546
Mexico	2,696	626	91	3,231
USA	103,893	28,135	9,434	122,594

Source: FAO, 1993.

board. Data are not provided on its production of roundwood because
it does not produce for international markets, though it does purchase
logs and lumber (see chapter 5). Other Asian countries have similar
net import balances. By contrast, Latin American countries produce
more than they consume and are net exporters. While southern hemisphere producers are not yet major exporters on world markets
(except for Indonesia in plywood), their share of markets has steadily
grown. This is especially true for market pulp.

Canada exports a large part of its forest produce, with dollar returns
three to four times those for the United States, twice those for Sweden
and Germany, and six times those of France. It exports chemical wood
pulp, principally to the United States and Japan, but as the world's
largest exporter of sawn lumber and a major exporter of newsprint,
it relies much less then the United States on the export of sawlogs.
Canada neither produces nor exports substantial quantities of fine
papers.

Figure 3.2
Paper and paperboard mills: capacity, consumption, and production by geographical region, 1992

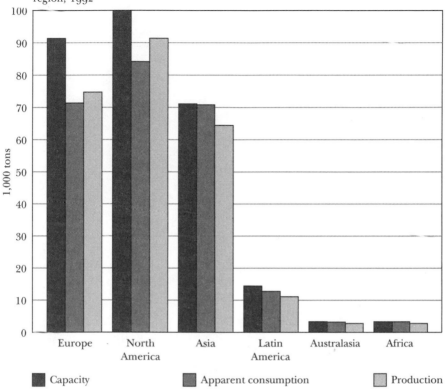

Source: *Pulp and Paper International*, July 1993: annual review section, 23–5.

The United States produces more of virtually everything than any other country. It exports more than European countries, but most of its produce enters its own high consumption market, and most of Canada's exports enter that same market. Measured by the dollar value at the end of the 1980s, the principal U.S. forestry exports were chemical wood pulp, fine papers and paperboard, and softwood sawlogs and veneer logs. Trade in hardwoods from the United States to Asia-Pacific markets has emerged since 1975. Prior to that time, Canada purchased most U.S. exports and Europe the remainder (Araman, 1988). This increase in trade with Asia is largely a function of the decreased availability of hardwoods from Indonesia, Malaysia, Singapore, and the Philippines, as one by one these countries imposed logging bans or log-export bans for tropical hardwoods. Japan, China,

Table 3.4
Number of pulpmills, capacity, consumption, and production by geographical region
and selected countries, 1992 (1,000 tons)

Region	Number of mills[1]	Capacity	Apparent consumption	Production
EUROPE	391	51,702	46,342	40,675
Nordic[2]	116	22,891	15,955	20,123
Germany	42	2,473	5,894	2,240
France	22	3,345	4,177	2,609
Italy	22	930	3,009	511
Spain	20	1,750	1,350	1,530
Portugal	8	1,670	621	1,592
East Europe	130	15,200	9,049	9,035
NORTH AMERICA	233	86,692	71,685	82,123
Canada	26	27,343	14,392	22,841
USA	207	62,349	57,293	59,282
ASIA	551	33,179	34,462	27,310
Japan	55	15,130	14,306	11,200
China	176	11,000	12,588	11,985
Indonesia	13	1,230	1,158	821
South Korea	5	704	1,841	323
Taiwan	3	440	1,062	368
Thailand	3	158	400	158
LATIN AMERICA	166	10,784	6,663	8,785
Brazil	105	5,699	3,810	5,368
Chile	9	1,903	472	1,681
Argentina	20	871	604	620
Mexico	11	1,072	908	560
AUSTRALASIA	17	2,440	1,922	2,270
Australia	10	1,000	1,202	982
New Zealand	7	1,440	720	1,288
AFRICA	30	3,298	2,270	2,896
South Africa	9	2,300	1,770	2,320
Total World	1,388	191,095	163,344	164,059

[1] Number does not include integrated pulp, paper, and paperboard mills. These are included in
table 3.5.
[2] Nordic includes Sweden, Finland, and Norway.
Source: Adapted from Pulp and Paper International, July 1993: annual review section, 23–25.

Table 3.5
Number of paper and paperboard mills, capacity, consumption, and production
by geographical region and selected countries, 1992 (1,000 tons)

Region	Number of mills	Capacity	Apparent consumption	Production
EUROPE	1,581	91,242	71,346	74,602
Nordic[1]	115	22,190	3,671	19,209
Germany	237	14,616	15,646	12,930
France	151	9,000	9,092	7,697
Italy	307	6,660	7,631	5,961
UK	89	5,730	9,568	5,128
Spain	149	3,800	4,869	3,448
Portugal	81	1,139	856	958
East Europe	316	17,936	9,671	10,072
NORTH AMERICA	656	99,898	84,074	91,323
Canada	112	19,365	5,317	16,594
USA	554	80,533	78,757	74,729
ASIA	1,620	71,046	70,635	64,156
Japan	444	33,071	28,318	28,322
China	250	16,000	19,464	17,251
Indonesia	53	3,304	1,845	2,263
South Korea	132	6,008	5,383	5,504
Taiwan	167	4,200	4,178	3,977
Thailand	35	1,245	1,562	1,245
LATIN AMERICA	446	14,388	12,837	11,007
Brazil	181	6,066	3,962	4,915
Chile	15	558	541	508
Argentina	80	1,340	1,427	1,027
Mexico	67	3,935	3,519	2,825
Columbia	23	774	843	628
Venezuela	14	817	801	659
AUSTRALASIA	26	3,400	3,330	2,804
Australia	20	2,500	2,755	2,072
New Zealand	6	900	549	732
AFRICA	89	3,398	3,395	2,615
South Africa	19	2,200	1,554	1,814
Total World	4,418	283,371	245,617	246,507

Source: Adapted from Pulp and Paper International, July 1993: annual review section, 23–5.
[1] Nordic includes Sweden, Finland, and Norway.

Korea, and Taiwan became major purchasers through the 1980s. Most of this trade takes the form of lumber, destined for the production of furniture and interior trim. The Taiwanese furniture makers prefer oak, because their chief market consists of U.S. consumers, whose preference for oak cabinets is established. The wood thus moves full circle.

During the late 1970s and through the 1980s, logging companies in the Pacific Northwest states sold increasing quantities of softwood logs to Asia. The sellers were companies with private landholdings, such as Weyerhaeuser. Profits were greater from the sale of logs than from finished lumber in the then-depressed American housing market. According to one analyst, Keith Ervin (1989), Weyerhaeuser enjoyed a 62 per cent profit on log exports, compared to a potential 10 per cent on lumber in 1979. A ban on the export trade in timber culled from federal lands reduced the total U.S. export, but did not stop the export from private lands. Since quantities of exports from private timberlands in the U.S. escalated after 1979, the profits must have been considerable. The log trade from the United States has a bearing on the trade war between Canada and the United States, described below. It also has an impact on the sustainability of Japanese rural villages and is reconsidered from their point of view in chapter 5.

Border Disputes in a Changing Market

Canada and the United States had a symbiotic relationship, albeit an unevenly balanced one, for the better part of a century. As supply and market conditions changed, the advantages of the symbiosis also changed for the United States. The first volley in a trade war came in the form of a 35-per-cent duty imposed by the United States government on cedar shakes and shingles imported from Canada during the early 1980s. British Columbia has 75 per cent of the remaining growing stock of western red cedar, the raw material for shakes and shingles. Unlike the lumber and pulp sectors, this relatively small industry was not integrated and was largely under Canadian ownership. The major market for the product was in the United States, but there was growing demand in Japan. The import tax appears irrational without the further information that BC had maintained an export ban on "baby squares," the raw material, and American producers could not obtain sufficient supplies to compete with BC producers on the Japanese or American markets. Pressure was applied to remove the export ban, but it was retained. American producers did not obtain new supply sources in consequence of the added cost for Canadians of doing business in the United States. Indeed, Canadian producers shifted more of their exports to the Asian market.

The second change came in a protracted dispute over lumber exports, beginning in 1983 and still continuing in 1994. Throughout postwar history, Canada was so integrated into the U.S. economy that American data sources and publications often included Canadian produce as part of the U.S. domestic output. Outsiders wondered why Canadians acceded to their own take-over and branch-plant status, but Canadian governments and branch-plant managements generally found the continental integration rewarding. Relying on what was deemed an inexhaustable supply of timber in the forestry regions and gaining quick profits on sales of relatively unsophisticated wood and pulp products, Canada as a whole and BC most particularly were naively confident that the trade would never end. So it came as a shock when, in 1983 and again in 1986, the United States Senate sought to take countervail action against Canadian lumber imports.

The instigators of the action were U.S. lumber interests known as the Coalition for Fair Canadian Lumber Imports. The lobby's argument was that Canadian lumber was heavily subsidized by provincial governments through low stumpage rates on crown land. The first round in 1983 went to the Court of International Trade, where Canada succeeded in fending off the countervail. The second attack in 1985–86 came about because Canada's share of the U.S. lumber market had increased from 28.1 per cent in 1980 to 33.2 per cent in 1985; this increase supported the U.S. producers' argument that imports were causing injury to their industry. The major culprit in this trade picture was the overpriced American dollar, with exchange rates favouring Canadian imported lumber. Other arguments in favour of Canada's position were that Canadian mills had modernized, Canadian productivity had increased, and the lower Canadian stumpage levels were attributable to differences in species composition, size, quality, density of timber terrain, and similar factors. These arguments notwithstanding, the U.S. International Trade Commission radically altered its 1983 position, deciding in 1986 that the American argument was valid. The United States was granted the right to impose tariffs within six months if Canada did not in the meantime launch a successful appeal. Just before the deadline the two country governments signed a memorandum of understanding by which the countervail action ceased, but the Canadian government agreed to impose an export tax of 15 per cent on lumber shipped to the United States. The tax was subsequently transferred to higher stumpage rates in the provinces.

The chief battleground was the Pacific coast, which provided the larger share of lumber exports in competition with U.S. producers of the same coastal region. The British Columbia government agreed to impose higher stumpage after October 1987 and to include in the price of exports the full costs of forest renewal. Responsibility for reforestation was transferred to the timber licensees. Thus the cost of doing business in the coastal province increased substantially in response to the U.S. threat. Exports to the United States thereafter declined to 1980 levels, and one largely unanticipated benefit was an increase in higher-value production intended for more diverse markets. At the time, however, benefits were less evident than costs; profits plummeted and much of the Canadian forest industry fell into a slump (Hayter, 1992, 1993).

There were still further rounds to this dispute, and when, in 1993, a tribunal of the General Agreement on Tariffs and Trade (GATT) declared the American interpretation of subsidies to be inaccurate, no one assumed that the end of the border war had been reached. Even when, that same year, a U.S.–Canada trade panel ruled there was no "substantial evidence on record" that Canadian softwood exports were harming American lumber producers, Canadians geared up for another round, while the U.S. International Trade Commission prepared a new brief for GATT.

One might well ask why, if imports from Canada were in American interests prior to the mid-1980s and so much of Canadian production was undertaken in American branch-plant or affiliated firms, the countervail duty was ever promoted in the United States. There are several explanations. The huge postwar construction boom had passed; many smaller, medium-sized, and even large companies were experiencing a decline in domestic market demand. Canadian imports, previously supplementing American products, were now actively competing in a smaller market. Further, American companies had, in the interim between 1980 and 1986, considerably reduced their ownership shares in Canadian industry; there were fewer companies straddling the border and defending their right to tariff-free imports from Canadian subsidiaries. American lumber producers recognized that import controls would permit them to maintain higher prices than would have been the case with more copious and cheaper supplies coming in from Canada. Hayter (1992) suggests that the move toward the Free Trade Agreement (FTA) was also a factor, particularly since it influenced the federal government and precipitated a split in the bargaining positions of the provincial and federal governments. It seems ironic that just as the FTA was in the works, the

American government chose to show its still powerful capacity to impose penalties on a weaker trading partner.

Another view is that the high price of lumber from American suppliers was due in substantial part to the scarcity of logs and that the scarcity was due in large measure to their export from private timberlands. Canadian companies took advantage of the shortage to undercut the Americans' domestic market for lumber. The log scarcity has another dimension, and this may be the strongest explanation for the American position. U.S. producers could obtain logs from Canada only if Canadian export restrictions were lifted. Without new supplies they were unable to compete on lumber markets in Japan, as well as in their domestic market. If the import duty imposed on cedar shakes and shingles was an attempt to force BC to remove export restrictions on raw material, the countervailing duty might well be seen in the same light.

Without doubt, Canadian producers enjoyed low stumpage, whatever the explanation or excuse for it, and some writers (including this author) have argued that the countervailing duty was a painful wake-up call for provincial governments. It is possible that further withdrawals of American and other companies in the late 1980s were connected to their loss of a preferential resource cost in Canada, though other escalating costs – including increasing transportation costs from remaining timberlands to mills, high wages throughout the BC industry, high infrastructure costs, and increasing energy costs – would have provided disincentives to stay in any event. If the explanation for the tariff war does involve a covert attempt to have log export restrictions removed in BC, one might suppose the companies that left for other reasons shared an interest in obtaining resources without having to maintain manufacturing plants in the region.

Comparative Prices

One final note on changing market conditions for northern companies: the south can produce pulp at much lower cost. Ultimate consumers may be individuals who buy books, newspapers, and magazines or lumber and other wood products. But importers and producers are companies, and they are concerned with the relative prices of like-quality raw materials. Although eucalyptus pulp was regarded as inferior in quality even in the mid-1980s, this valuation has rapidly changed. Indeed, for some purposes it is superior to traditional pulps from conifers. Quality, always somewhat subjective, is now also influenced by concerns about ecology, and in this regard, the plantation fibres may well have the ecological edge, provided the plantations do

Table 3.6
Comparative mill cost for bleached kraft pulp (softwood) by selected country, 1992
(US$ per tonne)

Country	Wood	Energy	Chemicals	Labour	Other	Total mill
Chile	120	20	55	45	62	302
Argentina	120	20	70	50	50	310
US South	130	25	54	60	63	332
New Zealand	135	15	45	88	60	343
BC interior	161	30	55	61	64	371
US Northwest	195	40	44	50	70	399
BC coast	191	49	54	85	64	443
Canada east	210	41	61	83	64	459
Finland	362	9	43	46	33	493
Sweden	345	15	40	77	56	533
France	285	45	75	80	90	575

Source: Wright, 1992.

Table 3.7
Comparative mill cost for bleached kraft pulp (hardwoods and eucalyptus) by selected
country, 1992 (US$ per tonne)

Country	Wood	Energy	Chemicals	Labour	Other	Total mill
Brazil	115	20	30	50	90	305
US South	112	30	50	60	63	315
Portugal	195	20	51	45	54	365
France	186	25	57	90	70	428
Finland	246	16	38	68	52	420
Canada east	151	42	54	80	78	405
Sweden	246	18	43	77	55	439
Spain	232	22	46	91	52	443

Source: Wright, 1992.

not displace tropical forests. Eucalypts are blond and require little
bleach to produce white paper.

Without question, plantation fibres have the edge in price in the
early 1990s, as demonstrated in the data provided by Roger Wright, a
partner in Hawkins, Wright of London, for 1992. Although exchange
rates would alter the specific comparisons from time to time, these
data show such substantial differences that changing exchange rates
would not likely produce reversals in a short time. Production costs
were lower in the southern countries. Labour costs were lower, but
the differences were greater for the fibre than for labour, so the
advantage to southern producers is not contingent simply on a temporary wage advantage. The cost advantage for hardwoods was more

marked than for softwoods. Energy, chemicals, and freight costs varied by region according to other economic considerations, but overall, the southern producers were able to provide pulp to world markets at better prices.

Comparisons are less useful for sawlogs and lumber because softwoods and hardwoods have different applications. The constraint for northern producers in lumber markets is imposed more by the dwindling availability of good sawlogs than by competition from southern producers of tropical woods.

MASTER PLANS: NEW ROLES FOR NORTHERNERS

Everyone is suddenly concerned about the ecological issues associated with tropical forests and plantations. "Everyone" includes environmentalists, villagers, tribals, academics, and journalists, many of whom oppose plantation economies, as well as predations against tropical forests. It also includes companies, governments, and consultants with international firms, international banks, and international aid agencies, most of whom are seeking ways of overcoming the resistance they encounter throughout the world. One of their ways is the master plan.

The most popular of the master planning agencies is the Finnish pulp and paper and consulting firm, Jaakko Pöyry, which has made a speciality of providing evaluations and recommendations to underdeveloped economies regarding potential and methods for the creation of plantations. Through the 1970s and 1980s Jaakko Pöyry undertook studies for governments and companies; its reports formed the basis for most of the forestry operations established over the past two decades. There are other consulting firms as well, including four Canadian companies: H.A. Simons, Sandwell, Stadler-Hurter, and SNC Cellulose. Finland has a second firm that has gained more international business with the current restructuring – Ekono.

Now master planning is taking the route of environmentalism. The World Bank, the FAO, the United Nations Development Program, and the U.S. World Resources Institute have all supported the idea of an international coordination system to stop the destruction of tropical forests. Thailand, the Philippines, Nepal, Bhutan, Bangladesh, and Sri Lanka are among the 79 of 125 countries involved in the Tropical Forestry Action Plan (TFAP), which has Jaakko Pöyry as its planning agency, funded by the Finnish development agency Finnida.

By 1990 two plans were completed in Southeast Asia, those for the Philippines and Indonesia. Both have been roundly criticized by environmentalists and non-governmental organizations in those countries;

in both cases, environment groups have called for a stop to international funding for TFAP. The World Rainforest Movement, a coalition of non-government tropical forest groups, has likewise criticized nine other completed plans in Latin America, Africa, and Asia. Among their complaints are that the master plans ignore social impacts, encourage intensified logging of natural forests, and fail to recognize the links between poverty, landlessness, and deforestation: in short, they are programs suitable for corporate logging.

Jaakko Pöyry was the firm selected to assess the findings of the Environmental Impact Assessment Review Board in Alberta following public hearings into the Al-Pac proposal discussed earlier. It said while there were risks, they were tolerable (Pratt and Urquhart, 1993:29). Jaakko Pöyry was also the firm listed by Crestbrook Forest Industries as an adviser when it submitted its contribution to the Al-Pac proposal process in 1988. Crestbrook, owned by Mitsubishi and Honshu, is one of Mitsubishi's subsidiary firms with shares in the Al-Pac venture (Pratt and Urquhart, 1993:18).

In addition to international engineering and consulting firms, the industrial world benefits from the sale of machinery to developing countries. Throughout these countries where pulpmills have been established, Voith (Germany) and Beloit (U.S.) paper machines are standard equipment. The alternatives include Black Clawson (U.S.), Escher Wyss (Germany), Karlstads MW (Sweden), and three Finnish equipment manufactures in the TVW group: Tampella, Valmet, and Wärtsila. The chief British exports in forestry today are more likely papermaking machinery than paper, including various components by Beloit Walmsley, Spooner Industries, Atlas Converting Equipment, Simon Engineering and its offshoot Simon-Holder, Black Clawson International, and some fifty other manufacturing firms.

Such a listing suggests a restructuring for the north and a new division of international labour, whereby the southern countries increase their participation as producers of staples, and the north becomes the supplier of advanced technology and expertise. This characterization is fair in reference to the Nordic countries, but less true for the North American sector. The difference is partly attributable to the remaining old-growth timber stocks in the Pacific regions of North America – forestry companies are reluctant to abandon them to environmentalists. In addition, the United States has new fibre sources in its southern states, and Canada, in its boreal forest. The more significant difference, however, lies in the role adopted by governments and industry: the Nordic regions have more readily accepted some central planning, and as we have seen, they began the restructuring process in the 1970s. As well, the forest industry in Finland was always closely integrated with the engineering industry, and Finland extended its industrialization by developing and exporting machinery and supplies. By contrast, Canada allowed foreign branch plants to dominate the production of machinery and other supplies; it imported these even when the original inventions were made in Canada. Finland ran into difficulties as early as the 1960s, but by the late 1970s it was restructuring its industry toward higher value-added products and machinery in the export markets of Europe. Canada exploited its timber and sold staples well into the 1980s, before the decline in timber reserves and changes in market context entered into policy deliberations. Changes in the international division of labour, then, are related to earlier development in these various countries.

In noting the persistent presence of one Finnish consulting firm, one might wonder if some world conspiracy is at work. While that company does have enormous influence, its success lies rather in two simple features typical of market economies: it happened to have the right skills to offer at the right moment, and its notions of global development match those of most governments, companies, and aid agencies. There is no conspiracy, but a world-view that has currency at the turn of this century.

SUMMARY: RESTRUCTURING INDUSTRIES AND MARKETS

A regional analysis indicated that Nordic companies have restructured, some have moved pulp-sourcing activities to southern climates, and the industry as a whole has become more concentrated; Swedes and Finns have moved into Europe as paper producers. Germany still produces substantial quantities of paper, but despite its success in regrowing stock since the Second World War, much of the raw material for its industry is grown elsewhere. Other European regions have increased growing stock since 1940, but like Germany, they import wood and pulp. The Iberian Peninsula is a major plantation region, and several large European companies are procuring pulp from subsidiaries in Portugal or Spain.

The U.S. industry has moved much of its investment to southern states and pine plantations. The Pacific coast region has suffered overcutting and understocking, though the eastern states have more standing timber than they had half a century ago. A major study of the U.S. Pacific region has resulted in promises to improve rehabilitation efforts and federal government support for new approaches to forestry. Canada is an example of a resource-rich country that lived off its inheritance. Even in the face of the new global challenges, its response has been to open up some 200,000 hectares of boreal forests for exploitation on precisely the same dependency model characteristic of past development in Canada. In this case the chief beneficiaries are Japanese companies, which have been granted harvesting rights to vast areas.

Global changes in investment patterns affect the ranking of companies, and while United States and European firms are still at the top of global rankings, Japanese and other Asian companies are moving up very rapidly. The lumber-trade dispute between the United States and Canada is one consequence of changes in market context. Northern industrial countries have taken on a new role as "master planners," which provides them with markets for more advanced manufactured products in developing regions, while those regions gradually become the pulp producers of the world.

4 The Profligate Century and Its Aftermath in British Columbia: A Case Study

British Columbia, Canada's Pacific coastal province, is an example of a region that has played out the captured-state model of contemporary history. It is also a prime example of a traditional northern forestry region struggling to emerge from its past and deal with a dramatically changed global marketplace.

Its 60 million hectares of softwood forest (just under 3 per cent of the world's total) gave it an export crop to be envied. Geography cut off easy east-west trade but made the much greater U.S. market accessible. Large American and central Canadian companies were part of the logging scene by the 1880s. Tariffs at the U.S. border were rigged to fit the interests of American lumber and newsprint buyers, as well as American branch plants in BC: no-cost entry for construction wood, pulp, or newsprint, but penalties for any more advanced manufactured products. Thus BC produced standard-dimension lumber, pulp, a small amount of newsprint, and not much else. Rewards for this simple trade in staples were so great that labour unions negotiated some of the highest production worker salaries on the continent.

The lush green gold covered so many hectares that British Columbia's population came to believe they owned an inexhaustable supply of timber that required no major investment in reforestation. A former president of Rayonier Canada described the province's attitude toward its forest thus: "In most of the 120 years since timber was first exported from B.C., both industry and government have paid little regard to regeneration" (James Buttar, 1980). Much the same sentiment was expressed by the founder of MacMillan Bloedel in testimony before

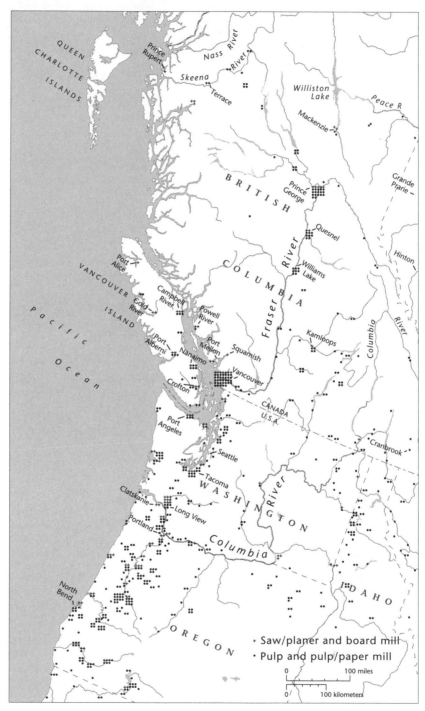

Map 3 Pacific coast regions, United States and Canada: saw/planer, pulp, paper, and paperboard mill locations

the Sloan royal commission in the mid-1950s, when he spoke disparagingly of "bureaucrats who have never had rain in their lunchbuckets." So thick and valuable were the stands that BC accounted for 40 per cent of Canada's total forestry export dollars in the 1970s, when some three hundred sawmills produced about 60 per cent of the country's lumber and twenty-four pulp and paper mills produced 28 per cent of its pulp and about 16 per cent of its newsprint (Marchak, 1983; Sopow, 1985). Ten per cent of the labour force was directly employed in the industry, and of all BC's exports, forestry products accounted for over 60 per cent.

When the crash came in the early 1980s, the industry assured the population it was one of the cyclic downturns that were inevitable in a resource industry. When the downturn became a depression, the industry blamed labour costs and environmentalists. When, in the 1990s, the depression became a long-term condition, structural changes were undertaken, and finally there emerged some recognition that forestry as it had been practised in the profligate century was no longer profitable or possible. All participants were damaged in the process, but unions and labour took the brunt of both depression and restructuring between 1980 and 1995.

THE PROFLIGATE CENTURY

Crown Ownership, Company Control

The provincial government ("the crown") owns about 95 per cent of forest land. The remaining 5 per cent is at the coast and has provided valuable harvests to the few companies that gained or bought it from the original railway owners, who themselves obtained it through government grants when British Columbia joined Canada. Provincial policy is directed toward the larger portion under crown ownership. Commercially exploitable land is allocated as forest licences of varying kinds and duration, which allow their owners to harvest what is known as "annual allowable cuts" (AACs). These licences have an intricate history of rechristenings, but under whatever name, they have been more beneficial to large, integrated companies than small independents. As the industry moved into the interior of the province in the 1950s, the government created "public sustained yield units" (PSYUS) for pulpwood harvesting, accessible only to companies with the capacity to construct or already possessing pulpmills. Government also created rules that obliged small sawmills to sell wood chips to specific companies in their regions. Further rules governing technological change in the sawmill sector obliged many small, family-owned mills to close. Over the period from the late 1960s to 1980, some 1,200 small, competitive sawmills shut down, and of the remaining 300, most were either part of the large corporate operations or directly feeding into them.

Crown ownership of forest land was never contested by large private companies. They enjoyed secure harvesting rights and had no responsibility for replanting and management. Resource rents (stumpage) supposedly funded management, but the rents were insufficient to cover the cost of reforestation and management or maintain a forest service sufficient to supervise and monitor the system. Where companies undertook replanting, costs were deducted from their stumpage payments, but they were not obliged to monitor the success rates for the seedlings or undertake the lengthy follow-up silvicultural treatment and thinning to ensure that growth actually occurred. Further, the stumpage system was based on felling rates, so that during downturns in the economy, when less timber was harvested, companies did not pay a guaranteed base rent as they would have in the United States (some had "negative stumpage" during the early 1980s, obtaining credits against future harvests).

In lieu of ownership, companies insisted on tenure security. A royal commission of 1976 advocated security of timber supplies to licencees, and this was provided in the 1978 Forest Act, if rather more lavishly than the commissioner had intended (see Pearse, 1980). Even the Ministry of Forests was worried about the level of concentration, but it was a natural outcome of continued preference for large, integrated operations, low stumpage rates, and the continued high rates of harvesting first established in the 1912 Forest Act. The 1978 act named sustained yield as an objective, but its definition rested on economic and political, as much as biological, criteria. During the depression of the early 1980s, the ministry was managing the resource, stumpage, and allocations with reference to such factors as the companies' cashflow problems (Sopow, 1985:66–7). When an entire economy rests on a few large corporations, governments are unlikely to apply rules that could damage or even inconvenience the "stakeholders."

Comparisons between resource rents on public lands in Canada and the United States have generally foundered because of inconsistent definitions and requirements, though by whichever method of reckoning, the conclusion is that stumpage rates are substantially lower overall in Canada. University of British Columbia forestry professor David Haley concluded that "the evidence strongly suggests that appraised stumpages in B.C. fail to reflect the full value of the province's timber resource and that, in the absence of competitive markets for stumpage, public revenues from the production and sale of timber are much lower than they should be" (Haley, 1981; see also Haley, 1985). A study conducted by Simon Fraser University students found BC rates per cubic metre to be Cdn$10.11 at the coast and $7.65 in the interior, compared to Cdn$51.84 for the U.S. Pacific Northwest coast and $41.13 for the U.S. Pacific interior (1992:82).

The British Columbia Royal Commission on Forest Resources (1976) reported that in the mid-1970s, some 59 per cent of all harvesting rights in the province were controlled by ten companies. A study of the situation a decade later indicated that four interlinked groups of companies controlled 93 per cent of allocated public forest cut and 84 per cent of the total provincial timber cut in BC (Marchak, 1983; Wagner, 1987). In 1988 two forms of harvesting licences, differing in specific management requirements and terms but similar in design, accounted for nearly 80 per cent of all AAC allotments in BC. These were called, respectively, tree farm and forest licences. Two parent companies, through various subsidiaries, held 54 per cent of allotment in tree farm licences; four companies held 72.6 per cent. The same four companies controlled 34.1 per cent of forest licences; eight controlled half of all forest licences. Combined allotments, then, provided two companies with control of 31.4 per cent, and five companies with 53.7 per cent of these cutting rights. Since many of the smaller companies holding AAC allotments were partially owned by the same larger companies or controlled by them via contracting and marketing arrangements, the real control by few companies was greater than the conservatively compiled data. As well, the large companies were not altogether independent of one another (Marchak, 1990).

Throughout the long postwar boom, the dominant integrated companies were MacMillan Bloedel (MB), Canadian Forest Products (Canfor), Northwood, and Canadian Pacific (CP), all majority owned in Canada; BC Forest Products (BCFP), jointly owned by Noranda (Canada), Mead (U.S.), and Scott (U.S.); Crown Zellerbach (CZ), Weyerhaeuser, Weldwood, Columbia Cellulose, Rayonier–ITT, Canadian International Paper (CIP), and Scott Paper, all subsidiaries of U.S. companies; and Eurocan, owned by Enso-Gutzeit of Finland. A second string of fair-sized comers began with West Fraser (U.S.), Crestbrook (Honshu of Japan, plus Canadian ownership), and Doman and Whonnock (B.C. companies). These companies were fairly stable in relative positions and steadily growing in production capacities throughout the quarter-century from 1950. Changes began in the early to mid 1970s, and then tumbled onto one another as the 1980s depression set in (Marchak, 1983, 1986, 1989).

Timber Supply Estimates

Allocations of timber harvesting rights, though supposedly embodying sustained-yield principles, resulted in an overall decline in timber stocks. The capacity of current operating mills exceeded the AAC by about a third by the 1980s. In various ways, with special government allowances because of pest infestations or economic conditions, the AAC turned out to be a bureaucratic, rather than an actual, indication of the harvest.

For most of the province's history, companies had argued that nature would do the replanting, and three-quarters of cut-over areas were left to nature (Association of B.C. Professional Foresters, 1984:17). Unfortunately, nature takes several centuries to restock a commercial forest, and conditions on cut-over lands are often beyond its regenerative capacities. The result was a growing expanse throughout the province of what are now known as "not satisfactorily restocked" (NSR) lands, defined as "land on which stocking is below a specified minimum or is not stocked with species that optimize the site's commercial productivity." NSR lands already constituted between 1.3 and 3.4 million hectares in the mid-1980s, according to Ministry of Forests estimates (BC, Ministry of Forests, 1984).* The Association of B.C. Professional Foresters estimated in 1984 that 1 million hectares fell into the category, of which at least half could be restocked. It would take thirty to forty years to eliminate the backlog at the rate of Ministry of Forests activity at that time. The association noted that without further management beyond planting, the effort would be wasted.

This and other reports identified forests situated next to cut-over areas where they succumb to winds and other forces (the "fall-down" effect), clogged stream beds, and silting of streams and rivers. Scarcely any management beyond basic restocking of denuded lands following logging and fires was undertaken prior to 1985. According to a 1984 report by Woodbridge, Reed, "Only a modest amount has been spent on, and even less appears available for, intensive silviculture." Every commentator since the early 1980s has urged provincial governments to double or triple investments in reforestation and management. Wood shortages were noted in numerous reports commissioned by both the federal and the provincial governments during the 1980s. The 1984 report by Woodbridge, Reed and Associates confirms earlier warnings dating back to the 1970s. It notes that the more optimistic

---

* The range in estimates reflects different terminology and methods used by branches within the ministry.

Table 4.1

British Columbia: productive forest land in timber supply areas and tree-farm licences by region,[1] 1991–92 (1,000 hectares)

Region	Mature timber[3]	Immature timber[3]	Not stocked[2]	Total
Cariboo	3,634	2,179	136	5,949
Kamloops	2,446	1,763	229	4,438
Nelson	1,320	1,889	239	3,448
Prince George	9,552	5,800	2,069	17,421
Prince Rupert	6,396	2,083	916	9,395
Vancouver	3,293	1,456	196	4,945
Total	26,641	15,170	3,785	45,596

Source: BC, Ministry of Forests, 1991–92.

[1] Includes only productive forest land where timber harvesting is restricted partly or wholly by factors such as timber quality, accessibility, and environmental sensitivity. Excludes crown forest land in timber licences, all private land, all federal land, provincial parks and miscellaneous reserves, and non-productive and non-forest lands. The province's total area is 94.8 million hectares.

[2] Not stocked productive forest land. Includes areas (current and pre-1982 backlog) that were disturbed by harvesting, wildfire, or other causes and were not satisfactorily restocked with sufficient trees of acceptable commercial species. Anticipated regeneration delay may be between one and seven years on current lands that are not satisfactorily restocked. Also includes non-commercial areas, either covered with commercial trees species or 60 per cent or more with brush one or more metres high.

[3] Stands with lodgepole pine and whitebark pine or deciduous species as the leading species are immature if the stand age is 80 years or less. When the stand age is greater than 80 years, the stand is mature. Otherwise, all stands having conifers, other than lodgepole pine and whitebark pine as the leading species, are immature when the stand age is 120 years or less. When the stand age is greater than 120 years, the stand age is mature.

projections of timber supply in the early 1970s suggested production in excess of 100 MCM in perpetuity, yet by 1980 the Ministry of Forests was able to project a timber supply of not more than 76 MCM annually, and that only under improved management conditions.

While reports on the availability of the resource were somewhat inconsistent with reference to the amounts of standing timber, there was general acknowledgment that a pulp-fibre supply deficit was already evident at the coast in the early 1980s (Sterling Wood, 1984), and it was apparent that lumber manufacturing capacity exceeded both the available supplies and the market demand. A 1985 analysis by Swedish economist Sten Nilsson predicted that by the turn of the century, the harvest would be small (one estimate was 44 MCM, compared to 77 MCM in 1985) and that the remaining resource would be so far distant from mills that transportation costs would be much higher. Also, the logs would be smaller in diameter and of inferior quality. As the industry approaches the target date of 2000, Nilsson's predictions are well within the range of probability.

Log Exports

Log volume increased by over 80 per cent between 1963 and 1978. Because integrated companies control the bulk of harvesting rights, there is no effective domestic log market. This precludes the growth of an independent manufacturing sector. On the other hand, restriction on log exports ensures that preliminary processing of BC wood is done in the province. During the early 1980s, these restrictions were modified because U.S. lumber markets were depressed, while demand for logs in Japan increased. From an upward limit of just over 2 per cent of the total harvest (constituting 776 thousand CM in 1979, for example), exports of logs increased to nearly 4 per cent of the total harvest (3.4 MCM) in 1987. Japan took three-quarters of these exports. In 1989 the British Columbia government imposed a "fee in lieu of manufacturing" on the export of logs by way of reducing the exports. By that time the recession was over, and the next downward slide had not started.

The arguments in favour of log exports were that overmature stands should be cut anyway, sales to the United States would be reduced by the cost increases, and lumber exports were down. There was another reason, noted by one forestry analyst: the cash-flow deficiency experienced by many companies during a general recession (Widman, 1984). This problem underlay the push to sell off logs from inventory; it also explained some of the sales of assets. The sale of logs had never been, and even during this period was not, a major part of BC's forestry export economy. By comparison, U.S. exports in the early 1980s amounted to 25 per cent of the total harvest from privately owned land, with half of it shipped to Japan.

Community Participation in Resource Decision-Making

Since no provisions for public hearings or community controls on resource management were included in the 1978 Forest Act in British Columbia, there was also no obligation to listen to public opinion. Most members of the public would have found it impossible to obtain accurate data on harvesting practices and allocations. This deficiency became more widely noted in the 1980s, when both native and non-native communities, together with environmentalists, began to experience the impacts of dwindling resources, prolonged poor management, and employment decline. While there were concerns in the 1970s and early 1980s about the health hazards of pulpmill effluents and there were small environmental-protection groups arguing for

Table 4.2
British Columbia: production of logs, lumber, and pulp and paper and employment
per sector, 1970–90

	Production				Employment		
Year	Logs 1,000 m³	Wood 1,000 m³	Pulp 1,000 t	Paper 1,000 t	Log	Wood	Pulp and paper
1970	54,726	18,067	4,100	1,675	18,581	38,329	17,089
1975	50,078	17,569	3,998	1,477	17,500	38,655	20,225
1980	74,654	28,207	5,727	2,165	24,270	49,708	21,540
1981	60,700	24,547	4,835	1,874	19,561	46,627	20,660
1982	56,200	23,406	4,823	1,872	16,371	40,309	18,458
1983	71,400	30,778	5,702	2,148	19,906	40,392	17,390
1984	74,500	30,862	5,358	2,085	20,586	38,901	17,433
1985	76,868	32,119	6,024	2,486	19,488	39,603	16,850
1986	77,503	30,996	6,346	2,601	19,848	37,204	17,254
1987	90,591	37,336	7,035	2,753	20,449	42,425	17,662
1988	86,807	36,376	7,251	2,878	19,871	42,283	18,207
1989	87,414	35,952	7,181	2,846	20,316	42,416	18,643
1990	78,300	33,514	6,799	3,002	18,106	40,312	18,427

Sources: Forestry Canada, *Selected Forestry Statistics Canada*, 1991: Table II–2 for roundwood production
1970–89, and 1990 data (derived from Statistics Canada cat. no. 25–201): table II–7 for lumber
(derived from Statistics Canada cat. no. 35–204, 35–250, 35–002, and 35–003): table II–17 for paper
and paperboard production (derived from Statistics Canada cat. no. 25–202, 36–204, and Canadian
Pulp and Paper Association monthly survey, 1980–90).

conservation policies in the woods, these were not high-profile features of the forest economy before the mid-1980s.

Unions and Employment

The British Columbia section of the International Woodworkers of
America (IWA) had 50,000 members in logging and sawmills at the
beginning of the 1980s and comprised half the membership of an
international (U.S.–Canada) organization. It had considerable clout
in the BC economy, and its leaders were the strongest voices in the
union movement. It bargained on a regional basis with an alliance of
employers, but it strongly supported its employers and their harvesting
rights (Marchak, 1983). The IWA's reasoning was that large companies
could plan on long-term horizons, could pay decent wages and benefits, and would invest in training for the labour force, safety measures,
and conservation of the forest (Marchak, 1983:61). The number of
accidents in the woodlands was high, though their frequency would
not necessarily have been lower under other policy regimes, and in

this respect the union was probably justified to be wary of small logging operations. With regard to integrated companies, it argued that pulpmills had to be huge to be profitable, and monopolies were inevitable. Pulpmill workers were represented by two pulp and paper unions, both much smaller than the IWA and generally cooperative with it. Strikes in forestry, considering the number of workers involved, were not numerous and were rarely prolonged. Wages moved upward with company profits, and company towns such as Port Alberni and Powell River, both dominated by MacMillan Bloedel mill complexes, were among the most affluent communities in Canada during the 1960s and 1970s.

Production volumes increased throughout the long postwar boom, but employment did not keep pace. Volumes of roundwood produced increased by about 80 per cent between 1963 and 1978, while the number of hourly workers in woodlands increased by just under 30 per cent. Lumber and pulp production volumes each increased by over 120 per cent in the same period, but hourly worker employment increased by only 71 per cent in pulp and newsprint mills and by 42 per cent in sawmills. These figures are all the more remarkable since in that same period, pulpmill capacity increased with the construction of some ten new mills (Marchak, 1983: charts 6.1–6.3). The rise in production volume, combined with the reduced rate of increase in employment, was due to technological changes introduced in woodlands and mills prior to 1980.

Aboriginal Communities

The fate of the Sekani (Tsekani) bands of the Ingenika and Mesilinka people of the Mackenzie district is typical of aboriginal conditions during the postwar boom. The Sekani were few in number and so isolated that their existence was not recognized in 1922, when Treaty No. 8 was negotiated in Alberta. Smallpox and other diseases had almost eliminated them. Both tribal relationships and geographical proximity might otherwise have ensured their inclusion in the treaty. Those who survived hunted and trapped in a region about 150 kilometres northeast of Prince George. When the original timber forest licence for the region was awarded in the mid-1960s, coincident with the building of the Peace River dam and the flooding of thousands of hectares in what is now known as Williston Lake, the Sekani bands were unceremoniously evicted from traditional lands. They were not informed of the events about to transform their way of life, and they did not surrender aboriginal rights to the land that was flooded. Indeed, it is not clear that outsiders were even aware of their existence.

The establishment of mills and logging operations at Mackenzie provided no livelihood for them; they were dispersed to two marginal locations north and south of the flooded area. Their land was occupied by predecessor companies to BC Forest Products and then by Fletcher Challenge.

Elsewhere, aboriginal peoples who survived the onslaught of industrial society and smallpox emerged during the 1970s as organized communities intent on reclaiming ancestral lands and ownership rights. Land title was never ceded to the Canadian government. Native peoples were exiled from their land as they were elsewhere, and the history of European settlement and industry is replete with appalling injustices, but when various native bands finally gathered sufficient strength and organizational capacity to mount land claims, they had many supporters who argued that European property law was on their side. Prior to 1992 the British Columbia government took the position that aboriginal title, if it had ever existed, had been extinguished by accumulated events over the century. Further, it argued that any responsibility for changing land title would rest with the federal government under terms of BC's entry into Canadian confederation. Land-claims cases thus fell between two levels of government, and industrial development of aboriginal land – no longer marginal as forestry and mining moved further north – continued without land-title settlements.

Major cases were mounted by the Nisga'a (Nishga) in the Nass Valley and the Gitksan Wet'suwet'en in the tributary valleys of the Skeena River near the town of Hazelton in northwestern British Columbia. The Nisga'a entered a land claim and lost the case in the BC Supreme Court in 1970; on appeal to the Supreme Court of Canada three years later, they gained a split decision, with the seventh judge arguing against the aborginal title on a legal technicality rather than on the substance of the case. They have since tried repeatedly to obtain a timber-harvesting licence and to negotiate solutions with the several successive companies that have controlled logging rights to the valley.

RESTRUCTURING: THE 1980S AND 1990S

Some restructuring after 1980 was imposed by border-war conditions. Stumpage was increased, and companies were required by revisions to forestry legislation to undertake more forest-management tasks. As well, they were obliged to accept greater costs for infrastructure and reforestation. At the same time, they obtained even more tenure security, and the capacity of a downsized Forest Service to regulate the yield was much diminished. Most harvesting licences now ran to between twenty and thirty years and were renewable. Where scaling of logs had previously been done by government employees, companies were now permitted to hire their own scalers. Further changes were subsequently proposed. A particularly controversial proposal involved conversion of tenure forms that would have resulted in enormous regions transferred to single-company control. This proposal elicited mass protests from environmental groups and many others, and it was dropped.

Other changes were induced by rising costs of production, together with greater competition in markets. Technological change benefited companies in the lengthy boom period, and as long as the larger economy was flourishing, unemployment was kept in check. Changes due to technology continued through the 1980s and 1990s, but now they occurred in a different context. Both companies and workers experienced setbacks. The immediate response in the early 1980s was to downsize management and production crews and curtail production. This brought temporary relief in the mid-1980s, but by 1990 the recession had settled in again. Some companies exited the province; others put mills up for sale or closed them. A few invested in the construction of new plants to produce higher value-added wood products and began to diversify their product mix to attract new markets in Asia. Success from the perspective of companies now depended on creation of a very different, small but flexible workforce.

Changing Conditions in BC's Central Interior Regions

Quesnel, a town situated in the centre of the province with 62 per cent of its employment derived from forestry in 1991 (Horne and Penner, 1992: table 1), provides an example of employment impacts related to a diminishing resource. Daishowa built Quesnel's two large pulpmills, Quesnel River Pulp and Cariboo Pulp and Paper, in 1971 and 1981. The mills have no harvesting rights, but in each case their large lumber partners – West Fraser and Weldwood (both U.S. firms) – provide the main raw material input, wood chips. Other, smaller mills supply the remainder. As well, these mills ship lumber to the United States and Japan. In the heyday of forestry in this region, the raw material consisted of varied softwood species. By the late 1980s, logging operations were far from the mills, harvesting lodgepole pine that was once regarded as unworthy for commercial use. An infestation of the mountain pine bark beetle provided the reason for a 40-percent incremental increase in the region's AAC over the basic 2.3 MCM established in 1984. Five years of hectic overcutting ensued, justified as necessary to overcome the infestation. The five-year battle ended in 1989 in the Quesnel area. More temporary licences were allocated in 1992 for a two-year period in the Williams Lake area south of Quesnel. As everyone expected, Quesnel went into a slump: the beetle had allowed the town to pretend that it could go on acting as if it were still in the 1970s.

The impact of technological change had been foreseeable in the 1970s. Modernization of the sawmills and the first pulpmill could have been anticipated to lower labour input; there was no reason to suppose the second modern pulpmill would provide substantial new employment for unskilled workers. The granting of an incremental harvesting right was a stopgap measure that would provide temporary employment for loggers and some sawmill workers, but townsfolk knew it was temporary. Elsewhere, beetle infestations were met by substitute cutting rights, so that at least when the infestation was controlled, the remaining forest would be available; in this case, both the healthy and unhealthy forests were cut in a short five-year period. Production increased, but employment did not, and in some sectors it decreased. The unemployment rate during the heavy cutting period was already over 10 per cent. A 1987 community study and questionnaire, conducted by a group of residents, concluded that some 400 direct jobs had been lost since 1979 in the processing sectors, even though production levels were higher than they had ever been before (North Cariboo Community Futures, 1987).

The larger region of Cariboo–Fort George was studied by Savage and Associates in 1993; they reported that "a 33% increase in the volume of timber harvested and processed in the Cariboo Forest Region between 1979/80 and 1992/93 created no net job growth (employment decreased by 1.5%)" (1993:1). The chief reduction in employment was found in sawmilling and plywood manufacturing, where both plant and management levels decreased by a total of 1,200 jobs, even though the volume of timber processed increased by 31 per cent. Production per worker in this sector grew by 79 per cent between 1980 and 1993. Savage and Associates also found some areas where employment had increased, specifically in silviculture, speciality or value-added manufacturing operations, logging, trucking, road building and maintenance, forestry consulting, and pulpmills. Although employment increases since 1980 had replaced most of the jobs lost in sawmilling and plywood manufacturing, the lost jobs had carried higher average wages, benefits, and security than the new ones in silviculture and speciality wood products. Many jobs throughout the region, and particularly in Quesnel, were supported on a temporary basis by the "beetle wood." Overall findings on unemployment indicated that 18.5 per cent of the region's working-age population was available for work and employable, but receiving Unemployment Insurance benefits or Income Assistance at the end of 1992. Increases in the number of Unemployment Insurance clients ranged between 51 and 106 per cent in various communities between January 1981 and December 1992, and other measures indicate the same rising levels of unemployment. Population, which had grown rapidly between the 1950s and late 1970s, was declining during the 1980s and early 1990s.

These data may be viewed in terms of the number of jobs supported by volume of timber at different periods. Savage and Associates estimated that "in gross terms, 966 cubic metres of timber supported a job in 1980, but 1300 cubic metres was required in 1993. In the major employment area, sawmilling and plywood production, 48% more timber is required to support a job in 1993 than in 1980," including jobs in management, plants, production, transportation, and infrastructure (1993:4). The growth in jobs in silviculture, an increase of 105 per cent, or 371 jobs, over the 1980 base of 181, was attributable to an increase in activity by the BC Forest Service; that is, these were public sector jobs. An increase of 132 jobs in forestry consulting employment, for a total of 150, was attributable to a pattern of contracting out by forest companies and the BC Forest Service to small, private firms. Road building and maintenance jobs are largely attributable to the public sector as well or are contract jobs paid for by the public purse. Thus much of the increase in employment was at public expense, while the decreases occurred in the private sector.

Quesnel is not unusual in its dependence on forestry; indeed, in the 1970s, before the pulpmills and incremental cuts, this town was shown to have a greater diversity of economic underpinnings than many others in BC (Byron, 1976). Throughout the Fraser–Fort George region, there are 7 pulpmills and 127 sawmills. Nine of the sawmills are large independents with harvesting rights. All logging operations are moving farther north to find sufficient wood to keep these huge sawmills and pulpmills in operation. Transportation costs are now impediments to continued growth; more problematic for the province, however, is the loss of wood throughout the region and very far north of it. In simple fact, the capacity of the pulpmills and sawmills far exceeds the availability of raw material, even were the population prepared to allow every tree to be cut.

Flexible Labour

Volumes cut and manufactured continued to increase through the 1980s, though the growth was less dramatic, except in pulp production.

Figure 4.1
British Columbia production and employment, 1980–90, in logging and wood:
percentage change relative to 1980

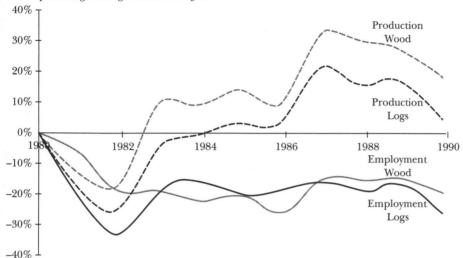

Source: Statistics Canada. *Canadian Forestry Statistics.* 1986, 1990, and 1991.

Employment, however, no longer increased at all. The total number of
employees in all forest sectors was close to 97,000 in 1979; it was down
to 70,000 in 1991. IWA membership in BC dropped to 27,500 by 1995,
a decline of about 43 per cent in fifteen years. The union severed ties
with the American IWA in 1987 over the countervail border war.

For the IWA, one of the contributing factors in membership decline
was an increase in contracting out to small companies. The integrated
firms began to phase out woodlands operations during the 1970s; the
practice of contracting out was extended by 1978 legislation requiring
that half of all logging be contracted, thus relieving large firms of
labour obligations and reducing the capacity of unions to organize
workers. Small firms competed for large company contracts, which
ensured their loyalty. Subsequently, a small business enterprise program was introduced that permitted independent logging companies
to bid on timber-cutting rights over limited territories. This reduced
by 5 per cent the harvesting rights already granted to large companies,
but these ostensibly independent small companies sold most of their
produce to the large companies dominant in their territories because
transportation costs inhibited alternatives. For the rest of their business, they competed for company contracts.

Powell River is an example of an old coastal mill town that flour-
ished during the postwar boom. MacMillan Bloedel (MB) and its

Figure 4.2
British Columbia production and employment, 1980–90, in pulp and paper: percentage change relative to 1980

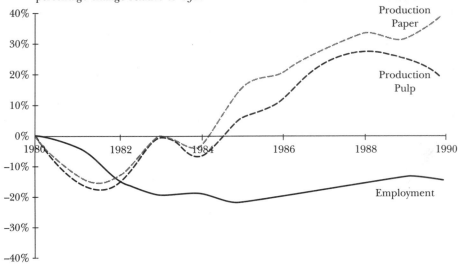

Source: Statistics Canada. *Canadian Forestry Statistics.* 1986, 1990, and 1991.

predecessor companies sold the town's high-quality lumber and newsprint in increasing quantity to the United States between 1915, when the town was established, and the late 1970s. Then came the downturn. Hayter and Holmes (1993) found that its lumber production declined by 58 per cent between 1980 and 1992, due in part to reduced supply of sawlogs and reduced capacity, as well as to reduced sales. Newsprint output in 1980 was already 66 per cent below capacity, and fibre costs were no longer competitive for U.S. markets. Production in 1992 was 75 per cent below capacity, and pulp production was 21 per cent below the 1980 level. Employment declined by 32 per cent between 1981 and 1992, with losses especially marked during the two major recessions of the early 1980s and early 1990s. Salaried employment in the papermill dropped by 45 per cent between 1980 and 1994; regular hourly employment by 50 per cent (Hayter and Holmes, 1994: table 1). Relief workers were nearly doubled, though their actual numbers were relatively small.

Seeking new markets and changing the product mix is standard strategy for coping with global change. In the Powell River case, MB was trying to do both. As well, it began to restructure its labour force at both Powell River and Port Alberni. Its stated objective was to develop a smaller, but more flexible workforce. Flexibility was defined as greater capacity of the employer to move workers from one job to another, to have open shops, to contract out work, to hire more part-time and temporary workers, and to waive credentialism for tradeswork if the job could be safely performed by unlicensed workers (Hayter and Holmes, 1994). Inch by inch, MB succeeded in gaining these conditions at Powell River, but when the company attempted in 1994 to bring in a splinter union labour force to construct a new coated-paper mill at Port Alberni, workers went on strike. The thin edge had been traversed; the company had reached too far.

Companies in the non-union southern states or in new "greenfield" mills starting from scratch (such as Al-Pac) are able to create smaller, lower-cost, flexible labour pools, and they then become the models pointed out to union leaders and aspired to by management. Not being competitive is a threat to everyone dependent on exports in a highly competitive market, but when all the basic factors of production – land, resource costs, and labour – are lower elsewhere, reducing only the cost of labour is unlikely to win the contest.

The Companies

Despite the enormous expansion of the postwar period, royal commissioner Peter Pearse found in 1976 that BC companies did not have a high return on invested capital. Richard Schwindt in a study in 1985 also found that the industry was not particularly profitable. If the resource rent was low, then the companies were dissipating the rents, or labour costs were consuming them. Their dissipation might have come about through poor wood utilization and wastage or the "misallocation of both capital and fibre" (Schwindt, 1985). By the 1990s the companies that remained were restructuring their plants and utilizing fibre more efficiently because stumpage had increased and resource supplies were decreasing; their balance sheets, however, showed little growth and many losses.

Foreign ownership in firms, as measured by shares exceeding 50 per cent, declined between 1975 and 1990 (from 29 to 18), and its share of timber rights declined from 35.5 to 30.4 per cent. However, as documented by Zhang, Leitch, and Pearse (1992), its share of manufacturing capacity in the pulp and paper sectors increased considerably over this period, while shares in manufacturing capacities for lumber and plywood declined. Majority foreign-owned firms controlled 45.2 per cent of pulp and 45.3 per cent of paper manufacturing capacity in 1990, compared to 37.4 and 17.6 per cent respectively in 1975. The number of separate firms had declined; concentration had increased. American ownership had declined; Asian ownership had increased.

Among the early changes – Columbia Cellulose, International Paper, Rayonier–ITT, and Crown Zellerbach (CZ) pulled out of Canada altogether. They sold mills and in the case of CZ some privately owned woodlands. The provincial government established a crown corporation, Canadian Cellulose (CanCel), to take over Columbia Cellulose properties and the antiquated Ocean Falls mill from CZ. Discovering that the Ocean Falls mill required complete renovation and that its timber supply was depleted, CanCel finally closed the operation. Eventually, it succumbed to the privatization process of the 1980s and was taken over by the Belzberg company.

The chief beneficiary of the CZ withdrawal was Fletcher Challenge (FC) of New Zealand, which picked up the more valuable of CZ private lands, public harvesting rights, and mills at low cost. Like Daishowa, FC had a relatively fluid financial condition, while Canadian and American companies were in straitened circumstances in the early 1980s. It subsequently purchased majority shares in BC Forest Products and Finlay Forest Industries. It also bought shares in Latin American companies, and by 1988, it had become the largest producer of market pulp, the second largest newsprint producer, the third largest lumber producer, and, with all forms of forest production combined, the fourth largest forest company in the world. FC owned forest land or had harvesting rights in Canada, the United States, Australia, New Zealand, Chile, and Brazil. However, its dominant position was temporary. Depressed markets and exchange rates battered the company as they did many others, and before the end of the 1980s, it had eliminated a third of its holdings in lumber, plywood, particle board, and other solid wood products throughout North America. In BC, Fletcher Challenge Canada had encountered native land claims and public rebuffs over its expansionist plans in the north. The plans of that period seem wildly out of keeping with the financial situation the company actually faced in the early 1990s. Quarterly losses mounted, and by the end of 1991, it posted a net loss of $26.3 million for all Canadian operations. It sold several mills elsewhere and put two on the block in British Columbia. In early 1992, FC announced further restructuring plans; later that year, it sold its shares in Finlay Forest Industries. It was still a principal company in BC in the early 1990s, but not as large either there or on the world stage as it had been in the mid-1980s.

Besides Fletcher Challenge, the chief remaining integrated companies in the 1990s were MB (Canada); Canadian Forest Products (Canada) and its joint venture with Oji Paper of Japan, Howe Sound Pulp and Paper; Weyerhaeuser (U.S.); Crestbrook (Japan); Western Forest Products (Canada); Northwood (Canada); and Scott Paper (U.S.).

Daishowa (Japan) was effectively integrated via its connections with West Fraser (U.S.) and Weldwood (U.S.). As the nationality of owners indicates, American ownership had declined somewhat and Japanese investment had increased.

MB began restructuring in the early 1980s, and while slimmer and less top heavy in the early 1990s, it posted net losses of $29.5 million in 1991 and $14.2 million in 1992. In addition to market problems and some antiquated mills, it was facing persistent conflict with environmentalists over its coastal properties. Such hot spots as Meares Island, Clayoquot Sound, and Carmanah Valley were all located on MB logging lands. In 1993 Noranda sold its shares in MB, and the shares were widely dispersed.

Originally owned by British Columbia residents, Canadian Forest Products (CanFor) operations in the province came under the control of holding companies during the early 1980s, but the company later re-emerged and began hunting for successful sawmills to feed its two large pulpmills, now amalgamated, at Prince George in northeastern BC. It already owned several of the region's major sawmills, together with independent wood supplies. In 1989, CanFor entered a joint venture with Oji Paper of Japan to renovate its Port Mellon mill under the new name, Howe Sound Pulp and Paper. The Howe Sound venture provides Oji with a guaranteed newsprint supply for its expanding Japanese market; CanFor, with a guaranteed buyer. These restructuring moves, however, have not saved the company from continuing financial decline. It reported a $22.2 million net deficit in 1992. While Howe Sound seemed to be successful, the Prince George operations were more problematic. In January 1992, CanFor announced it would substantially reduce production capacity in newsprint and other paper.

As a railway company, Canadian Pacific had obtained land grants on Vancouver Island when BC entered Confederation. Although its forestry holdings in the province were not extensive, it was one of the world's largest producers of newsprint, pulp, and paperboard, with locations in five Canadian provinces and in Washington State. It was reported to have access to over 11 million hectares of forest when it announced in August 1993 that its forest subsidiary, Pacific Logging, was up for sale. This company had been a wholly owned subsidiary of Canadian Pacific Forest Products until June 1993, when minority shares were put on the market. Pacific Logging continues to be blessed with a secure supply of wood in excess of its annual sawmilling needs, yet the parent company reported a $57.9 million net loss in 1992 (*Globe and Mail*, 5 June 1992: B7).

Weyerhaeuser, the largest private timber owner in the world (with 14.6 million hectares in the United States) also has harvesting rights, sawmills, and a huge pulpmill in the central interior of BC. Daishowa, owner of the two mills at Quesnel, opened another mill in northwestern Alberta in 1990. West Fraser, majority owned by the Ketchum family in the U.S. and co-owner of Quesnel River Pulp, is also the co-owner of Eurocan (with Enso-Gutzeit, Finland) and of Babine Forest Products and Houston Forest Products with Weldwood. By linking its sawmills and timber holdings to the Japanese pulp market via Daishowa, West Fraser found a market for pinewood that was not of high enough quality for lumber, as well as for its chips. Its lumber also gained markets in Japan. Weldwood, the other co-owner in some of these ventures, is owned by the giant American company Champion International. Honshu and Mitsubishi of Japan hold majority shares in Crestbrook, situated at Cranbrook in southeastern BC. Honshu has been involved in this mill for many years, originally in a joint-venture arrangement with a BC family. The Crestbrook interests are linked with Daishowa in proposed mill operations in Alberta. .

In the mid-1980s, BC Forest Products (BCFP), together with two smaller BC companies, Doman (Western Forest Pulp) and Whonnock (International Forest Products), created Western Forest Products as a joint venture. With BCFP disintegrating in the late 1980s, Doman eventually acquired controlling shares in this company even while the owner, Herb Doman, was fighting a court battle over an insider trading deal that incidentally involved the then premier of the province. International Forest Products is a major logging operator at Clayoquot Sound. These companies are not integrated forward to pulp and paper, but they are important because they hold extensive timber-cutting rights.

Federal Government Initiatives

Under Canadian constitutional agreements, provinces retain control over land; thus the provincial government is the custodian of forests. However, the federal government became increasingly active during the 1980s in attempts to improve the condition of the country's forests. Its revenues are affected by the decline of the industry, just as are those of provincial governments. In 1986 it entered into a partnership with the British Columbia government to restock NSR lands and provide funds for research. According to its publicity, under the agreement at least 150,000 hectares of the estimated 738,000 hectares of good- and medium-site backlog NSR lands would be replanted in the first five years. The agreement was renewed in 1991 for another five years, but other priorities came into play in 1993, when the final budget of the Conservative government dropped the agreement.

Under neither agreement were the companies that benefited from the exploitation of old-growth forests required to contribute to the funding of reforestation; indeed, some of those companies – such as Rayonier, whose president was quoted at the outset of this chapter – had left the province. The program offered some benefits to municipalities, native Indian bands, and small woodlot owners. Its stated philosophical objective was the advancement of sustainable development, by which was apparently meant sustenance of the industry within guidelines deemed appropriate by governments and until such time as budget deficits overcame concern for sustainability of either the region or its environment.

NEW VOICES AND NEW CLAIMS

Restructuring is a process undertaken by companies, with or without the support of governments and unions, to cope with global change in their raw material sources, technology, and marketplace. But it is never a containable process. Once it is set in motion, there are unintended consequences, one of which is the opening up of debate about the uses of land, resources, labour, and government subsidies. This secondary process is well under way in British Columbia.

The new voices are those of environmentalists and indigenous peoples: the first to claim the remaining forest against both companies and unionized labour, the second to claim it for development or preservation by native bands. With the industry in decline and companies in some disarray, the government is listening to these voices for the first time in the province's history, and real changes have been initiated. As well, there are voices at both ends of the traditional political spectrum in BC arguing in favour of structural changes in the tenure system.

Indigenous Peoples' Land Claims

Land claims in BC go back to the 1920s, but they became much more prominent in the 1980s. Organized native groups launched court cases in defence of their rights, and they challenged the rights of forestry and mining companies to exploit traditional hunting and fishing lands. In 1985, Nisga'a chief James Gosnell launched a public complaint with the provincial ombudsman about the Westar company. A study commissioned by the band showed that most of the area logged by this company had not been satisfactorily reforested. Soil degradation and loss of fish and game habitat had prevented band members from hunting and fishing, and very little employment was provided band members in local logging operations (BC, Office of the Ombudsman 1985). The Gitksan Wet'suwet'en also launched a case in 1984 relating to some 50,000 square kilometres. It became a *cause célèbre* when the presiding judge declared in 1991 that there was no extant aboriginal culture to preserve. In another case, the Sekani, located at MacLeod Lake, south of Mackenzie, blockaded a logging access road in 1986 to bring public attention to their claim for some 80,000 hectares in the region. Other aboriginal cases have affected virtually every major watershed along the coast and several regions in the interior.

Eight cases, including the Gitksan Wet'suwet'en, were heard by the BC Court of Appeal in 1993. A panel of five judges ruled 3 to 2 against claims for unqualified aboriginal title to 57,000 kilometres of traditional territory for the Gitksan Wet'suwet'en, but all five ruled that aboriginal rights exist and are protected by the constitution, thereby overturning the 1991 judgment of the chief justice of the BC Supreme Court. The 1993 judgment, while not granting ownership rights to the native bands, provides them with enhanced bargaining capacities over land use. In several other cases, hunting and fishing rights were upheld, but not the right to sell on a commercial basis fish caught on reserves or contrary to fisheries laws.

The forest industry expressed consternation over the 1993 decision, less because it recognized aboriginal rights than because it created ambiguity and will lead to further negotiations over forestry practices and harvesting rights. Prior to the 1993 ruling, the BC Council of Forest Industries (COFI) had complained that the economy was adversely affected by outstanding claims. "Since 1985, a growing number of companies in BC have faced disruption of logging operations, suspension of logging activities, the need to pursue costly litigation to protect harvesting rights, and growing uncertainty about the security of long-term tenure and access to Crown-owned timber" (COFI, 1988:20). Some land-claims cases involve coalitions between native groups and environmentalist organizations. COFI charges, for example, that environmentalists began a particularly hostile battle over logging on Meares Island, off the west coast of Vancouver Island, and that the land claim by the Clayoquot and Ahousaht bands was launched later, when logging was about to begin (1988:15–16). The BC Court of Appeal issued an interim injunction in 1985, preventing MacMillan Bloedel from logging the island. A similar history is told of the Stein Valley and the Lytton and Mount Currie native bands. According to a 1989 Price-Waterhouse study, uncertainty about aboriginal rights has been costly in foregone investments; the 1993 court decision has not cleared up the uncertainty.

A number of major land claims were under discussion in the court system when, in 1992, the British Columbia government established a quasi-judicial body to investigate and rule on land claims. This approach may reduce the pressure on the judicial system and improve the likelihood that negotiated agreements will be arrived at through essentially political processes. In 1992 also, a commission was established by the federal and provincial governments, together with First Nations associations, to deal directly with land-claims issues. The objective was to remove the process from both politics and courts and to attempt to arrive at negotiated settlements in shorter time periods than could be achieved through legal channels. In the land-use decisions of the 1993–94 period, the BC government set aside some lands and established limited rights to other lands or their produce for established native groups in the affected regions. In mid–1994, the Gitksan Wet'suwet'en announced that they would not pursue their case through the courts because the process might now be negotiated directly. By February 1995, the Nisqa'a claim was close to settlement as well. Other bands are likewise hopeful of arriving at settlements. Although these settlements may not be all that the participants wish, for the first time in the province's history, land claims are being dealt with directly.

While the forest industry expresses its frustration at the intrusion of "outsiders" in the competition for forest land, there is one aspect of aboriginal cultures that forces the rest of the society to reconsider its own values. This is "community," or an emphasis on collective well-being. In spite of a history of degradation, oppression, isolation, and discrimination, despite alcoholism and high rates of suicide, native communities are struggling to ensure that as communities – not simply as individuals in a larger society – they survive. Further, these communities claim land, and they claim land not solely for economic use but as well for spiritual reasons. The forest is a spiritual home, as Chief Simon Lucas, who was quoted in a previous chapter, has noted; it should not be defiled. If it is to be enjoyed by people, then their role has to be that of stewards, not exploiters. Cynics may anticipate that if native bands succeed in claiming territory, they will simply become the forestry entrepreneurs of the twenty-first century. Since native people are neither more saintly nor more sinful than other folk, they are unlikely to re-create the world totally as self-sufficient village communities, but there is reasonable evidence that some bands at least are fully capable of changing the forest-exploitation patterns established by other settlers in the profligate century.

Community Control

Advocates of change in BC include many who believe that community control over resource management is the solution. A community plan known as the "Slocan Valley study" advocated local control in 1975; it has become the prototype for grass-roots plans throughout the province. A detailed plan for decentralization of the entire economy, including a "devolutionary contract," was proposed by participants at a conference in 1983 (D. Wilson, 1985:128–39). One of the current advocacy groups is a coalition under the label "Tin-Wis," which purports to represent aboriginal people, organized labour, and environmental, educational, and social-justice advocacy interests. It has proposed a forest stewardship act based on the same ideas. They are repeated again in a community proposal by the Village of Hazelton in 1991. Decentralized decision-making by elected resource boards are the proposed alternative to centralized government allocation of resources and the concentrated industrial units using them. Other proponents of these concepts include Herb Hammond, who was involved in the original Slocan Valley study and has since published a major treatise on community forestry (1991), and university professor Michael M'Gonigle (1989–90), whose work champions environmental concerns.

Two pioneer ventures in British Columbia are the North Cowichan Municipal Forest and the Mission Tree Farm Licence, examined in a study by Daowei Zhang (1992). Although the first had more problems than the second, Zhang concluded that both were economically viable. The Mission TFL in particular generates revenue and employment for local residents and has allowed the municipality to increase economic size and diversity in recreation, tourism, and education. In Zhang's opinion, the success of the project is due primarily to the contribution of committed and competent individuals who created a sound management plan. As well, though not primary reasons for success, the project benefited from low taxation and from start-up and other funds provided by the provincial government. Other studies of the Mission operation have also indicated its success (e.g., Reid, Collins and Associates, 1987).

To create a community project, there must first be a socially cohesive community, and Mission appears to fit the requirements. Since many of the forest-dependent towns now are artificially created locations for a resident labour force to a large company, they are not, as a matter of course, socially cohesive. Transience rates are high if the economy is sufficiently buoyant to sustain job changes. When it is in a downturn, these towns experience various other signs of social tensions, such as domestic violence, high suicide rates, alcoholism, and family breakups. The nature of the community, then, would have to be established prior to allocations of management rights.

As well, the definition of "community" becomes a hazardous undertaking as soon as practical implications are attached to it. Is a community the one closest to the resource? Is it the region of the resource? Should it include the more distant town where the manufacturing facilities are located or the city where the head office resides? Is it a watershed or a single valley? Is its basis economic or environmental? Would the government establish contractual agreements to govern decision-making, and would these limit the options or leave the fate of forests entirely to democratic processes? Would "stakeholders" have any particular role, and if so how would they be identified? These are questions that have already confronted – and confounded – decision-makers in BC when they tried to determine who should be on advisory bodies to develop sustainable-forestry plans on Vancouver Island (see Darling, 1991; Taylor and Wilson, 1992). There are no obvious answers, but in the 1990s the questions are at least on the agenda. The objective is to decentralize a range of resource-allocation decisions, or at least to bring into existence a consultative process that is, and is seen to be, accountable to residents of the province.

Non-Timber Values and the Allocation Process

The emergence of dedicated ecologists altered the dynamics of the forest industry and public acceptance of it. Some of the opponents to industrial forestry held advanced degrees in relevant subjects, and indeed some were trained foresters; they knew how much was being logged and what ecological damage logging had caused to watersheds and mountainsides. Their protests became stronger through the 1980s and gained widespread support in the 1990s. Dependent communities had high unemployment or had lost population by the 1990s, and there were more people with enough time to mount protests against logging in old-growth forests.

These new voices posed an impertinent question: How much land should be allocated to the forest industry? Since that question did not previously exist, conflict with the industry was inevitable. Companies put out many brochures about "multiple use," and some companies established camp-sites for the recreational use of land in their tenure area, but they were not legally obliged to protect old-growth forests as habitat for wildlife and non-timber flora. According to one study (Luckert, 1988), proposed mandatory expenditures on rehabilitation of forests would cause tenure holders to minimize effort and spend less than they might voluntarily invest in silviculture. Whether industries minimized effort or not, the more basic issue was that they had no incentive to expend any effort in the first place. The objection to voluntarism as a means of sustaining a forest is expressed by Paul M. Wood: "There is no apparent reason why a just democratic society should put itself in a position where it is dependent on an irrational, altruistic motive held by a private corporation ... it is the responsibility of government to avoid this situation" (1989). All very well, but neither did governments have any incentive to push companies in a direction that promised no revenues.

The first attempt by the BC government to defuse public hostility to the allocation process then in place was the establishment of a Permanent Forest Resources Commission, named in mid–1989. Its guidelines allowed it to hold public meetings. The first round of such meetings was concerned with proposed tree farm licences on Vancouver Island and in the Fraser – Fort George region, where public objections were so strong that the government stepped back from intended legislation.

Manipulating Public Opinion in the Conflict over Land Use

Following the initial round, companies and the government of the time mounted public-opinion campaigns and promised reforms to their harvesting and management practices – promised much, but far short of a cessation to logging. Employers in the forest industry, together with banks, insurance companies, and other industrial interests, had much earlier created a financial support system for the Fraser Institute, a right-wing think-tank that specialized in publishing anti-union, anti-Keynesian, and anti-welfare material. As the land-use conflict became more heated, the same groups sponsored the Forest Alliance, a public-relations enterprise on behalf of the large companies. The Forest Alliance hired Burson-Marsteller, a huge public-relations firm with a history in Latin America, to advertise the companies' case. The former head of the IWA became the president of the Forest Alliance, signifying the close relationship between company and union in opposition to the environmentalists.

Jeremy Wilson has suggested that the extraordinary countermeasures taken by industry in the late 1980s were due more to fear of change in the allocation system than to concern over the actual land already allocated. Parkland and ecological reserves amounted to 5.4 per cent of forest land in BC in 1975; by 1988, only 1.4 per cent had been added (Wilson, 1990:142). Government foresters argued that upwards of 60 per cent of the forest land was protected, but this argument lost weight when the figure was understood to include all land too remote or rugged to be of any value to the forest industry.

Environmentalists and native peoples were questioning a system that had always allocated timberland on long-term leases to companies, had charged too little for the resource, and had resulted in land either not adequately restocked or no longer salvageable, together with silted stream beds and damaged ecosystems. The rules were at issue at least as much as the particular valleys and mountainsides. While various wilderness groups fought hard battles to preserve an island, a valley, or a particular coastal area, the industry actually increased the area clearcut each year. Wilson estimated an annual increase between 1976 (as documented then in a royal commission report) and 1985 in the order of 54 per cent, roughly 70,000 hectares per year. By 1985, over 200,000 hectares per year were clearcut (Wilson, 1990:162; see also BC, Ministry of Forests and Lands, *Annual Report,* 1986–87:52). Each time an area was preserved, industry intensified its extraction rate in other areas. In Wilson's view, "environmentalism imperils the unwritten code of speculative rights at the heart of the B.C. capitalist ethos, threatening a system that has long legitimized a profitable traffic in rights to Crown resources" (1987:3).

Companies were not alone, however, in trying to manipulate public opinion. When the basic allocation rules are at stake, as they undoubtedly were in this time of restructuring, all contestants use whatever means may achieve their ends. The environmental movement also became a sophisticated user of media and publicity stunts. In 1990 the *New Yorker* carried a long article by Catherine Caufield based on a visit to BC's coast. She noted: "It is in British Columbia, where critics of the province's forest management have little access to information, no right of administrative appeal, and no resource to the courts, that civil disobedience has been most effective. One reason is the alliance between environmentalists and the province's native groups, most of whom, never having signed treaties with the government, maintain claims to their traditional lands, which compose the bulk of the province" (14 May 1990:72). Other world "names" were recruited to the cause, including Robert F. Kennedy Jr, who toured the Clayoquot Sound region on the west coast of Vancouver Island and produced copy for the news media in February 1993.

Proposals for Change

Among the numerous proposals for changes in the allocation system was one by graduate students at Simon Fraser University, who assessed the cost of a comprehensive wilderness protection program as advocated by a group known as the Valhalla Society (SFU, 1990). The SFU study estimated that a reduction in the provincial AAC of 3.5 per cent could be expected if the proposal were implemented in its entirety and without offsetting policies; variation would run from 1.9 per cent in the Prince George region to 5.2 per cent in the Cariboo region. Probable employment losses were estimated at 2,554 jobs, plus an equivalent number of jobs indirectly related to forestry. Provincial revenues would decline by approximately 53 million dollars in the first year, an amount below 0.5 per cent of provincial revenues from all sources in the fiscal year 1988–89. The SFU study concluded that the job and revenue losses could be sustained by the provincial economy and that offsetting increases in tourism and other industries would ameliorate the impact. It also urged a reallocation of commercial timber rights to industries engaging in better utilization and value-added manufacturing. This recommendation was one of many inputs into the debate; it was important because it moved beyond condemnation and moralizing into specific proposals with cost-benefits calculations. While rhetoric dominated the public forum, behind the stage, participants were now aware that solutions would involve trade-offs, and the trade-offs could be measured in jobs, income, and revenue.

Proposals for Log Markets

Arguments in favour of log markets are becoming more insistent. Apart from the advantages for those who would sell to Japan and other Asian producers, there is a potential advantage in discovering the true market value of the wood. The rationale for public ownership and private control of harvesting rights is also under scrutiny. Whether land were privatized or timber sold in small lots as on public lands in the United States, logs could be sold on the open market, with anticipated benefits to small manufacturers as well as increases in the valuation of wood. On the other side of the argument, apart from the interests of integrated companies in the status quo, is a concern about a possible "cut and run" sequence that would soon denude the Canadian forest. The question is related to the matter of ownership and the control of harvesting rights: if harvesting rights were divorced from manufacturing and allocated in larger part to small woodlot owners, log markets would be essential mechanisms in the process and the market price of logs would become established.

Government Responses

The combination of harvesting rights in enormous land areas, large sawmills, and large pulpmills, integrated and owned by relatively few companies puts any government in a tight spot. A new government is obliged to honour long-term contractual obligations and cannot abruptly change the system. If it were to withdraw its concessions or reduce its allocations, the entire provincial economy would experience the effects. As well, the government would likely fall, since governments depend on the financing and support of such corporations where they dominate the economy. None the less, the increasingly strident demands of environmentalists, with their strong popular support base and their alliance with First Nations tribal councils where land claims are at issue, and the globalization of the conflict have imposed new and unavoidable pressures for the recognition of non-economic values and more diverse stakeholder rights. The New Democratic Party government arrived in 1992 with obligations to very diverse constituencies – the IWA, public-sector unions, First Nations peoples, and environmentalists – and its first test was to find a solution for Clayoquot Sound.

The Clayoquot Sound battle pitted logging companies against ecologists, and it was fought out in the heart of small towns on the coast. No decision would be a winner; the government's own supporters were at war with one another. In April 1993 the government made a decision to give most of the forest to the loggers, though it did reduce logging area by 95,000 hectares and log volume by 33 per cent (to 600,000 CM per year). It estimated that 400 direct and 600 indirect jobs would be lost. There would be revenue losses at $7.6 million per year, with a total impact on the provincial economy of $46 million per year (BC, 1993b). The objections of the industry were muted; indeed, at least one company informed its managers that the decision was in its favour but that they should not "overtly" support the government over it. Further and more militant action was proposed by environmentalists. Robert Kennedy Jr returned; the Hesquiat Band vowed to initiate land claims.

It was some time before the government's strategy became evident. While the government had accepted the contractual claims of industry, it would use its discretionary powers in allocating AACs and enforcing regulations toward improved harvesting practices and environmental protection. An "interim measures agreement" with the Nuu-Chah-Nulth Tribal Council was announced by the government in December 1993. This agreement gave the First Nations a broader management role and veto power over land-use decisions in Clayoquot Sound while land claims are under consideration. About the same time, the government established a "scientific panel" to investigate and make recommendations in line with the objective of sustainability of the region. Then in the spring of 1994, the Ministry of Forests quietly reduced the AAC for Clayoquot and imposed more stringent rules for forest management. The BC Forest Service imposed maximum penalties on MacMillan Bloedel for illegal cuts in the sound, sending a clear signal to all companies that the decision to accept industrial claims on the forests was not "business as usual." Contractual rights were sustained, but new conditions and the enforcement of old, as well as new, rules would fundamentally change the way the business of forestry was conducted in British Columbia.

By 1994 the industry was in a defensive posture against a Greenpeace campaign to discredit BC forestry in Europe, with the Clayoquot issue at the forefront. Consumers demanded "proof" that paper purchased from Canadian mills was not contaminated by fibres crudely clearcut in old-growth forests. Companies began developing a form of credentialism to show how well they were avoiding pollutants and unsafe logging practices, and the premier toured Europe with assurances of moral probity in the BC forests.

A new system for determining land use was established by the BC Government in 1992 known as the Commission on Resources and Environment (CORE). It included representatives of all "stakeholder groups," ranging from loggers to the most militant environmentalists in every major forest region of the province. They were to negotiate outcomes in round-table-style regional meetings and make recommendations to government. This was a bold and high-risk strategy since it erased the corporatist arrangements by which land-use decisions were made by governments on the advice of companies and advanced a more consultative approach. Although CORE had a rocky start and faced opposition from both ends of the political spectrum, the process gained momentum and legitimacy over the next two years.

While CORE groups were meeting toward the end of 1993 and in the spring of 1994, the government initiated a Forest Sector Strategy and created a Forest Renewal Plan. These established an agency with representation from all vested interests mandated to take responsibility for managing investment in forest renewal. Companies would pay higher stumpage, and the increased revenue would pay for the retraining of loggers to work in silviculture and for more intensive forest-renewal practices. As well, new royalties on timber harvested from timber-licence areas were introduced, and these were much higher than industry had bargained for. The renewal plan is linked to a Forest Practices Code introduced in the legislature in May 1994. This legislation would impose stringent new conditions on companies, stronger enforcement, and substantially higher penalties for non-compliance. Overall, these various measures will reduce the profitability of logging on marginal timber lands, increase the penalties for logging beyond strict limits, and reduce the amount of logging allowed on an annual basis.

CORE came forward with recommendations on Vancouver Island land use in February 1994. It was labelled "devastating" by the forest managers and manufacturers sector of CORE (FMM, "Update #13," 15 Feb. 1994). However, the premier announced a land-use plan for Vancouver Island based on CORE's proposals the following June. Twenty-three provincial parks were to be established; total park area would then be 13 per cent of the island's land base. Although the premier said that 81 per cent of land would be allocated to industrial forestry, about half of this is not harvestable forest and about 8 per cent can be logged only under such stringent rules that large companies would likely choose to withdraw (horse logging is possible). The government anticipated a loss of 900 jobs; opponents on the industry side estimated a loss of 3,500, but the government argued that retraining will change the labour force and alter the entire context for forestry.

While no changes will be universally praised, the process of change in BC's forestry is seen by the general public, if not the wilderness groups and the companies, as an improvement over the arbitrary and secretive allocation of property rights that has characterized the coastal province since its beginning. For some opponent groups and companies, any concessions are viewed as terminal, either because the forests, once cut, are gone or because the tenure system, once abandoned, ends company control of property rights in BC's forests. But these groups do not speak for everyone, and the prevalent reaction to the premier's announcements in 1994 was relief that a decision had been made and that it embodied recommendations achieved through a consultative process (*Vancouver Sun*, 25 June 1994).

A Note on Comparative Land Costs

Land costs are a central factor in determining comparative advantages between regions or countries. In the 1980s and 1990s, southern regions had lower land costs, but as the value of plantations increases so will the cost of land. In British Columbia there is some land in the northern interior which has no value for agriculture and is not of interest to the tourism industry. Forestry and mining are currently the only industries likely to make a living there, and forestry on a long-term basis only if it can grow a crop of trees that reach commercial size in under forty years. For some boosters, that potential is sufficient to create optimism: BC forestry may yet be saved.

SUMMARY

British Columbia provides a case study of a region dependent on the extraction of timber by large, integrated companies in a provincial economy that has been a client state over most of its history. The pay-off for selling timber in the form of dimensional lumber, pulp, and newsprint was affluence; the downside was a truncated economy and seriously depleted timber reserves. Even during the prolonged postwar boom, employment did not keep up with production growth. Techno-logical change steadily reduced labour input. The boom came to an end in the early 1980s; some restructuring in the form of downsizing took place then, and employment levels sharply declined over the next decade and into the 1990s. The state of the forests was found to be fragile: large areas of insufficiently restocked land, together with land badly managed and damaged through logging practices, were docu-mented in the 1980s. Resource supplies dwindled because of overcut-ting, and the price of supplies increased both because of scarcity and because of changes in legislation pursuant to the softwood-lumber dispute with the United States.

Restructuring began in the 1980s. It included company mergers, closures of older mills, renovation at other mills, the seeking of new markets, the creation of new product mixes, and finally, attempts to create small, flexible work groups for mills facing rapidly changing market conditions. Opposition to logging practices and allocation rules grew during the 1980s and joined with protests by native groups. Companies, unions, and environmentalists engaged in struggles over land use and public-relations manipulation.

There are clear signals in the early 1990s that the BC government is prepared to change the allocation rules and process and that it is attempting to retrain the labour force for higher value-added jobs. Struggling to break out of the deadlocks between environmentalists and loggers, since 1992 it has introduced new forms of decision-making that take into account values other than the economic and that provide for greater public participation in the allocation process. These are significant changes, and they signal the end, not only to monopoly by integrated forest companies, but to an economy too long dependent on mining the forests.

5 Japan and the Creation of the Global Forest Industry

Japan's construction companies buy wood in neighbouring tropical countries and North America. Its pulpmills devour wood from every forested or plantation region. Its papermills are major consumers of imported pulp from a dozen or more countries. Its newsprint and fine-paper producers are now becoming established in offshore locations. The companies and mills of Japan, more than any other participants, have created the global forest economy, and they have been major forces in the deforestation connected to that industry. This chapter considers them in the Japanese context. They are engaged in so many ventures elsewhere that we meet them again and again in following chapters.

Tariff barriers in Japan had to drop to allow these companies to import wood supplies. While free trade benefits them, however, it harms small-scale tree farmers, whose product cannot compete with imports in domestic log markets. Ironically, their plight is related to that of villagers in neighbouring countries, whose trees are cut and exported to Japan, and of families in North American lumber communities, whose jobs disappear as logs are shipped to Japan. Thus do globalization and restructuring affect all comers: the dominant countries are not above experiencing deleterious impacts.

FORESTS AND FOREST HISTORY IN JAPAN

Japan has few natural resources on its four islands, and forests, which cover 67 per cent of the land area, are the exception. Among temperate-

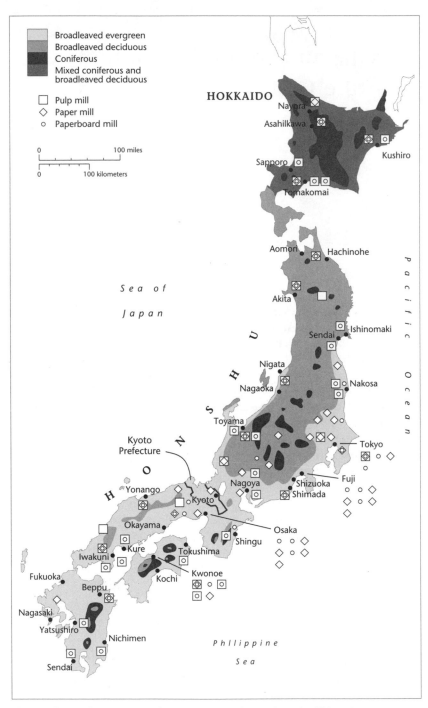

Map 4   Japan: forest types and pulp, paper, and paperboard mill locations

zone countries, only Finland has a larger proportion of its land in forests. Hokkaidō, the northernmost island, and the northern half of Honshū are mountainous territories with subalpine coniferous and broadleaf forests. A mountain ridge running north to south along Honshū provides a temperate subalpine climate where conifers are grown in plantations. A warm-temperate zone characterizes the more southern islands of Shikoku and Kyūshū, distinguished by indigenous hardwoods (Cox, 1988:164).

Hokkaidō still has natural forests with such species as hiba or false arbor vitae or cedar (*Thujopsis dolabrata*), tsuga or northern hemlock (*Tsuga diversifolia*), sugi (*Cryptomeria japonica*), Aomori todomatsu or Todo fir (*Abies sachalinensis* or *A. mariesii*), and Japanese larch (*Larix leptolepis gordon*), as well as some less valuable in commercial terms, such as the Japanese black pine or kurumatsu (*Pinus thunbergii*). Nearly three-quarters of the northernmost island is forested, occupying some 5.6 million hectares in all and comprising about 22 per cent of the total forest area of Japan. About 27 per cent of the forests on Hokkaidō are private plantations, most in larch trees, and these average 5 hectares in size. The remainder of the forests are under government or other public jurisdiction.

The forests of the main island, Honshū, have been destroyed at least twice, first during the Tokugawa period (1590–1868) and more recently during and immediately following the Second World War (1939–45). In the earlier era, land was used primarily for agricultural production, wood being of lesser value than rice, and was owned either by hamlet communities or by wealthy shoguns and feudal clans (Totman, 1986). Renewal eventually took place, and a growing national timber market increased the relative value of wood toward the end of the feudal period. During the Meiji era (1868–1912), hamlet forests were divided into national and non-national components. Parts of the national forests were then sold to ex-samurai and other high-ranking persons, while non-national forests became the property of communities (now municipalities). Lands with obscure ownership, together with property previously owned by the shogunate, were redistributed. An enlarged forest-land owning class was established toward the end of the nineteenth century under the National Forest Law (1899) and other legislation, although some lands continued to be under the jurisdiction of municipalities or the central government. A forest service was established in 1879, and its management role was enhanced through subsequent legislation. Both reforestation and afforestation were undertaken, with planting in private and public sectors reaching a peak around 1910 (Totman, 1985; Kumazaki, 1988).

Forests were again destroyed in the Second World War or by fuel-gathering citizens in its aftermath (Kumazaki, 1988:12). Afforestation projects, encouraged by the central government, replaced former oak and pine forests with hinoki (*Chamaecyparis obtusa*, yellow cedar) and sugi (*Cryptomeria japonica*, or cedar). Oak, used in earlier times for making charcoal, lost much of its economic value when charcoal burning was banned in the 1960s. Japanese red pine or akamatsu (*Pinus densiflora*), beechwood, and other less valuable species are also found on Honshū and the more southern islands.

Prefectural governments were required under the 1959 Forest Law to develop regional forest plans in non-governmental forests. Government decrees ensured planting on private lands and prohibited cutting except for thinning of trees under the standard rotation age (about 35–45 years). Low-interest loans and subsidies for new plantations were provided in return for this cooperation (Handa, 1988a: 22–4). Under a 1958 share-renting law, forest owners, capital suppliers, and groups in charge of afforestation were encouraged to work together and share the harvest revenue. Pulpmill owners did not cooperate, having cheaper offshore fibre sources, and capital was finally provided by the central government.

Forestry has long been organized in the form of cooperatives. Their legal origins go back to a 1907 amendment to the Forest Law. This legislation obliges all forest owners in any defined community to belong to a cooperative if among them two-thirds, or enough owners to make up an ownership portion of two-thirds of the land, agree to join (Funakoshi, 1988:20). Cooperatives became the basic management units, buttressed by the forest service, and were charged with carrying out national forest policies. In 1939 the forest cooperatives were obliged by new legislation to draw up silvicultural plans for all members and to engage in afforestation. Subsidies were provided for planting. The tradition of cooperatives carrying out national forestry directives continued after the war and is still the core organizational form in rural areas. These cooperatives still tend government, shrine, and municipal, as well as private, woodlots. As this fact suggests, the hinoki and sugi forests, like the oak and pine forests that preceded them, are embedded in complex cultural traditions that tie people into communities and communities into a natural world. Local wood is always present in community rituals, and the aesthetic taste of communities is expressed in wooden furnishings and decorations. Thus changes are not merely economic when the forest economy fails.

Mountain village economies began a long decline in the early 1970s. Domestic timber constituted some 70 per cent of total fibre supplies in Japan in 1965, but between 35 and 44 per cent by 1985 (Handa,

1988a:23 and appendix table 6-2; the lower figure is given by JFTA, 1985). Although some of the decline was due to a general economic slump after the oil crisis, longer-term trends were already in evidence. Substitutes were by then in general use for construction, but even where wood was the main material, many sawmills in Japan were buying imported Douglas fir and cedar. Both quality (Douglas fir and red cedar are not grown in Japan) and price, where species and quality are equal, were cited in field interviews with sawmill managers as reasons for this trend (my field notes, 1988), but both Japanese and American economists have different explanations, as discussed below.

LOG MARKET DEBATES

Log markets are controversial, and Japan is at the centre of the controversies. One debate is about logging in Southeast Asia. The issue is introduced here, but in part II of this book we return to it often, as we consider the social and ecological impacts of tropical deforestation. A second debate is about Japanese logging operations (or contractors to them) which have recently moved into the boreal regions of Russia and Canada. Of those in Russia, insufficient information is currently available; those in Canada have been discussed in previous chapters. A third debate is about the log trade between the western regions of the United States and Japan. This debate rages in the rural regions of Japan and is discussed below.

Logging in Southeast Asia

Most tropical woods destined for Japan have been used in construction as plywood, veneers, or sawnwood. The Philippines and Malaysia were the major exporters of tropical hardwoods to Japan in the 1950s, selling lauans, apitongs, merantis ramin, and keruing (all *Dipterocarpus* and *Shorea* species). Indonesia became an exporter by the mid-1960s. By the early 1970s, these three regions accounted for two-thirds of the volume of all tropical hardwood exports, much of it going to Japan. Kenji Takeuchi, under the imprint of the World Bank, which provided funding for deforestation, forecast an average annual rate of earnings from these exports of up to 12 per cent during the 1970s (1974:xv). Korea, Taiwan, and Singapore developed processing facilities and became intermediaries for logs from their neighbouring supply countries, producing plywood, sawnwood, and veneers subsequently shipped to Japan and the United States.

Sources of tropical hardwoods declined between the early 1970s and late 1980s as supply regions became exhausted. One by one,

Figure 5.1
Japan log imports (hardwoods and softwoods) by country, 1991

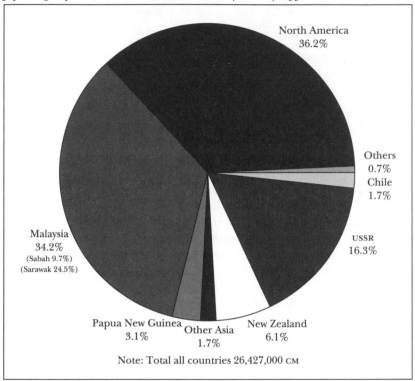

North America
36.2%

Others
0.7%
Chile
1.7%

Malaysia
34.2%
(Sabah 9.7%)
(Sarawak 24.5%)

USSR
16.3%

Papua New Guinea   Other Asia   New Zealand
3.1%   1.7%   6.1%

Note: Total all countries 26,427,000 CM

Source: Japan Paper Association, 1991; as reported in Penna, 1992.

following massive floods and public outcries, Southeast Asian tropical-wood suppliers banned log exports. A few countries – Indonesia in particular – were gradually developing manufacturing capacities and for this reason were resisting log exports. During the 1970s, Malaysia captured much of the Japanese market, but by the end of that decade, it too was facing denuded lands where forests once stood. None the less, it continued to allow large-scale logging in the states of Sabah and Sarawak, most of the logs going to Japan for the construction industry.

Some logging of tropical forests is done under the aegis of pulp and paper companies. Catherine Caufield described the logging in the mid-1980s of a 330-square-mile area of rainforest in Papua New Guinea by Jant, a subsidiary of Honshu Paper Company. This operation had been in place since 1975 and had an option on another 260 square miles when the first area was deforested. She quotes a Jant prospectus as saying, "Every day one hundred various types of heavy vehicles and their operators are working and making their presence known in the green jungles of the Gogol area." Twenty thousand tons of wood are shipped out each month, but Jant, according to Caufield, pays neither dividends or income taxes in Papua New Guinea "because it sells the wood to its parent company at such low prices that it never makes a profit" (1985:152). These trades and practices are examined for their social, as well as ecological, impacts in the chapters on Southeast Asia.

Boreal Forest Supplies

The former USSR exported coniferous logs and hardwoods to Japan throughout the twentieth century, with interruptions during various conflicts. Japan became a major contributor to foreign currency in the USSR through its imports of logs in the 1970s and 1980s. This history in itself is not controversial, but current moves by Japanese companies into former USSR territories and the exploitation of boreal forests there are becoming matters of concern amongst ecologists and forest-ers elsewhere. The desperate search for immediate cash returns by Russians and their former subject peoples may denude northern for-ests; it may also flood world markets and lower prices everywhere. Both fears, held by diverse interest groups, were current in the early 1990s.

North American Logs on the Japanese Log Market

In Japan itself, the more controversial topic is North American log sales. These, like sales from Russia, were not an issue until the late 1980s. Both logs and lumber were shipped from North America to Japan in the late nineteenth century, though they were not major export items. Shipments of "baby squares," or wood cut in blocks suitable for construction of Japanese houses, along with Douglas fir and cedar logs, became more frequent in the 1930s. The log trade increased considerably between 1929 and the onset of the Second World War. Thomas Cox traces the impact of this trade on the inter-national trade in plywood, for which the Douglas fir peelers were the raw material. Japan captured much of the market, while U.S. plywood plants in the Pacific Northwest were idle (1988:177–8). But the log trade became a major issue only in the 1980s, and it became an issue in America and in Japan for opposite reasons.

Log sales from Washington and Oregon states (and a small quantity from Canada) increased during the 1980s, when exports could fetch higher prices than the same wood manufactured as lumber on the

American market. It is these log sales that have elicited the wrath of Japanese forest owners and the attention of Japanese forest economists. Y. Iwai (1989) argues that American companies, unconstrained by lifetime labour contracts in the fashion of Japanese employers and enjoying virtually free resources, began exporting logs to Japan instead of processing them through sawmills and selling lumber in the North American market because they chose to avoid high American labour costs and had no necessary ties to communities. (The export of logs harvested in U.S. national forests was prohibited in 1973, but private landowners such as Weyerhaeuser continued to export volumes of logs into the high-return yen market.)

Iwai and other Japanese forest economists argue that Japanese producers must pay high taxes for land, high wages on a lifetime contractual basis, and high and obligatory costs for replanting and silviculture. They cannot compete under these conditions. Their children are leaving the countryside for urban jobs because they cannot make a living as forest owners, even if they inherit timberland. The forest owners have repeatedly called for import restrictions or tariffs to remedy their economic situation.

There are also complaints about lumber imports. As Japan has changed, more North American–style houses are being constructed. Japanese sawmills do not, in general, produce dimensional lumber; thus the American producers have another edge that was not there in an earlier period, although this advantage should not be overestimated since there is insufficient space in Japan for extensive North American–style housing.

The Japanese side of the argument on logging is countered by an American one. It points to inefficient production methods in Japan. Logging is labour-intensive. Average wage rates for silvicultural workers rose steadily between 1961 and 1985 (estimates provided in my 1989 interviews varied from between five and fourteen times), while the average stumpage price of (privately owned) sugi rose by only 67 per cent (Kumazaki, 1988:12). The escalation of the exchange rate on the yen after 1973, high costs of internal transportation in Japan, and the very high value of land itself are all factors in maintaining the price differential. As well, the plantation forests are not yet of an age-class that makes them fully competitive with mature timber from outside. As a result, Japanese trees are uncompetitive, but that is not the fault of producers of the more competitive North American product.

Thomas Cox argues that a major cause of problems in the Japanese forestry sector is the inefficient organization of forestry. In his view, "attitudes as well as ownership patterns stood in the way of adoption of forest management practices that would maximize production"

(1988:170–3). He attributes the cooperatives to Occupation authorities and sees community traditions as generally hampering the efforts of the forestry ministries to plan rational operations "compatible with present-day realities and needs." One must treat this argument with some caution. Cooperatives predated the war, and they are not creatures of American contrivance. Indeed, they are totally inconsistent with American world-views and typical social organization. They are certainly different from integrated companies. These reservations aside, Cox is correct in noting that the sawmill industry in Japan consists of a plethora of small mills that cannot capture economies of scale. There are some 18,000 sawmills, most producing small quantities of custom-sized lumber. During my own fieldwork in 1989, I noted that an average sawmill in Kyoto Prefecture in the late 1980s might use 1.5 to 3 thousand CM annually, consuming both imported and domestic wood. By comparison, an average sawmill in BC might produce anywhere from 30,000 to over 1 million CM in the same period. On the other hand, small Japanese sawmills often produce highly specialized and aesthetically satisfying items, while BC mills produce standard-sized (dimensional) lumber.

Another possible line of argument is that the Japanese government made a policy decision to save the country's forests while obtaining industrial raw materials from elsewhere. When the hinoki and sugi forests reach maturity, they will be more valuable because alternative sources for these species will be scarce. No evidence is available to test this speculation. The less sinister and more probable explanation is that other regions value their forests so little, they sell the trees below the costs of regeneration. As long as Japanese companies can obtain cheap fibre sources elsewhere, they have no reason to turn to the much costlier domestic sources. One might note that a similar argument has been mounted from time to time about the United States, which reduced its rate of deforestation, reforested or afforested some sections of the country, and imported large quantities of wood from Canada throughout the entire twentieth century. The evidence of a national policy in the U.S. during the 1950s is considerably more persuasive than any so far presented with reference to Japan (see Clark-Jones, 1987, for an argument along this line).

Japan's reserve supplies are not on the same scale as those in the United States. The largest company, Oji, owns 120,000 hectares of forest land (primarily on Hokkaidō), but unlike American counterparts – Weyerhaeuser, Georgia-Pacific, and others – most Japanese firms own no land and have no long-term harvesting rights; they buy wood on local and global markets. The Japanese pulpmill industry and other sectors required supplies from elsewhere because local supplies were insufficient and, with the rising exchange value of the yen, too costly; the market had to be opened up. While forest owners' cooperatives are important, they are less influential in Japanese politics than rice growers' cooperatives, and the government was able to relax tariffs on wood where it was unable to do so on rice. The saving of forests, then, was an incidental benefit, and the damage to mountain villages an incidental cost. Japan's selective adaptation to free-market principles had earlier protected the sawmill sector, with its thousands of small mills. But the needs of other industries, including the paper sector, finally obliged the government to reduce remaining tariffs.

Impacts on Rural Communities in Japan

In 1989 I conducted fieldwork in the Tanba Highlands of the Kyoto Prefecture, concentrating on tree farmers whose income had gradually eroded with the decline of their domestic market in villages that had lost population (Marchak, 1991–92). The area contained three municipalities (chō), each with populations of under 8,000. In one town, 5,000 people remained in 1985 from a base of 10,000 in 1960, and of those remaining, 30 per cent were over 60 years of age. Since 1980, 95 per cent of high school graduates moved to cities. The employment profile had changed since 1965; then, 60 per cent were in agriculture and forestry, 16 per cent in manufacturing, and 24 per cent in services. In 1985, 20 per cent were in agriculture and forestry, 40 per cent in manufacturing, and 40 per cent in services.

In the 1920s and 1930s, these municipalities produced rice, matsutake and shiitake mushrooms, charcoal (from oak), and some construction timber. Isolated from urban centres, they were scarcely involved in the cash economy. New road links reduced the isolation, legislation outlawing charcoal reduced that source of cash income and fuel, and forest products gradually became important sources of income for a rural society now more integrated into the market economy. However, traditional land ownership patterns prevailed, with the majority of private forest holdings being under 5 hectares. Families, communities, and shrines owned land, and all were connected with one another by cooperative practices and ancient communal obligations. Municipalities as a whole functioned as cooperatives, where decisions affecting others were made communally by both elected and informally recognized leaders.

Following the war, the same communal organization of small-lot farmers re-created a forest economy. They replanted, and they sold the still immature trees thinned from new plantations. In the 1950s, local construction used local wood, grown and milled by members of the district community. But by the mid-1980s, with the planted forest requiring expensive silviculture, the market had collapsed. Even national forests had accumulated a deficit. Plans for future afforestation were being shelved. The central government revised its harvesting plans and reduced its employment from about 70,000 in 1964 to about 43,000 in 1987; another 20,000 were scheduled for reduction by 1997 (Handa, 1988a:31). A sawmill owner in Wachi-chō, the smallest of the three villages in the study area, informed me that 50 per cent of the lumber used in construction in the region was imported. Douglas fir is not grown in Japan, so no domestic product competed directly with imports; however, with growing affluence, construction companies were purchasing the exotic Douglas fir in preference to local construction timbers. Where the species were identical and quality similar, construction companies were buying imports because they were cheaper than domestic wood.

In Keohoku-chō, the largest of the three communities and the one closest to Kyoto, a formerly independent community was gradually transforming itself into a commuter satellite for Kyoto. The population was nearly 11,000 in 1955, just over 7,000 by 1985. Of those continuing to live in the rural area, many travelled regularly to urban jobs, and some of these were forest owners. That is, they owned some 5 hectares of land, had invested considerable family funding and time in replanting and silviculture, and may well have carried out traditional rural roles of ancient standing for their family and community. But they could not make a living out of tree farming. Many such farmers had university degrees (indeed, several university economists were also tree farmers), and their children attended urban schools with a view to gaining entry in top universities. The economic conditions of the rural area, then, should be understood within the context of rising affluence and inextricable ties to urban life for forest owners.

Although owners and cooperatives are expected to provide lifelong employment to workers, in fact, there is a labour shortage in the rural regions of Japan, and employers have difficulty attracting labour to silviculture and logging. In one logging operation that I witnessed, the felling crew were in their late fifties and early sixties. Young people are not being trained as loggers because few seek work in the woods. Combined with other conditions, this provides an incentive to sell land to urbanites for summer cottages. Some owners have succumbed to this pressure, though there is considerable social disapprobation of such decisions, and a large real estate company would not be able to obtain extensive holdings. If a tree-farm owner sells the land to a developer, he (rural land in Kyoto Prefecture is owned almost exclusively by male heads of families) might take on the building task, using his own logs for construction.

Several new industries were getting started in this municipality. One of the effects of economic conditions was that cooperatives and individuals were becoming more entrepreneurial in their interactions with the external marketplace. Individuals whose families did not own land established real estate businesses catering to wealthy urban Japanese who wanted country homes; a cooperative of forest owners created a successful polished cedar pole business, and some of the cedar was imported. Another business in prefabricated cedar homes, using only imported wood, was thriving outside the confines of traditional cooperative organization. Cheaper imports caused the decline of the traditional rural village, but they also created the rise of industries in what used to be rural Japan. The municipal government tried to attract industries, providing community land as an incentive. The Kyoto Prefecture government, one municipality, and local business organizations, brought together by the local banker, purchased 6,000 square metres for 1 million yen from a local forest owner and then sold it to the Panasonic corporation at 80 per cent of the cost, in an attempt to bring an employer into the region.

Similar situations in northern and central Hokkaidō are reported by Japanese researchers. Yutaka Ishii and Hitoshi Arai (1986) surveyed small-holding forest owner-managers in Furen-chō, northern Hokkaidō, discovering that a majority had no plans for planting because of low prices for Japanese larch (used mainly for pulpwood). Forest land was being turned over to farming and grazing land. Another study in central Hokkaidō by Shigeru Shimotori and Yukio Akibayashi (1989) provides further evidence of the decline in rural forest villages. The township of Shimokawa had a peak population in 1960 of 15,500. By 1985 the population was a third of that figure. Almost all land in the township is forest or uncultivated field, 88 per cent of this in nationally owned forest. The same pattern of decline characterized this region as in others. Various levels of government and the local population were attempting to revitalize the region through establishment of small-scale industry.

Villagers in rural Japan obviously care about wood. As we surveyed plantations, they talked about each tree as if it had its own personality. They worried over any defect. Sawmills produced highly crafted furnishings and centre poles for shrines and decorative purposes. They wasted little. Public-opinion polls in Japan show high regard for the environment (Handa, 1988a; Ishii, 1985; Mitsuda and Geisler, 1988; but see Cox, 1988, for a contrary opinion). One might note that this attitude does not seem to extend to the forests of neighbouring countries. The apparent shift from conservationist attitudes (expressed in the rural villages) to hard-nosed market attitudes (expressed with

respect to foreign operations) is not a peculiarly Japanese characteristic, however. The same change seems to affect people of all national
origins once they become competitors on global markets.

THE PULPMILL AND PAPERMAKING
INDUSTRIES

The pulp and papermaking industries were established before the
turn of the century, were re-established in 1948, and have been
expanding steadily since then. In 1960, Japanese consumers used 4.4
million tons of paper and paperboard; in 1993, 28.1 million tons (JPA,
1994: table 4). Every paper product enjoyed stronger market demand,
including newsprint and advertising, packaging, and office supplies.
Japanese paper companies maintain the world's second largest production (the United States is first). They supply some 96 per cent of
the domestic market, but export very little.

Pulp is the essential ingredient for their products, but with modern
transportation and new plantation capacities in many countries, there
is no longer a necessity to produce it in Japan. Declining external
supplies of logs and wood chips, together with increasing concerns
about energy costs and air pollution, underlie new government regulations that should lead to reduction in mill numbers and greater
rationalization of the industry. The Ministry of International Trade
and Industry (MITI) imposed a moratorium on the installation of new
machinery and took further actions to reduce overcapacity in the early
1980s. Further regulations require detailed plans for production
increases (Julian, 1988; as reported in Penna, 1992). The restructuring is evident in the number of mills. A census of manufacturers in
1980 tallied all pulp, paper, and paperboard mills at 30,479; the same
census in 1992 counted a total of 20,196. MITI data for 558 (apparently larger) companies in 1963 numbered 655 mills, of which 44
produced only pulp; in 1990 the same survey counted 375 remaining
companies with 452 mills, of which 8 produced only pulp. Integrated
companies (pulp and paper/paperboard) were also fewer in the MITI
survey, down from 93 to 54 in the same three decades. Non-pulp-
producing paper companies, likewise, declined from 518 in 1963 to
290 in 1990 (JPA, 1994: tables 29, 30).

Pulpwood Supplies

Japanese papermakers own very little forest land in Japan itself. Oji
Paper holds the largest land area at 120,000 hectares on Hokkaidō.
That island and northern Honshū are the primary regions for natural

Figure 5.2
Japan: sources of material for paper supply, 1990

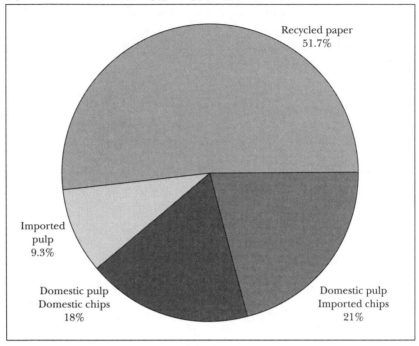

Recycled paper
51.7%

Imported
pulp
9.3%

Domestic pulp
Domestic chips
18%

Domestic pulp
Imported chips
21%

Source: Japan Paper Association, 1991; as reported by Penna, 1992, figure 3.

beechwood or conifers grown in pulpwood plantations on public lands and put up for bid. Although domestic pulpwood sources have slightly increased since the 1960s, consumption has increased faster, and imports of pulpwood, especially hardwoods, have increased substantially. In 1987, 13.9 MCM of pulpwood was imported, constituting 43 per cent of all pulpwood supplies; roughly half of this was in softwoods. By 1993, a total of 20.8 MCM was imported, constituting 59 per cent of the total supply; a third was in softwoods. Imported pulpwood in 1993 derived from the United States and Australia (each about a third of all imports), Chile (about a fifth), and South Africa (a thirteenth), with Canada and New Zealand distant competitors.

Wood Chips

Wood chips were not traded on international markets in any substantial quantity before the early 1960s. Japanese companies introduced specialized chip carriers in 1966, after which trade increased from 523,000 CM (1961–65) to 1.8 MCM and then to 7.4 MCM by 1970.

Table 5.1
Foreign direct investment by Japanese paper companies in pulp and paper
manufacturing ventures, 1994

Venture name	Location	Japanese companies	Product
Alaska Lumber & Pulp Co. Ltd	Sitka, Alaska, USA	Alaska Pulp Co.	dissolving pulp
Crestbrook Forest Industries Ltd	Skookumchuck, BC, Canada	Honshu Paper Mitsubishi Corp.	bleached sulphate pulp
Cariboo Pulp and Paper Co.	Quesnel, BC, Canada	Daishowa Paper Marubeni Corp.	bleached sulphate pulp
*Carter Oji Kokusaku Pan Pacific Ltd	Napier, New Zealand	Oji Paper, Sanyo-Kokusaku Pulp	thermomechanical pulp
Cellulose Nipo-Brasileira SA	Minas Gerais, Brazil	Japan-Brazil Pulp and Paper Resources Development	bleached sulphate pulp
Quesnel River Pulp Co.	Quesnel, BC, Canada	Daishowa Paper Co.	thermomechanical pulp
Tenma Paper Mills Co. Ltd	Bangkok, Thailand	Tenma Paper, Sumitomo Corp.	whiteboard
North Pacific Paper Corp.	Longview, Washington, USA	Jujo Paper Co.	newsprint
New Brunswick International Forest Products	Dalhousie, NB, Canada	Oji Paper, Mitsui & Co.	uncoated groundwood paper
Kanzaki Speciality Papers Inc.	Ware, Mass., USA	Kanzaki Paper Mfg. Co.	thermosensitive & tack paper
Caraustar Industries Inc.	Austell, Ga, USA	Settsu Co.	recycled paperboard
*Daishowa America Co. Ltd	Quebec, Canada	Daishowa Paper Co.	directory paper
Daishowa Forest Products Ltd	Quebec, Canada	Daishowa Paper Co.	newsprint
Matrena-Sociedade Industriale de Papers SA	Lisbon, Portugal	Settsu Co.	printing paper
Daishowa Canada Co. Ltd	Peace River, Alberta, Canada	Daishowa Paper Co.	bleached sulphate pulp
Howe Sound Pulp and Paper Co. Ltd.	Howe Sound, BC, Canada	Oji Paper Co.	pulp & newsprint
Forestale Anchile Ltda.	Chile	Daio Paper Co.,	newsprint
Pan Pacific Forest Industries	Napier, North Is., New Zealand	C. Itoh & Co.	thermomechanical pulp (240,000 mt/y)
Kanzan Spezial Papiera GmbH	Neumuhl, Germany	New Oji Paper Co.	thermosensitive paper (25,000 mt/y)

Table 5.1
(continued)

Venture name	Location	Japanese companies	Product
Daishowa-Marubeni Industrial de Papeis SA	Peace River, Alta, Canada	Daishowa Paper; Marubeni Corp.	bl .sulphate pulp (340,000 mt/y)
Alberta Pacific Forest Industries Inc. (Al-Pac)	Athabasca, Alta, Canada	New Oji Paper, Hokuetsu Paper, Mitsubishi & Co.	bl. sulfate pulp (500,000 mt/y)
Jujo Thermal Oy	Kauttua, Finland	Nippon Paper Industries Co. Ltd, Mitsui & Co.	thermosensitive paper, others (25,000 mt/y)

Source: Japan Ministry of Finance; published in JPA, 1994.
* These were not included in 1994 source but are noted in Penna, 1992: appendix 5.

Japan became the largest importer of wood chips, taking more than 80 per cent of the world's total through the late 1970s and early 1980s. The west coasts of Canada and the United States were major supplying regions, and Weyerhaeuser was the largest single supplier in the 1970s.

In 1979, U.S. and Canadian suppliers of chips increased their price by 67 per cent over a six-month period, causing what the Japanese called "chip shock" among the pulp and paper producers in Japan. MITI, together with members of the industry, introduced guidelines to prevent any recurrence of the dependence on single supply sources and of surprise escalations in prices. The guidelines involve an expansion of domestic pulpwood supplies, an increase in the percentage of waste paper in the total fibre furnish, and general diversification of overseas fibre supply sources. New sources for chips are hardwood or radiata pine regions: New Zealand, Papua New Guinea, Australia, Chile, and Brazil. The USSR was considered a likely new source for softwood chips before its political breakup. The chip trade nearly tripled to 19.2 MCM by 1980 and has been constant at or slightly below that level since then (Schreuder and Anderson, 1988:162).

By the mid-1980s, the Japanese fleet consisted of sixty-nine carriers with an average capacity of 55,000 CM of chips (Kato, 1985; Schreuder and Anderson, 1988:163; see also Ishii, 1990). Further expansion of the chip trade is not anticipated because as more countries develop their domestic industries in pulp and paper, their wood will be used internally. For Japanese producers, this will be a distinct limitation on growth.

Table 5.2
Foreign direct investment by Japanese paper companies in wood-chip supply ventures, 1991

Venture name	Location	Japanese paper companies	Product
Harris-Daishowa (Aust) Pty. Ltd	Eden, NSW, Australia	Daishowa Paper Co., C. Itoh & Co.	eucalyptus chips
Jant Pty. Ltd	Madang, Papua New Guinea	Honshu Paper Co.	hardwood chips
PT Chip DECO	Tarakan, Kalimantan, Indonesia	MDI (Jujo, Kanzaki, Sanyo-Kokusaku, Oji)	mangrove chips
California Wood Fiber Corp.	Sacramento, Calif., USA	Daio Paper Corp., Marubeni Corp.	chips
Oregon Chip Terminal Inc.	Coos Bay, Oregon, USA	Daio Paper Corp., Kanematsu Ltd.	chips
Primex Fiber Ltd	Delta, BC, Canada	Sanyo-Kokusaku Pulp Co. Ltd.	alder chips
Forestal Anchile Ltda	Chile	Daio Paper Corp., C. Itoh & Co.	plantations
Forestal Tierra Chilena	Chile	Mitsubishi Paper, Mitsubishi Corp.	plantations
Volterra SA	Chile	Nippon Paper Industries Co. Ltd, Sumitomo Corp.	plantation
Southland Plantation Forest	South Is., New Zealand	New Oji Paper Co., C. Itoh & Co.	plantation
Albany Plantation Forest	Albany, Australia	New Oji Paper Co., C. Itoh & Co.	plantation

Source: JPA, 1991; as reported in Penna, 1992, appendix 6.

Waste-Paper Utilization

More than any other country, Japan has utilized recycled paper. It became a source for domestic pulp in the 1960s, and usage increased following the oil crisis of 1973. Recycled paper in 1993 constituted 53 per cent (11.7 million tons) of the total fibre supply for pulp (JPA, 1994: table 14). In a discussion note to the FAO *Proceedings* in 1987, I. Ushiba says that "the collection ratio and utilization rate of waste paper in Japan are already nearly 50 percent, and we can say that a certain saturation has been reached. We think that waste paper imports from the United States and other countries will remain relatively easy" (FAO, 1987b:175). That prediction was given before major U.S. state legislatures decreed that newsprint must contain up to 50 per cent recycled paper. These new laws will increase the marketability of waste paper in North America and reduce its supply on external markets, but by 1993 the impact was not yet affecting Japanese imports.

There is also an upper limit imposed by the nature of paper products now demanded by Japanese consumers. No longer content with products suitably furnished by waste paper, they demand fine-quality papers for artistic applications. Ian Penna noted in 1992 that cheap wood-chip supplies were also reducing the use of waste paper in Japan. It "appears that the industry has the technology at its disposal to substantially increase its use of recycled paper," he says, but "1990 industry demand forecasts expected [Japan's] dependency on foreign wood supplies to increase to 70% of total raw fibre requirements by the year 2000" (1992:1).

Wood Pulps

Pulp produced in Japan for papermaking (using domestic or imported fibre) has continued to increase through the 1980s and early 1990s, but imports have also increased in that time. Domestic pulp consumed by papermakers totalled 9.5 million tons in 1987 and 10.5 million tons in 1993; imports totalled 2.1 million tons in 1987 and 2.6 million in 1993. (A total of 3.3 million tons of pulp was imported in 1993, but the remainder was not consumed directly in the production of paper and paperboard, and a small quantity was wood-free pulp.) In 1993 the chief source countries were Canada and the United States, each providing about 1.2 million tons, and Brazil, New Zealand, and Chile, providing 247,500, 246,500, and 174,200 tons respectively. South Africa, Sweden, Portugal, Finland, and Indonesia accounted for most of the remainder (JPA, 1994: tables 14, 19–21).

Paper and Paperboard Products

Japan absorbs most of its produce on the domestic market and neither imports nor exports much paper. Since about 1988, however, its imports have increased, primarily in printing, writing, and newsprint grades. Its chief sources for newsprint imports are the United States, Canada, Finland, and Sweden; for printing and writing papers, these countries plus Norway and Southeast Asia (JPA, 1994: tables 8 and 9). As Japan's appetite for papers continues to grow, imports are increasingly from Japanese-owned mills located elsewhere or from mills working in contractual arrangements with Japanese merchant companies. Investment abroad from about the mid-1980s was finally including newsprint and even papermills. As mentioned in the last chapter, Oji, together with the Canadian company Canadian Forest Products, has

Table 5.3
Pulpwood imports by origins and species, 1987–1993 (in 1,000 CM)

Year	USA	Canada	Australia	New Zealand	Chile	South Africa	Other	Total	Per cent of total pulpwood supply
1987 total	5,449	1,306	4,503	722	103	690	1,105	13,878	43
Of which softwood	4,346	1,246	86	495	103	–	568	6,844	44
Of which hardwood	1,103	60	4,417	227	1	690	536	7,034	42
1988 total	6,918	1,740	4,583	468	667	566	1,407	16,439	47
Of which softwood	4,919	1,650	144	311	322	–	623	7,969	47
Of which hardwood	1,999	89	4,439	157	344	566	786	8,380	47
1989 total	7,652	2,427	4,585	518	1,579	527	1,642	18,929	51
Of which softwood	4,955	2,096	182	336	506	–	691	8,766	48
Of which hardwood	2,696	331	4,403	181	1,073	527	952	10,163	52
1990 total	8,427	1,998	4,411	516	2,352	504	1,731	19,939	53
Of which softwood	4,989	1,875	173	397	477	–	714	8,625	49
Of which hardwood	3,438	123	4,238	119	1,875	504	1,017	11,314	56
1991 total	9,263	1,922	4,864	814	3,348	672	1,728	22,611	57
Of which softwood	4,947	1,714	469	736	682	–	664	9,212	51
Of which hardwood	4,316	208	4,396	78	2,666	672	1,063	13,399	62
1992 total	8,547	1,721	4,549	573	3,173	670	1,815	21,048	57
Of which softwood	4,113	1,663	607	508	455	–	544	7,890	47
Of which hardwood	4,434	58	3,942	65	2,718	670	1,270	13,157	64
1993 total	7,848	1,369	4,947	605	2,842	1,110	2,070	20,791	59
Of which softwood	3,662	1,182	757	528	216	–	700	7,045	46
Of which hardwood	4,186	187	4,190	77	2,626	1,110	1,370	13,746	69

Source: JPA, 1994.

invested in a newsprint mill at Howe Sound in Port Mellon, British Columbia. This mill uses the cheaper energy and wood costs in Canada to supply the Japanese market. *Pulp and Paper International* quoted the administrative general manager of Oji's Tomakomai mill as saying that "many other industries, like automobiles or electronics, went off-shore quite a while ago in search of lower costs, whether it's raw material, energy or labor. Now I think the paper industry is following this trend and I expect it to grow in the future" (O'Brian, 1988:68). This prognosis is supported by investment flows. Canada and the United States are among the recipients; the limits on fibre supplies for Canadian mills are temporarily overcome by tapping the boreal forest supplies. But if the Japanese are planning to move substantially greater proportions of their newsprint industry offshore, they are more likely to place it where the coniferous raw materials are cheapest while still of high quality and that would mean in Paraná, Brazil, and in southern Chile. Both produce pulp from plantation pines.

Employment

Unlike the sawmill sector, the pulp and paper industry in Japan is fully modern and rationalized. As a corollary of that process, employment declined while production expanded. In 1965 the pulp sector employed 24,335 workers; in 1990, 9,516. Corresponding figures for the paper sector were 58,807 workers in 1965 and 38,270 in 1990. Employment in the paperboard sector in 1965 was 23,581; in 1990, 11,921. Ian Penna calculated the changes in these statistics, using tons per person for production measures. For pulp there was a 61 per cent reduction in employment and a 461 per cent increase in production per person (1992:16). In paper there was a 35 per cent decrease in employment and a 499 per cent increase in production per person. In paperboard the decrease in employment was 49.9 per cent, and the increase in production per person was 683 per cent. These employment-production ratios are similar to those in North America and Europe. As the pulpmill industry became automated, the number of workers everywhere declined (see also JPA, 1994: table 30*).

Dominant Paper Companies

The first paper company in Japan was Shoshi, which later became the Oji Paper Company; it was founded in 1873 and used cotton rags as

---

* While trends are clearly evident in the directions discussed in the text, precise data provided by JPA vary considerably.

a fibre source. Oji and another company, Fuji, began to produce sulphite pulp from spruce and fir in the late 1880s. Unlike other forested regions of Japan, harvesting rights in public forests on Hokkaidō were allocated on short-term contracts to large companies, including Mitsui Bussan and Oji Paper, as early as 1907 (Ishii, 1985). Oji's Tomakomai mill was established on Hokkaidō in 1910. By that date there were eleven paper companies, twenty papermills, and a total production of 86,000 tons of paper in Japan (Ishii, 1990). In 1933, Oji merged with Fuji and another major firm, Karafuto Kogyo, to form the Oji Paper Company.

The loss of the Sakhalin islands after the Second World War depleted a major source of timber for the pulpwood industry. As well, the *zaibatsu* (large industrial and financial houses) were dissolved by the Allied Command (though most were restructured and continue to this day). Paper companies under Mitsui and Oji were broken up. Oji was obliged to divide its operations into three groups, becoming Jujo, Honshu, and Oji, all of which became major companies. All operate globally; all three, for example, are active in British Columbia and Alberta and have been discussed in chapters 3 and 4. With the Korean War spurring economic recovery, investment in modern pulp and paper plants began around 1950, and production steadily climbed. In short order, Japan became the world's second largest paper and paperboard producer and the third largest pulp producer.

The Japanese pulp and paper industry, like other maturing industries everywhere, has fewer, but larger companies today than in its formative years. The major part of production is actually controlled by very few of these. Table 5.4 provides data on the percentage shares of total production held by the largest companies in 1960, 1970, and 1990. Four major companies, three of them the 1949 successor firms of Oji (Oji, Jujo, and Honshu), plus Daishowa, founded in 1938, are clearly dominant. They are followed by Sanyo-Kokusaku, Daio, Rengo, Mitsubishi, Kanzaki, and Hokuetsu as ranked by sales (JPA, 1994: table 32).

Oji controls 9.5 per cent of total production in 1990 and has the leading share of the newsprint market at 29.8 per cent and of A-grade printing paper at 20.3 per cent. It is the second largest producer of coated papers, with 26.5 per cent of production. Its giant mill at Tomakomai on Hokkaidō has been regularly updated and expanded over the entire twentieth century, and now it has moved into the world market. Daishowa, the second largest paper and paperboard manufacturer, has built its strength by creating joint-venture pulpmills with American companies that already have extensive harvesting rights in Canada and the United States. Among them are the two pulpmills at

Table 5.4
Japan: Leading pulp and paper manufacturers in sales for the March 1993 term

	Company	Sales in 100 mill. yen	Profits after tax in 100 mill. yen	Total assets in 100 mill. yen	Number of employees	Profit to sales ratio (%)	Profit to total liabilities & net worth ratio (%)	Net worth ratio (%)
1	Oji Paper Co. Ltd.	4,500	78.1	6,604	6,181	1.7	1.2	37.2
2	Honshu Paper Co., Ltd	3,903	32.5	4,592	5,972	0.8	0.7	20.1
3	Jujo Paper Co., Ltd	3,789	35.0	4,908	4,409	0.9	0.7	32.6
4	Daishowa Paper Mfg. Co., Ltd	3,100	3.1	6,690	4,489	0.1	0.0	16.1
5	Sanyo-Kokusaku Pulp Co., Ltd	2,876	31.9	3,879	4,433	1.1	0.8	33.1
6	Daio Paper Corp.	2,706	8.1	4,373	2,970	0.3	0.2	19.4
7	Rengo Co., Ltd	2,428	33.1	2,071	3,456	1.4	1.6	35.0
8	Mitsubishi Paper Mills, Ltd	1,867	12.1	2,724	4,038	0.6	0.4	39.0
9	Kanzaki Paper Mfg Co., Ltd	1,405	7.1	1,934	3,018	0.5	0.4	45.0
10	Hokuetsu Paper Mills, Ltd	1,013	13.6	1,557	1,082	1.3	0.9	28.4
11	Chuetsu Pulp & Paper Co. Ltd	913	27.0	1,474	1,388	–	–	30.2
12	Settsu Co. Ltd	690	31.6	2,361	978	–	–	43.1
13	The Japan Paper Industry Co., Ltd	556	5.3	530	1,098	1.0	1.0	34.3
14	Nippon Kakoh Seishi K.K.	546	55.7	943	1,156	–	–	22.5
15	Tokai Pulp Co. Ltd	518	2.1	771	655	0.4	0.3	19.5
16	Kishu Paper Co., Ltd	512	7.7	548	1,322	1.5	1.4	52.7
17	Tomoegawa Paper Co., Ltd	461	35.5	506	1,333	–	–	12.6
18	Sanko Paper Mfg Co., Ltd	368	2.7	384	640	0.7	0.7	11.2
19	Takasaki Paper Mfg Co., Ltd	290	0.3	430	387	0.1	0.1	25.6
20	Tokushu Paper Mfg Co., Ltd	259	19.6	401	618	7.6	4.9	75.6
21	Chuo Paperboard Co., Ltd	238	3.2	516	509	1.3	0.6	11.2
22	Mishima Paper Co., Ltd	166	2.0	204	600	1.2	1.0	45.1
23	Chuetsu Co., Ltd	162	3.3	121	488	2.0	2.7	19.9

Source: The Nihon Keizai; published in JPA, 1994.
Note: This table includes only listed companies.

Quesnel and planned mills in Alberta. Unlike the Canadian mills, those in the United States – a former James River mill in Washington State and Reed International's North American Paper Group in 1988 – are not designed to supplement Japan's supply of pulp; rather they are intended to provide an inroad into the American domestic market for newsprint.

While American companies remained dominant throughout the 1980s, the listing of major international companies by *Pulp and Paper International* provided in chapter 3 shows Oji, Jujo, Sanyo-Kokusaku, Honshu, and Daishowa among the top twenty-five and Rengo and Taio among the next fifteen. It should be noted that unlike American and European companies, few Japanese firms are listed as having locations in numerous countries. This is partly a function of inadequate reporting. Much of Japanese investment takes form within consortia and intricate joint ventures in developing countries and does not show up on company listings designed to accommodate American-style business. As well, until recently, Japanese activity was directed primarily toward obtaining raw material, while American activity was focused on establishing production units elsewhere. These patterns are changing as Japanese companies move production of paper, as well as pulp, to other locations and begin to sell in the large American and European markets. Even with inadequate reporting, these ranks indicate the relative size of companies that have moved into the international arena only since the mid-1970s.

Although individual companies have holdings throughout the world, much of the Japanese importing business is carried out by merchandising companies (the *sogo shosha*). The most significant of these in the forestry business are Marubeni, Mitsubishi, C. Itoh, Mitsui, Sumitomo, Nissho-Iwai, Kanematsu-Gosho, and Nichimen. Several of these are also investment companies, purchasing shares and creating marketing contracts in various materials-sourcing countries. Some of their investments are in fact consortia arrangements under the name of the *sogo shosha*, which provides the organizing and sometimes the engineering skills.

Marubeni, for example, the largest seller of pulp and paper, has half-ownership of a Canadian company, Cariboo Pulp and Paper, and it owns a mangrove wood-chip operation in Irian Jaya, Indonesia (Bintuni Utama Murni Wood Industry). As well, it imports paper from Finland and the United States. Mitsubishi owns half of the Canadian company Crestbrook Forest Industries, has controlling shares in Al-Pac, and imports pulp as well from Sweden and the United States. Mitsui owns two paperboard companies in Japan (Takasaki Paper and

Tohoku Paper), for which it imports groundwood paper from its Canadian subsidiary New Brunswick International Forest Products (co-owned with Oji). In several of these international ventures, Weyerhaeuser is an important link and supplier to the *soga shosha*, especially in connection with Marubeni, Mitsubishi, and Sumitomo. C. Itoh and Company is the largest importer of pulp and the second largest importer of wood chips. As noted in chapter 11, it has substantial shares in Cenibra of Brazil. It also imports from the United States, Canada, Norway, and West Germany. In addition to its imports to Japan, it exports paper on the world market.

The consortia formed to back Cenibra, the Japan-Brazil Pulp and Paper Resources Development Company, is made up of eighteen Japanese papermakers and administered by C. Itoh together with the Japanese government agency, the Overseas Economic Cooperation Fund (*PPI*, April 1987:42). Despite the shaky Brazilian economy and escalating interest rates, Cenibra, with cheap wood resources, low costs, and growing demand for eucalyptus pulp, is expanding its share of export markets. The Japanese companies take half the output of the mills, manufacturing it into high-grade papers in Japan. As well, C. Itoh has started expanding sales from this mill to China.

Mitsubishi formed a joint venture with Forestal Colcurra of Chile in 1987 to export chips to Japanese producers, but Japanese investment in Chile lags behind American and New Zealand investments there. Several joint-venture arrangements between Japanese companies and either governments or private groups in Indonesia provide both pulps and plywood for Japan. The Japan Paper Association announced in 1988 that it would grow eucalyptus trees in Thailand under a joint-venture arrangement with the Thai government. Plans call for the construction of five wood-chip plants and a plantation of 200,000 hectares within five years. The consortium includes Oji Paper (*PPI*, 11 July 1988:10).

As the foregoing discussion indicates, Japan is now a powerful force in global forestry; indeed, in many respects, it is Japanese companies that have globalized the industry. In the case studies reported in part II, the overwhelming presence of the Japanese market and Japanese investment is shown to affect the establishment of plantations, sawmills, and pulpmills throughout the southern hemisphere. New Zealand and Australia have courted Japan's trade; Canada and the United States have invited Japanese investment; Chile, Brazil, Indonesia, Thailand, and Malaysia are all deeply involved in the Japanese wood trade.

ASIAN COMPETITORS

Japanese companies broke the northern monopoly, but they themselves are now competing with companies and consortia throughout Southeast Asia, Oceania, and Latin America, and they are extremely dependent on fibres sourced in countries that are beginning to produce their own manufactured wood products.

Ten Asian countries – Japan, Singapore, Malaysia, Thailand, Indonesia, the Philippines, Korea, Taiwan, Hong Kong, and China – have a combined population 4.4 times that of western Europe and representing about 30 per cent of the world's total population. Japan accounted for 64 per cent of the region's total consumption of paper in 1987, yet consumption growth between 1983 and 1987 in Japan, at 10 per cent, was dwarfed by a growth rate of 25 per cent in the other nine Asian countries (Patterson, 1989). The potential market demand for paper and wood is enormous. Japan cannot supply these markets; its companies have all they can do to supply their domestic market. Companies from other countries are moving into the competition.

Large companies have formed in Korea, Taiwan, and Singapore, each of them processing wood purchased from less-developed neighbours. Korea's paper companies are now venturing into the global marketplace. They have already substantially reduced that country's imports of paper and increased its fibre imports from elsewhere. The Taiwanese company Yuen Foong Yu is one of the applicants to build a pulp and paper mill in northern Alberta. It works on the same principle as other successful Asian companies: process the raw materials of others and become the market intermediary. Taiwanese capital is moving into neighbouring countries and Latin America, as the following chapters indicate. Total Asian newsprint capacity is rapidly rising, with new mills throughout the region in Korea, China, and Malaysia, as well as in Indonesia (Sedjo, 1989). Tied supply for Asian companies (from mills with joint-venture ownership and market contracts) makes much of the region self-sufficient. Such arrangements include the joint venture between the Singapore Newspaper Services and the CIP Gold River project, the Howe Sound CanFor–Oji joint venture, and the proposed China-Westar project, all in B.C.

Thus we come full circle. Japan initiated many of the changes that undercut traditional northern market arrangements, including those in the Pacific Northwest and British Columbia. Now it has become an important component of the economy of these regions through both purchases and investments. Its competitors are in Southeast Asia, and they too are becoming producers of pulp in the Pacific regions of

North America. Meanwhile, those regions have lost much of their unique forest inheritance, a fact that will not be overlooked by Japanese forest owners currently bedevilled by low-priced North American logs.

SUMMARY

Japan has reforested its mountainous regions since the end of the Second World War, but both construction and papermaking companies purchase logs on external markets. Rural forest farmers complain that their traditional villages are being depopulated because the domestic log market is dominated by cheap imports. While there is some evidence to support the claim that North American logs are sold at less than replacement cost and without reference to social costs, there are other explanations as well for the plight of Japanese forest owners and sawmills, including the more general effects of growing affluence, urbanization, industrialization, and inefficient production methods. Suppliers of wood chips, logs, and more recently pulp for the papermaking companies of Japan include all the regions of the Pacific. Softwoods are purchased from northern countries, hardwoods from all available sources, including the tropical countries of Asia. Japanese papermaking companies are investing in pulpmills and in pulpwood plantations to obtain their fibre, and they use high quantities of recycled waste paper. The links to North America were described in this chapter; linkages to other Asian countries and Latin America are examined in those that follow.

PART TWO

Forests and People in the Southern Hemisphere

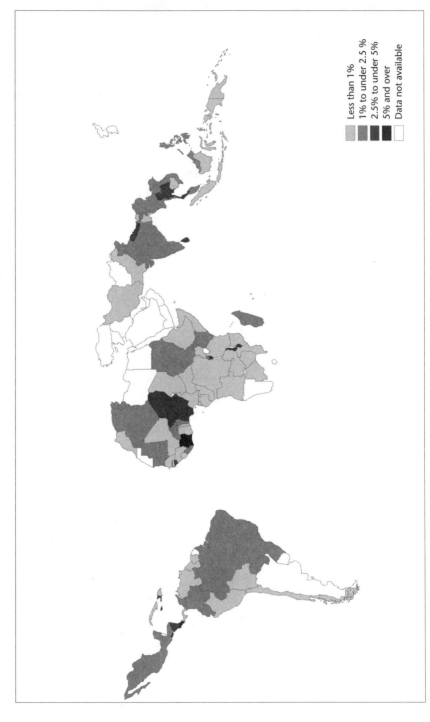

Legend:
- Less than 1%
- 1% to under 2.5 %
- 2.5% to under 5%
- 5% and over
- Data not available

Map 5 Southern hemisphere: average annual percentage rate of deforestation, 1980–89
Source: World Resources 1990–91 (New York: Oxford University Press).

6 Tropical Forests
and Forest Dwellers

Tropical forests are fragile. In many places they are irreplaceable. Before encountering the particular nature of industrial forestry in tropical and subtropical regions, we will consider some of the characteristics of the forests themselves and some data on the extent of deforestation and its ecological implications.

People of the forest are also fragile, or at least vulnerable to the predations of other people and organized states, companies, and groups who exploit the forests. Exploiters disperse, harm the bodies, or destroy the spirit of forest dwellers. Many who live in forests are indigenous peoples whose tribes or bands have subsisted as hunters or as shifting cultivators for centuries. Others are poor people without land title who grow subsistence crops at the margins of the forest.

In the next several chapters, we consider assaults connected to industrial forestry against tropical forests and forest dwellers. Here we will concentrate on the nature of the forest itself, estimates of deforestation, the causes of deforestation other than industrial forestry, and the characteristics of indigenous forest dwellers in tropical regions. The intention here is to remind the reader that contemporary logging practices and plantation forestry were preceded, and even today are accompanied, by other destructive agents in forests. This chapter concludes with a discussion of some destructive agents in the Amazon prior to the growth of a forest industry in Brazil.

LOCATION AND CHARACTERISTICS OF TROPICAL FORESTS

Of the remaining tropical forests, about 45 per cent are located in South and Central America; Brazil alone has nearly 31 per cent of the world's total. Africa contains about 36 per cent, and Asia, Australia, and Oceania have about 19 per cent (Rubinoff, 1982). By country, Indonesia has about 10 per cent; Zaire, just over 9 per cent. Ranging from just under 6 per cent to 2 are Peru, Colombia, India, Bolivia, Papua New Guinea, Venezuela, and Burma. Mexico, the Congo, Malaysia, and other Latin American, African, and Asian regions have less than 2 per cent (Guppy, 1984: table 1; based on FAO and other sources).

Tropical forests may be subdivided according to climatic conditions and precipitation patterns into categories ranging from the moist closed-canopy forests, through various more moderate types (including monsoon, closed, semi-evergreen forests), to partially dry, open-canopy woodlands. At the equator, the forests have high rainfall – from about 160 inches up to 400 inches a year – and temperatures above 80 degrees Fahrenheit all year round. Subtypes of equatorial forests are due to soil and altitude differences, but their characteristic feature is a heavy canopy and relatively sparse underbrush. They are rich in both flora and fauna. To the north and south, as seasonal climates become more pronounced, the tropical moist or semi-deciduous forests have between 40 and 160 inches of rain a year. Their trees may lose their leaves during the dry season and thus have less canopy cover, though they may have more understorey.

All of these subtypes are threatened with extinction, but it is the moist canopied forests that have raised the greatest anxiety. These lush, wild forests spawn myriad life-forms and comprise approximately 1,500 million hectares (about 31 per cent of land in the tropics and 10 per cent of the world's land area). Tropical America has 45 per cent of these forests (mostly in Brazil); tropical Africa, about 30 per cent; and tropical Asia, about 25 per cent (Sanchez, 1981:348–9). The annual deforestation rate in the 1980s is estimated at just over 16,000 hectares, of which nearly 11,000 were in tropical America (mostly Brazil), and nearly 4,000 in Asia (WRI, 1991). Though most attention is given to these rainforests, Jack Westoby has pointed out that the dry, open forests which constitute about two-fifths of all tropical forests (some 900 million hectares by his estimation) are also extremely vulnerable to exploitation (1983).

Like temperate varieties, tropical rainforests have growth phases. A mature forest has an enormous variety of both flora and fauna. Natural causes and human interventions may create gaps, that is, small areas with fallen and decomposing trees. In time, herbs, vines, and tree seedlings appear in the gaps; then larger species of trees gradually create a new pioneer forest and canopy about 15 metres above the ground. Provided that the seeds have not been destroyed, the original mature-phase trees will gradually emerge, rise above the earlier second-growth trees, and eventually create a canopy some 40 to 55 metres above the ground. This process may take a century and a half or more, and it will be four centuries before all the original inhabitants of the forest are back in place. These include all the smaller trees and shrubs and the mollusks, insects, reptiles, birds, and mammals. A large rain-forest, such as those in Sarawak and Brunei, may contain over 2,000 tree species.

The middle-level trees – those occurring in the "building" phase between the gap and the mature forest – are fast-growing and light-weight. Because they grow quickly, they are targeted as commercial species. Their utility has increased with fertilizers, biotechnology, and technological change in pulping methods. But harvesting them is dangerous for many reasons: they provide the essential canopy for the growth of the major large varieties of the tropics, and they are neces-sary in the sequence from gap to mature forest. Further, they use soil nutrients and moisture, and while they are critical in the transition of the forest, to cut and regrow them on the same soil is a quick method of destroying the soil altogether. Although the growth phases are lengthy, a tropical forest can be regenerated if the gaps are small, the soil has not been degraded (as by heavy machinery and roads), the seeds remain in the vicinity, and no further intervention occurs. As the gaps increase, or as their proximity to one another increases, the regenerative capacities are correspondingly diminished. These conditions vary from one region to another.

Soils in tropical forests differ significantly, and what is true of one region is not necessarily true of another. Several writers have argued that humid tropics become a "red desert" once they are denuded of tropical canopy. The argument is that tropical forests typically survive on poor soils. The trees have evolved ways of capturing the nutrients. Tree roots sometimes rise above the ground, using fallen logs as nutrient sources. In some areas, the soils are pure silica, the fossil sands of ancient seas. Others are high in silicon content and low in such other minerals as calcium, potassium, and phosphorus. Once the canopy is gone and these soils are exposed to the blistering sun, the nutrients are leached out, the water-table is immediately reduced because its rainsource has disappeared; and the land dries up. Even selective logging can be highly destructive. Roads and machinery destroy the soil as well as the trees. Extraction of even 10 per cent of the foliage can cause massive losses in canopy cover, exposing what remains to further erosion. Climatic change follows from the disturbance, and the wastelands, if not transformed into caked deserts, become brushlands topped by the alang-alang grass already so evident throughout Thailand, Indonesia, and Malaysia. Nothing else grows there; the diversity of life is gone; the gene pools are destroyed. Dipterocarpaceae forests of Southeast Asia may have more capacity for regeneration than the denser and more mixed forests of Latin America and Africa, but their destruction is now so extensive that it may be impossible to undo the damage; even where regeneration is started immediately, human intervention cannot recapture the original complex ecological composition of the forest. (Among contributors to this theme are Guppy, 1984; Conway, 1983; and Proctor, 1983.)

Pedro Sanchez (1981) offers a less pessimistic view. He argues that some soils are fertile, and if deforestation is selective of soils as well as of trees, crops can subsequently be planted with success. He notes that shifting cultivators frequently have superior knowledge about the chemistry of the soils, and if their knowledge is harnessed, better results can be achieved. As well, he advocates burning rather than mechanical clearing of land because the ash provides fertilizer and the soil is not compacted or displaced as it is under mechanized land clearing. The crops he cites are the traditional ones produced by slash-and-burn shifting cultivators: yams, cassava, upland rice, corn, plantain, cowpea, peanut, and various vegetables. Perennial tree crop production is also possible and has a sufficient history to allow for judgment. The plantations of rubber, oil palm, cacao, coconuts, bananas, and other export crops, in Sanchez's view, mimic the rainforest. However, all of these crops – both traditional vegetables on areas larger than the shifting cultivator's plots and tree crops – require fertilizers; the amount required depends on acidity and other soil conditions.

The Gene Pool

Much of the difference between those who believe that tropical forest soils are infertile once denuded of original canopy and those who do not has to do with values, rather than technical facts. Some soils will support tree crops, some will not, and that matters, of course. But what may matter more is that deforestation destroys what some people value above export dollars: the diversity of gene pools and the multiplicity of life-forms. Others are more concerned with the viability of conditions for commercial monoculture crops. If a secondary forest grows at all, it is an impoverished shadow of the original, and its wildlife is diminished. Birds, mammals, insects, and plants alike disappear when the habitat is destroyed, and the loss is not only of individual members of species but of genetic reservoirs specific to regions. Overexploitation, even where it does not result in destruction of habitat, may result in diminution of the genetic quality of surviving populations. In some instances, so rare are living examples of species that erosion or loss of a single habitat may result in its extinction.

The gene pool for species already considered economically valuable, as well as for those not yet discovered by the market, has already been seriously depleted. This loss may occur by repeated selective logging as well as by clearcutting, where selections fell stronger varieties and diminish the potential for sturdy offspring. Overexploitation of the resource also reduces the genetic quality of the surviving populations. Among species so affected by 1980 were Cuban mahogany, Macassar ebony, Andaman padauk, and some rosewood species (Oldfield, 1981:320).

Coevolved species are also affected by the removal of any particular species. L.E. Gilbert (1980)) provides examples, including a canopy tree that supports some twenty-two species of fruit-eating birds in a Costa Rican rainforest. If the *Casearia* tree were removed, these birds (one of which is the Ramphastos toucan) would be deprived of an essential food. The same relationships bind together trees and insects, trees and mammals, trees and reptiles; though their full range of interaction is not understood by humans, it is known that many species would be affected by the logging of one species. As well, it is now recognized that rainforest species are not able to adapt quickly to disturbances in watersheds (Gomez-Pompa et al., 1972; Brunig, 1977; Jacobs, 1979; Plumwood and Routley, 1982; Bodowski, 1988).

Atmospheric Pollution and Its Effects on Climate

There is a considerable body of thought indicating that an increase in atmospheric carbon dioxide would result from deforestation in the tropics. Tropical rainforests and their soils hold about a fifth of the world's terrestrial carbon pool of 500 billion tons, nearly half of this in the living trees themselves. When the trees are destroyed, the carbon dioxide is released into the atmosphere, where it contributes to the "greenhouse effect" on climate. Some writers argue that this release would be sufficient to raise the level of the seas; among other impacts, the rising seas would inundate industrial lowlands and cities (Brown and Lugo, 1981; Guppy, 1984:931). However, there is not consensus on the impact. Robert Dickinson argues that only for microclimates, as contrasted with regional and global climates, can general assertions be made now about the atmospheric effects of deforestation. The microclimate is "especially influenced by the shading of solar radiation and the mechanical production of turbulence by the canopy of leaves and branches" (1981:415), and local microclimates, in his view, would be substantially changed by tropical deforestation. The uncertainty about larger changes comes about in part because undisputed information on base data is unavailable. Earlier measurements and inferences have been found inadequate. Dickinson notes that "there is currently no general agreement either as to what changes have taken place in tropical forests in the past due to human activity or as to what is now occurring" (1981:412).

In addition to the problem of agreement on base data, there are daunting questions about measurement and inference on a scale larger than the microclimate. Even with respect to the microclimatic risks, we need to discriminate between the widely different conditions of deforestation followed by secondary-growth replacement, agricultural crops, plantation crops, grasslands, or desertification. It may reasonably be assumed that if extensive deforestation occurred in a region, the microclimatic changes would modify the climate in nearby regions as well. In Dickinson's view, "extensive removal of tropical rainforests also would change the global heat balance significantly." However, the "global climate changes due to even complete tropical deforestation are expected to be no larger than either natural climate fluctuations or the changes that will result from past combustion of fossil fuels" (1981:436).

A. Henderson-Sellers, examining similar issues, also notes that predictions are difficult, and he cites particularly the several feedback mechanisms within the climate system itself and the still inadequate climate-modelling techniques, including those she uses. Despite the uncertainty of the techniques, she concludes that deforestation does have the potential for influencing climate at all scales. She states: "Almost certainly the increasing levels of atmospheric carbon dioxide will modify local and global climatic regimes before the turn of the century" (1981:444).

The extreme anxiety expressed in some of the literature regarding global pollution and rising ocean levels may be unwarranted, but with the scientific community so uncertain, the wisest course should be caution in deforesting tropical lands. At the very least, deforestation would have impacts on the local and regional climates that could be very serious for countries with tropical forests.

Pharmaceutical and Other Properties of Tropical Forests

Among the losses incurred when tropical forests are destroyed and the human inhabitants are dispersed or relocated is traditional knowledge about the pharmaceutical properties of plants. As noted later in this chapter, ethnobotanists in the Amazon region are attempting to recover some of this knowledge from remaining tribal peoples, but already, vast stores of human skill and cultural understandings have been lost. In addition to medicines, these forests produce many useful gums, rubbers, fibres, oils, foods, implements, and decorations. Small tribal populations have subsisted on Brazil nuts and minimal other produce of a tropical forest environment, yet these are being destroyed along with the forests in Brazil and elsewhere. There are regions where non-wood products are extracted for export markets. For example, people of the Peten region of Guatemala, a reserve of national parks along the country's border with Mexico and Belize, produce and sell spices, chicle, and xate palm for international markets. The sales amount to substantial cash income for these people and for Guatemala as a country, and extraction can be sustainable if it is not organized on an overexploitative basis (Redford and Padoch, 1992: Nations, 1992). Similarly, rattan is a source of income for tropical forest regions of Asia.

In addition to losses of knowledge and alternative subsistence or marketable produces when forests are destroyed, there are potential impacts on human health (assuming survival after the depletion of ozone supplies). Some reporters argue that the spread of malaria throughout Southeast Asia and of African river blindness, now found in Amazonia, are linked to deforestation (e.g., Goodland and Irwin, 1975). There are more diffuse claims as well – that epidemics emerge where forest cover is insufficient. Such claims are not proven, though credible scientists are taking them seriously. More to the point at this stage is a recognition that deforestation of the tropics has ripple effects on a global scale.

PEOPLE OF THE FOREST

Forests are the habitat for many species, one of them being humans. An estimated 696 million people dwelled in the tropical rainforests of the early 1980s (Lanly, 1982:34–55). These included peoples whose ancestors lived in forests for centuries, even for thousands of years, and whose tribal cultures were built around hunting and gathering or slash-and-burn agriculture. Hunting and gathering peoples inhabited most of the earth's forests at some stage, though few of them remain now in Southeast Asia and Amazonia. Carib and Arawak Indians were slaughtered or later died of disease when colonial powers penetrated Central America. Elsewhere, such as in northern Thailand, indigenous cultures disintegrated as their land base disappeared.

"Slash and burn" is the common description of tribal economies engaged in agriculture within forests; the cultivators were often descended from hunters and gatherers or practised both hunting and agriculture. The term is unfortunate, since it carries a negative connotation of carelessness and indifference to the forest, when in fact the cultures supporting the practice, also known as swidden or shifting cultivation, were protective of the land and capable of sustaining multiple species on it. Swiddening is an old English term with affinities to traditional Nordic expressions for a burned clearing in the forest and shifting agriculture (Raumolin, 1987; Arbhabhirama et al., 1988:175). The land in Europe, once tilled by this method, has been deforested for so long that tribal cultures ceased to exist many centuries ago. Tribal people of the Amazon regions survived into the late twentieth century. Their land was not so easily penetrated and not so obviously valuable, and they were left alone a little longer. In India, Southeast Asia, and Africa, the swiddeners have suffered in varying degrees as external cultural agents, external technologies and consumer products, and external claimants to their land imposed on them.

Swiddening involves the clearing of a small area of forest – half a hectare or so – and the planting of such crops as cassava, yams, maize (in highlands), and various vegetables. Highland soils are typically nutrient-deficient, but through the burning of trees, nutrients drawn through the roots are released into the soil and can be transferred to crops. The burning also neutralizes the acidic brush layer of the forest floor and destroys pests. In traditional swiddens, some trees were left in the centre of plots to encourage regeneration following cultivation. Planting may be undertaken for up to three years, after which the soil is exhausted, weeds take over, pests return, and crop yields are much reduced. Traditionally, the shifting agriculturalists move on to another area in the forest and begin again, but they do not entirely abandon the first area. They plant trees after they have moved their cropland, and they may continue for some time thereafer to intervene in the regeneration of the forest. Under ideal conditions, when the swidden agriculturalists are few in number and have little competition for the land, they will not recultivate the same patch for a half-century or more, by which time a second-growth forest will be well advanced (Conway, Manwan, and McCauley, 1983; Dean, 1983). In more fertile areas, they may settle into a permanent village, planting a small section for only one year and practising crop rotation over a several-year period.

These tribal peoples had an impact on forests: they possibly prevented tropical forest expansion; they were responsible for planting species outside the range destined for them by nature; they destroyed soils and patches of forest, at least temporarily. However, as long as population density remained low and rotations lengthy, their impact was not of great moment. Population growth was never characteristic of tribal groups actually living in forests; there is ample evidence that both social and biological limitations on growth maintained stable-sized tribal groups over many centuries. As the forest shrinks because of other assaults, settlement, industry, and logging, the area available for shifting agriculturalists likewise shrinks, lengthy rotations become impossible, and the cultivators themselves finally become destroyers, not because they so choose, but because they are so confined. In their current situation, they are frequently targeted as the major culprits in deforestation, especially in northern Thailand and Indonesia. Time-honoured cultural taboos have disappeared, along with hill-tribe cultures and means of subsistence, and population pressures are now endemic. As well, political turmoil and persecution may oblige these peoples to move beyond traditional territories and become competitors on land already occupied by others. It is not now swidden agriculture that is practised in forests, so much as slash-and-burn without plan or reasonable rotations, or else highly intensive irrigation agriculture.

Throughout the southern hemisphere, survivors are often obliged to settle in alien social and environmental shack villages, where they subsist on meagre welfare rations or scrounge for food in a cash economy. Those who manage to stay near the margins of forests compete with thousands of equally displaced and impoverished villagers, as they move in behind logging operations to establish subsistence plots on degraded land – a destructive activity, to be sure, but not the primary cause of deforestation. Tribal peoples and landless peasants alike lack fuelwood. By FAO estimates, two and a quarter billion rural people suffer fuelwood deficits; as Westoby puts it, "What this jargon means is that most of them will have to eat less in order to buy the wood or charcoal with which to cook their food." He suggests that the problem is not with shifting cultivators, it is with *shifted* cultivators, displaced people "obliged to penetrate the forest to clear a patch of land from which they can scratch a precarious living" (1983:4).

Most of our attention in this book is on shifting (or shifted) cultivators and landless peasants, but there are also forest dwellers who are not quite in either category. These include the *caboclos* (backwoodsmen) of Brazil and the *ribereños* (inhabitants of the flood plain) of Peru. They are detribalized Amerindians, offspring of European-Amerindian marriages, and descendants of immigrants of various origins. These groups are vital links between tribal and essentially European cultures in South America (similar intermediate groups exist elsewhere), bringing to the margins of forests or interfluvial zones where they live a vast reservoir of native knowledge (Hiraoka, 1992; Padoch and de Jong, 1992).

THE LOCATION AND EXTENT OF DEFORESTATION

Deforestation has been a global process for a very long time. There is nothing absolutely fixed about forests. They expand, they recede, with a vast variety of natural conditions; virgins they are not, but there is a limit to the damage a forest can sustain. There are conditions under which soils do not regenerate nutrients, or air becomes so overloaded with pollutants that it cannot sustain life. At some stage of deforestation, water-tables drop, watersheds fail, and silting, flooding, and general erosion of the land become commonplace. There are moments when the end is actually reached, when it is too late to reverse a trend. By the time the forest industry had developed the markets and technology to make profits by pulping hardwoods, such a stage was already reached in many tropical and subtropical forests of Latin and Central America, Africa, and Asia.

The FAO estimated annual tropical deforestation in 1981 in the range of 11.3 million hectares; revised estimates half a dozen years later totalled 20.4 million hectares annually (WRI, 1991). FAO estimates are conservative, are based primarily on timber values, and are restricted to the outright destruction of forests. Other estimates, such as those of Norman Myers, use broader criteria and include degraded lands; he estimated total destruction in 1980 at 24.5 million hectares per year. Annual deforestation by even the lower count is comparable in extent to whole European countries: Austria, England, and Belgium are appropriate comparisons. While estimates vary considerably, many scientists and other students of these forests fear that they face extinction by the early to mid twenty-first century if the current rate of destruction continues (see Lanly and Clement, 1979; Myers, 1980; Oldfield, 1981; Rubinoff, 1982; Lanly and Clement, 1982; Guppy, 1984; Panayotou and Ashton, 1992:16–26; WRI, annual reviews).

Estimated losses for Latin America are about 37 per cent; for Asia, 42 per cent; and for Africa, 52 per cent of the original tropical forest areas (Rubinoff, 1982:253; Lanly and Clement, 1979). The projected annual deforestation rates in 1987 put Brazil at the top of the list at 1.76 per cent, the rest of tropical America and tropical Asia ranging from 0.80 to 0.90 per cent. Tropical Africa has lower rates only because so vast are the areas already deforested (WRI, 1991; Panayotou and Ashton, 1992:21). Estimated average annual deforestation between 1981 and 1990 includes 1,315 hectares in Indonesia, 255 hectares in Malaysia, and 110 hectares in the Philippines (*Economist*, 14 Nov. 1992:40, based on 1990 data). In India, only about 11 per cent of the country's 328 million hectares of land remain forested; in

Thailand one-quarter of the total disappeared in the decade of the 1970s; in the Philippines one-seventh disappeared in a five-year period after 1975; in Nepal, what is left is fast disappearing; and in Malaysia, there is unlikely to be anything left by the year 2000 (*World Wood*, Oct. 1987:39–40). Bangladesh, Haiti, and Sri Lanka have already been substantially deforested; Sierra Leone, the Ivory Coast, Nigeria, Ghana, and Guinea are either deforested or seriously degraded; most of Central America is now degraded or deforested. The remaining forests of Brazil, Zaire, and Indonesia have become the focus of world attention in part because they are survivors in a global deforestation process.

CAUSES OF DEFORESTATION OTHER THAN INDUSTRIAL FORESTRY

Shifting cultivation and encroachment on land, together with the search for fuelwood, are major causes of deforestation everywhere. Some of the encroachment has the blessing of governments, as in Indonesia and Brazil, where governments have obliged poor people in urban centres to relocate as farmers on land formerly forested. In both cases, the resettlement is related to the inability or unwillingness of governments to reduce the privileged occupation of more fertile agricultural land by large landowners. Logging for commercial sale is one of many causes, and in some parts of Asia, Africa, and Latin America, commercial logging had a significant role before the late twentieth century, but most deforestation has other causes.

Contrary to popular opinion, overpopulation is rarely a direct cause of deforestation. Population pressures have indirect impacts, to be sure, and there is an upper limit to the carrying capacity of land. At the end of the twentieth century, that upper limit has possibly been reached in China and much of South and Southeast Asia. But deforestation is not a simple function of the population-land ratio. Cynthia Mackie, in a 1986 study of shifting cultivation in the upland rainforest of Indonesian Borneo, observed that the intensity of land-use practices varied with numerous socio-economic conditions, and some of the most intense practices were found where population density was lowest. That is one example of the need to pay attention to how the population is organized, how it is distributed, and how its activities impinge on the forest. Population increases, moreover, have tended to follow, rather than precede, deforestation; in blaming population pressures, we are often ignoring the history that created them.

Massive deforestation processes of the nineteenth century were not in areas of high population density; in the delta region of Burma, for example, the British administration destroyed forests and imported settlers to control the territory. The Philippines were scarcely populated where deforestation was greatest. Deforestation of the highlands of India had causes other than, or at the least in addition to, population pressures. Siberia has never been overpopulated, though its forests have been exploited. Overpopulation of many lowland regions of Asia occurred after, not before, deforestation. The most threatened forests today are not in the regions most afflicted by high population growth. The Amazon is one of the least densely populated regions of the earth; Malaysian Borneo and Indonesian Kalimantan also have very sparse populations. Likewise, the destruction of the temperate rainforests of North America cannot possibly be attributed to high population densities. Settlement in the nineteenth and early twentieth centuries in the United States and Canada destroyed forests on soils suitable for other crops, but the pressures were not high-density populations so much as constant frontier expansion and land greed. The boreal forests of the north are on less arable land, and these regions are sparsely populated, yet they are now threatened.

Large-scale deforestation in most tropical countries awaited colonial incursions. Cash crops were introduced by the colonists in lowland regions. These reduced the variety of forest species and the genetic variability, and deforested large-scale areas. Much of the remaining forests of colonized regions in India, Indonesia, and Africa are, in consequence of plantations, secondary successions. The plantations included local crops, such as cinnamon, coconuts, nutmeg, and oil palm, but also exotics, such as tea, coffee, cocoa, sugar, and rubber (Jordan and Herrera, 1981; Jordan, 1987b:73–5). Similar plantations and cattle ranching were common causes of deforestation in South America.

The creation of colonies was the first step toward the formation of nations in regions that had not hitherto developed the nation-state structure. With the conclusion of the colonial era, subsequent governments sought means of controlling their territories and sustaining the people. Where the forests were degraded, whether by agricultural encroachments, colonial plantations, selective logging, roads, or other uses, governments tended to classify the regions as industrial and give what remained to logging concessions. Alternatively, they encouraged the poor of their urban centres to settle and farm where forests once grew, a still-common practice. In the nineteenth century, land clearances became more frequent and extensive in connection with the construction of major transportation systems, new industries, and mechanized agricultural development. It is worth repeating that population increases followed, rather than preceded, such developments.

Wars have contributed to deforestation. Most have not involved herbicides, as in Vietnam, where 44 per cent of the cover was destroyed by Agent Orange or bulldozers and bombs. But trampling by armies, the use of wood for fuel and construction, and general abuse while wars rage on eventually destroy the forest. Strategic defence has motivated the destruction of forest land elsewhere: in Malaysia, in Indonesia, and in Brazil. The aftermath of wars and revolutions are also destructive of forests. In Japan and China, as in the USSR and Europe, poor people used up remaining wood for fuel.

Some deforestation does not have even such concrete causes; people destroy forests for reasons that cannot be articulated clearly – fear of them, perhaps. Forests stand in the way of human control over nature. They have their own, often mysterious rhythms; they are not domesticated, they are not tame. Roads, railways, airstrips, dams, even scientific research are not self-evidently designed to do something, so much as to displace something: to get rid of this dark and independent presence. In reviewing the history of forests, one is constantly faced with puzzles about human interventions. Railways, costing thousands of lives and millions of dollars, often last but a moment of time; dams fall apart and floods destroy the land; roads go nowhere. Yet new projects are always under way; new riches are to be reaped in the mines, in the waters, and in the trees of the few remaining forests.

Most deforestation has multiple causes, which change over time. The tropical and subtropical forests on the Atlantic coast of South America were virtually destroyed by the end of the nineteenth century. Warren Dean (1983) has provided a study of deforestation in the southeastern subtropical zone, the region that now includes Rio de Janeiro in the state of Espírito Santo, about a third of the state of Minas Gerais, and about four-fifths of the state of São Paulo. "Only a few patches of this primary forest remain," he says, "and all of them are to some degree disturbed." The causes of disturbance include aboriginal swidden agriculture, intensified agriculture undertaken by Portuguese and French settlers after 1500, the cutting of wood for firewood and charcoal, hunting with increasingly destructive weapons, gold prospecting and mining, sugar cane and coffee plantations, and finally commercial logging. As elsewhere, railways and roads contributed to deforestation.

THE AMAZON

The world's attention has been focused on the Amazon for the past several years because its huge tropical forest is believed to be essential for sustenance of the planet and its deforestation is believed to be

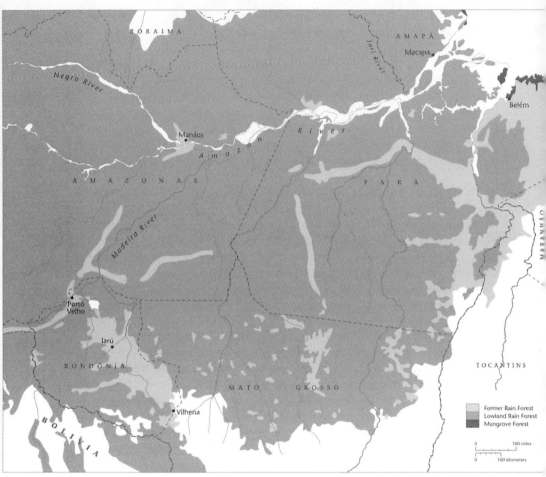

Map 6 Brazil, the Amazon region: forest types

especially dangerous. The Amazon River runs through northern Brazil, with a drainage basin extending over 5,000 square kilometres into Bolivia, Peru, Ecuador, Columbia, Venezuela, Guyana, French Guiana, and Suriname. The region contains diverse soil and forest conditions, classified as lowland tropical rainforest, tropical moist forest, and evergreen seasonal forests. Plant species are estimated to number between 250,000 and 1 million, and the vegetation is regarded as the most ecologically complex on the planet (UNESCO, 1978). The precise rate of deforestation may be in dispute (Myers, 1980; Lugo and Brown, 1982; Jordan, 1987a), but most scientists agree that the eastern region will be gone and much of the western Amazon will be depleted by the year 2000.

The causes of this deforestation are complex, and, as is true elsewhere, accusations are more often directed at the victims than at the chief beneficiaries. Deforestation has been going on for much of recorded human history. Shifting cultivators have both deforested and afforested the area for at least 3,000 years, but it was not they who brought about the destruction. Nor can the forest industry be blamed for much of it, though the industry's potential for destruction has greatly increased since the 1960s.

Population pressures are not the cause either. Brazil has high population density in its urban areas, but overall, and excluding the Amazon, its density is about 65 persons per square mile, the same as that of the United States and considerably lower than in most of Europe (Hecht and Cockburn, 1989, 95–128). The country as a whole is twice the size of Europe, and it has about a quarter of the population. If we count the Amazon as 60 per cent of the total territory, and we estimate that somewhat less than 10 per cent of the population lives in the region, the equivalent of half of Europe's population still lives in a space the size of that continent. Thus it is not population pressures that are destroying the Amazon.

Peasants in fact cleared less than a fifth of the deforested area of the Amazon between the mid-1960s and the mid-1970s; 60 per cent was destroyed by ranchers and the Brazilian highways program (Plumwood and Routley, 1982:7). There are other lands far more suitable for agriculture if the rich of Brazil were willing to give the poor access to them. Catherine Caufield estimates: "Taking potential farmland into account but still leaving aside Amazonia, each person in Brazil could have 10 acres. Instead, 4.5 percent of Brazil's landowners own 81 percent of the country's farmland, and 70 percent of rural households are landless" (1985:39). Westoby calculates that 1 per cent of Brazil's population owns two-fifths of cultivated land (1983:4). Myers estimates that every family could be given sufficient land for fuel and food without touching the Amazon (1984:989). The problem is not land and it is not population density: it is greed, and the most greedy appear to be those with the most land already.

Of some unsuccessful projects in the history of the Amazon, we need not provide details; each in its way has destroyed parts of the forest. There were railways that were never completed, roads that went nowhere, and industrial ventures that produced nothing of any value. Of other enterprises it is useful to learn more because they have a bearing on the current development of industrial forestry.

The Trans-Amazon Highway, Transmigration Policies, and Cattle Ranches

The construction in the early 1970s of the 5,000-kilometre Trans-Amazon Highway, by way of integrating the Amazon into the Brazilian economy and establishing sovereignty over it, was the most massive of the assaults on the forest and on the remaining tribal people who lived in it. Construction began in 1970, and connecting roads were linked to the new highway to provide a perimeter near the borders with Peru, Venezuela, and Guyana and routes through the Amazon from east to west and from north to south. By 1974 much of the construction was completed. The project gave employment to the Brazilian Army Corps of Engineers, and with funding from the World Bank, the Inter-American Development Bank, and other international agencies, it moved ahead at record speed. American and European companies contributed technical expertise, machinery, and mapping services, and the U.S. Agency for International Development provided an $8.4-million grant-in-aid to the government for participation in the Earth Resources Observation Satellite Program in the United States (Davis, 1977:64–5).

As the highway progressed, the government introduced its transmigration policy. Access to agricultural land in the southern states of Paraná and Rio Grande do Sul were closed to settlers in the early 1960s, and thousands of tenant farmers and sharecroppers had moved out of rural areas. They were encouraged to relocate in the Amazon, along with migrants from the drought-stricken northeast region. The migrant settlers were provided with a simple house, a small regular payment over a three-year period for the purchase of provisions and implements, and 100-hectare lots at low cost, payable over 20 years. Early expectations were of millions of settlers and thriving farms, but ultimately only 8,000 families took up residence. There were an estimated 30 million impoverished peasants in the northeast, but the Amazon apparently did not attract a large number of them. The most optimistic estimate is that 40 per cent established economically viable conditions for themselves (Smith, 1981; Moran, 1981; Jordan, 1987b:104–5). The reasons may include poor advice on crops from the government services, the settlers' inexperience with Amazon conditions, weed and pest competition, and transportation and storage deficiencies; but the most critical problem was low soil fertility (Jordan, 1987: 58–75; see also Smith, 1982; Sanchez et al., 1982; and Stone, 1985). Yet again in the tropical forest's history, would-be settlers learned that the nutrients are in the trees, not in the soils; once the trees are removed, the soil is infertile.

There have, however, been small successes in establishing agriculture in the Amazon. A relatively successful settlement was started at Yurimaguas, Peru, by Japanese peasant immigrants. The settlement, called Tome-Assū and established in the early 1900s, was less than flourishing at first, and then managed to develop black pepper plantations (even while the area was used as a Japanese internment camp during the war) and ultimately more diversified plantation crops. Working as a cooperative and said to be highly disciplined, the colony observed rigorous rules for land utilization, biomass conservation, rotation, ground cover, and soil conservation (Jordan, 1987b:70–2; Sioli, 1973). The settlement, as a consequence, is unique; other settlers had less discipline and much less patience. As well, they were caught in the shifting winds of government policies, for while the early proposal was to settle the landless poor, subsequent Brazilian government policy clearly favoured the growth of large-scale cattle ranches.

Remote forest lands are difficult to control, but the governments of modern nation-states apparently feel it necessary to establish their sovereignty right up to their remote borders. Venezuela, an oil-enriched nation in the 1970s, became concerned that shifting cultivators and landless peasants from Colombia and Brazil were migrating illegally into its San Carlos region. Since Venezuelans were less than eager to settle the region themselves, the government established a development project to provide families with low-interest loans and other aids if they would become cattle ranchers in the area. The project did not succeed, and the intended pasture, a burned-out forest, lay fallow. A scientific team obtained permission to study the pasture, borrowed some cows, and examined the effect of grazing. It concluded that a lightly grazed pasture may be less environmentally damaging than shifting cultivation. However, there was a net loss of cattle, and "from the aspect of cattle production, the project was a disaster" (Buschbacher, 1987:57). The climate in the Amazon basin is not suitable for cows.

Brazil, none the less, chose beef cattle as its instrument for colonization of the Amazon. Abandoning its attempt to settle the poor, the government established the Superintendency for the Development of the Amazon, with a mission to promote corporate investment. Tax incentives and land grants induced investment in cattle ranching. The impoverished peasants now stuck in the area provided the potential labour force. One of the first companies to accept the conditions was the King Ranch of Texas. In a joint arrangement with the Swift-Armour Company of Brazil in 1968, King Ranch obtained 176,000 acres in the state of Pará. Other major ranching and beef-packing companies followed, including firms, such as Liquigas of Italy (industrial chemicals) and Volkswagen of West Germany, which had not hitherto participated in the agriculture business (Davis, 1977:126–31).

Environmental scientists criticized these developments. They claimed that the low fertility of the soils would not carry the cattle load. Nutrients would be leached out, and overgrazing would cause erosion of topsoils and further loss of nutrients. Pasture grasses would be overcome by weeds better adapted to degraded soils (Goodland, 1990). The scientists turned out to be correct in their predictions. By the late 1970s, pasture degradation was widespread. By 1984, some 85 per cent of all ranches in the region were deemed failures, and many had been abandoned (Hecht, 1982). A subsequent examination of soils throughout the region suggested that although the abandoned low-intensity pasture sites might progress toward forest, the high-intensity sites would not. The differences depended on soil compaction, availability of sprouting trees, competition for growing space, microclimates, and efficiency of seed predators and seed carriers (Buschbacher et al., 1987).

Besides soil degradation, the ranchers had to contend with cattle diseases. These turned out to be so plentiful that meat could not be exported. Hamburger restaurant chains in North America did not benefit from cheap Brazilian meat as some popular writers claimed. In view of endemic hoof-and-mouth disease, the U.S. Department of Agriculture prohibited the import of fresh beef from South America (Tracey, 1990). Thus the "hamburger thesis," as the claim is called, is not valid for the Amazon (though it is true of Caribbean and some other tropical regions).

Shifting Cultivators

Shifting cultivation, widely regarded as an unmitigated evil by governments in both south America and Asia, is viewed in a positive light by many anthropologists and ethnobotanists, including William Balée, Janice Alcorn, Darrell Posey, and Suzanna Hecht and Alexander Cockburn. Hecht and Cockburn note: "Wherever one turns, the landscape almost invariably bears the imprint of human agency, starting with fire ... there are vast tracts of forests created by man" (1989:28–9). They point to contemporary interventions by the *caboclos* in the Amazon estuary, where intensively managed systems have given timber, palm hearts, palm fruit, cocoa, and rubber to local and international markets, and the *ribereños* near Iquitos in the Peruvian Amazon, whose cultivation strategies have created orchards of fruit and food trees.

Posey's extensive studies of the Kayapó Indians (1981, 1983a, 1983b, 1984, 1985a, 1985b, 1989) indicate a long history of deliberate and sophisticated reforestation in open grassland areas and the planting of medicinal and other species in selected locations. The Kayapó live on a 2-million-hectare reserve in the Brazilian state of Pará; though they are officially protected, Posey notes that Kayapó culture is persistently undermined by plantation owners, settlers, and gold miners, and their cultural knowledge is undergoing erosion. The uncontaminated culture involved replanting of species in order to attract or maintain wild animals or to provide food near hunting areas or gardens. Since the Kayapó lands cover acreage about the size of France, these interventions have substantially altered the natural forest over the centuries. Unlike the massive deforestation and unselective burning carried out by colonizers, cattle ranchers, mining companies, and road builders, however, the interventions of the tribal cultivators contributed positively to the long-term growth of the forest. Posey argues that "indigenous use and management of tropical forests are best viewed as continua between plants that are domesticated and those that are semi-domesticated, manipulated, or wild. Likewise, there is no clearcut demarcation between natural and managed forest: much of what has been considered 'natural' forest in Amazonia is probably the result of millenia of human management and co-evolution." Of the native peoples whose knowledge of plants and animals is extensive, he observes, "They are a living, human wealth whose loss our planet can ill afford" (1984:125).

As colonists and missionaries penetrated the forests of Latin America, tribal people lost their isolation and gradually became enmeshed in trade. They obtained more sophisticated implements such as steel axes and machetes in exchange for food, medicines, and labour (Miracle, 1973; Wilbert, 1972; Gross et al., 1979; Korten and Klauss, 1984; Jordan, 1987a, 1987b). Shifting cultivation, and the cultural traditions and rituals to which it is attached, adapted to these new technologies, but showed very little fundamental change in the century of contact up to the massive deforestation in the 1980s.

The Carib-speaking Ye'cuana (also called Makiritare) of the north-central Amazon basin in Venezuela near the upper Orinoco River are known to have inhabited the region for at least 3,000 years. They cultivated manioc or yuca; since contact with Europeans, they have also cultivated plantains, bananas, maize, sugar cane, pineapples, papaya, and some root crops and medicinal plants. Farther south and west, where the Casiquiare and Guainia rivers unite to form the Rio Negro, Arawak-speaking tribes supplement horticulture in low-fertility soils with fishing. To the east of them in Brazil, the Yanoama tribes continue to hunt and gather much of their food, and they apparently adopted horticulture only within the past century.

The shifting cultivation practices of these groups were studied between 1974 and 1983 as part of a scientific project to determine the impact on soil of such cultivation. The scientific team concluded that nutrient stocks in the soil did not decline greatly during the period of cultivation, but there was a decline in phosphorus, which may have been responsible for crop-productivity decline. Following abandonment, the productivity of native vegetation increased rapidly. Apparently, native plants were better able to adapt and acquire phosphorus than crop plants. Carl Jordan concludes: "Cultivation of Amazonian soils, when practiced in the traditional manner, does not appear to inhibit the beginning of natural secondary succession. Whether succession will continue, and a forest will develop similar to that existing prior to the cutting, is another question" (1987b:23; see also Herrera et al., 1981).

Juan G. Saldarriaga takes up that question in a second case study in the upper Rio Negro. This involved the investigation of twenty-three sites in a 200-square-kilometre area, where soils were comparable and cultivation chronology could be accurately determined. He concludes that "it takes about 80 years for the biomass to build up to a value of about half of that of the mature forest" (1987:33). The data suggested that a return to the mature primary forest is possible, but that it would take more than a century for the nutrient stocks to be fully rebuilt.

If population densities are low and the territory is large, tribal groups may not recultivate the same site in a single lifetime. But as pressures on the land are imposed and population densities rise, shifting cultivators are obliged to compete for land and begin cultivation in immature secondary forests. Geoffrey Scott studied a situation in the Peruvian Amazon where arable land was limited and an influx of colonists to the region imposed pressure on the available land. He concluded that although shifting agriculture appears to be sustainable without commercial fertilizers if the fallow is fifteen years or more and no fires occur, "a 15-year fallow does not build nutrient stocks back up to the level of the primary forest. However, it does appear to make available most of the nutrients necessary to sustain a 2–3 year cycle of agriculture" (1987:45). The colonists in the Gran Pajonal region of Peru were trying to use seven-year cycles or even less, and this study suggests that the soil nutrients would not permit such agriculture to be sustainable.

Acculturation of Indians to European technologies and habits was rarely a gradual or mutually respectful development. In Brazil, as elsewhere throughout the world, penetration of forest lands already occupied by tribal cultivators and hunters was a cruel process. Though the Brazilian government of 1890, prodded by international opinion, appointed an army officer to carry out peaceful expeditions into Indian territory with the objective of protecting the natives, the Indian Protection Service (SPI) ultimately succumbed to the general disease of non-native cultures. Under its first leader, Câdido Mariano da Silva Rondon, the SPI made contact with several tribes, offering the usual gifts and protection while taking their lands and incidentally introducing diseases that killed them. Following the army were the rubber collectors, nut gatherers, settlers, and ranchers, and following them were the much more destructive national policies for opening up the Amazon.

Catherine Caufield cites anthropological data to estimate that between 6 and 9 million Indians inhabited the Amazon basin in 1500. One million survived in 1900, and fewer than 200,000 by the 1980s (1985:86). About half of the 230 tribes known to have lived there in 1900 are now extinct. Brazilian anthropologist Darcy Ribiero published a statistical report on Indian tribes in Brazil in 1957, which showed that since 1900, over eighty tribes had been destroyed by disease and deculturation. Six tribes had become extinct in areas of agricultural expansion. Thirteen had disappeared in cattle-raising areas. Fifty-nine had been destroyed in areas where rubber collecting and diamond prospecting had taken place. Those few tribes that survived were depopulated and wretchedly poor (Ribiero, 1957; see also Davis, 1977:5–6).

A decade after Ribiero's studies, a government commission reported widespread corruption and sadism practised by the contemporary SPI. Dynamite attacks, machine-gun forays, arsenic mixed into sugar, sexual perversion, murders, robbery, embezzlement, and illegal sale of Indian lands were documented. Foreign observers who followed up on the Figueriredo Commission reported that germ warfare was among the weapons employed to exterminate tribes. One reporter claimed that in the Amazon basin between 1964 and 1965, the year the military junta took power, tuberculosis organisms were deliberately introduced. Another reporter numbered the Indian victims of genocide in Brazilian history at six million (Lewis, 1969; Kent, 1968a, 1968b; Davis, 1977:10–14). The plight of native peoples was not viewed as a high priority, however; a year after the Figueriredo Commission report, the new president announced plans to build the Trans-Amazon Highway.

The plantation and pulpmill industry bears some responsibility for decimation of tribal peoples in the Amazon. It is among the industrial interests for whom railways and roads are constructed throughout the tropical forests, and such massive installations as Jari (described in chapter 11) necessarily take over land previously used by indigenous peoples. The sawmill and mahogany trade (described in the next chapter) displaced people, but these trades are more consequences of the opening up of the Amazon than causes of it. The impact of plantations and mills in the more southern regions is less clear. Many of the tribal groups there had already been dispersed by ranchers and settlers. The araucaria trade on the southern Atlantic coast, described in the next chapter, undoubtedly displaced tribal peoples, but it predates the modern plantation industry.

SUMMARY

Deforestation of tropical lands is occurring at a rapid pace. The causes are multiple. These forests are fragile and generally cannot be regenerated. The impacts of deforestation include reduction of biodiversity, downstream flooding, desertification, and reduction in the world's oxygen supply. Indigenous and poor peoples are displaced by deforestation. Parts of the Amazon region have been deforested by highway construction, transmigration projects, and attempts at cattle ranching. Population pressures are not important factors in the destruction of the Amazon (population density in Brazil as a whole is much less than in Europe and in the Amazon region is exceptionally low). Shifting cultivators are often blamed for deforestation, but ethnobotanists who have long studied the region argue that the blame is not theirs; they, indeed, are the chief victims.

7 Industrial Forestry in the Southern Hemisphere

In response to a plea from a British schoolchild to save the tropical forest, the prime minister of Malaysia defended the logging industry in these terms: "The timber industry helps hundreds of thousands of poor people in Malaysia. Are they supposed to remain poor because you want to study tropical animals?" (letter to Darrell Abercrombie, 15 Aug. 1987; printed in Malaysia, INSAN, 1992:76–7). While the prime minister's retort reflects a self-interested policy that benefits the few – overwhelmingly politicians and their families – who control logging concessions in Malaysia, there is also a barb for developed countries in the response. The chief buyers of tropical logs exported from Sarawak and Sabah are Japanese forestry companies.

Southeast and South Asia, together with Africa, have been deforested by the same general agents as Latin America: colonization, settlement, agriculture, plantation crops, roads, and dams. But commercial logging has been a more significant agent in Asia and some parts of Africa than in Latin America (the Atlantic coast of Brazil excepted) over the last century. European colonizers in Asia and Africa used wood in construction and furnishings and also exported small quantities of the especially decorative woods, such as teak, to their home countries. Latin America was a long distance from European markets and was blessed by an extreme species mix in often inaccessible locations, thus it was somewhat more sheltered while Asian and African forests were plundered.

INDUSTRIAL FORESTRY IN NATURAL FORESTS

Industrial forestry has a range of guises: simple plunder by loggers who sell the wood or are on contract to cut it, through to companies carrying out huge clearcutting operations. Plantations, which we examine in the next chapter, are often the "succession" crop; indeed, much of industrial forestry in places such as Malaysia today is openly intended as a means of clearing the land for planting cash crops.

Decorative cabinet woods are not the primary hardwoods on post-war markets, except for those from West Africa. Major exports in the 1980s were construction timbers and pulpwoods. In 1950, hardwood exports from tropical countries amounted to 4.2 MCM; by 1973, 53.3 MCM; and by 1980, 67.6 MCM (Myers, 1980:5; see also FAO, 1979).* Most of these consisted of tropical forest hardwoods, since the plantations were not yet important producers or exporters. One by one the Asian source regions were logged over. Finally, in the late 1970s and early 1980s, log exports were banned in the Philippines and then in Thailand and Indonesia. As each source dried up, the loggers moved on to other areas. Thai loggers moved to Myanmar, and Philippine loggers to Indonesia; Japanese companies looked more to Papua New Guinea. Peninsular Malaysia banned sawlog exports in 1985, but this policy merely shifted intensive logging to Sarawak and Sabah. The following year, 65 per cent of the 29.1 MCM cut in those areas was exported, most of it to Japan. Nearly half of all trade in hardwood logs in the 1980s and early 1990s consisted of exports from Southeast Asia to Japan (see table 7.1).

When log exports were banned, production shifted to plywood, lumber, and occasionally pulp. While in some cases, this allowed the exporting countries to upgrade their manufacturing capacity, it did not reduce the logging of tropical forests, except where logging itself was banned. Sawnwood has domestic markets and is traded between Asian countries; as well, it has an export market in the European Community (see table 7.2). Plywood exports move from Asia to Japan, the United States, and the EC. Latin America exports a small quantity of sawnwood to the United States and increasing quantities of pulp to Japan, Europe, and the U.S. (pulp and paper production data in the early 1990s are shown in tables 3.4 and 3.5 in chapter 3).

---

* Historical data should be read with some caution: statistics prior to 1950 are estimated by the FAO. As well, variable definitions of tropical hardwoods are in use, thus the data base between regions may not be comparable.

Table 7.1
Africa, Asia, Central America, South America, and Oceania: industrial roundwood
production, imports, exports, and consumption in selected countries, 1991 (1,000 m³)

	Production	Imports	Exports	Consumption
WORLD	1,599,272	127,074	123,474	1,602,872
AFRICA	58,871	716	4,613	54,973
Nigeria	7,868	0	8	7,860
South Africa	12,601	7	498	12,109
ASIA	251,999	73,789	22,493	303,295
China	90,099	9,480	919	98,660
India	24,522	1,298	46	25,774
Indonesia	29,314	0	676	28,638
Japan	27,945	49,500	8	77,437
South Korea	1,994	9,006	0	11,000
Malaysia	40,388	0	19,383	21,005
Myanmar	5,426	–	702	4,724
Philippines	4,473	358	6	4,824
Thailand	2,875	1,747	0	4,622
Turkey	5,968	1,082	7	7,043
Vietnam	4,846	37	420	4,463
CENTRAL AMERICA	4,515	50	117	4,456
SOUTH AMERICA	102,979	122	6,843	96,259
Argentina	6,487	2	923	5,566
Brazil	74,478	119	4	74,593
Chile	12,060	–	5,896	6,164
OCEANIA	33,713	9	11,190	22,532
Australia	16,423	1	5,465	10,958
New Zealand	13,937	7	3,578	10,365
Papua New Guinea	2,655	–	1,594	1,061

Source: FAO, 1993.

Tropical timber values, like their temperate counterparts, rarely (if ever) reflect true costs of regeneration (where it is possible) or of the foregone advantages of letting the forest remain undisturbed. Panayotou and Ashton argue that tropical forests are undervalued, first, by low stumpage, subsidized logging, and low rates of taxation; secondly, by overestimation of benefits from the timber industries and the conversion of natural forest to other uses; and thirdly, by ignoring non-timber goods and other forest services (1992:61–9). They note: "In order to ensure the long-term sustainability of tropical forest production, benefitting consumers and producers alike over the long run, adequate investments in forest management, regeneration, and planting are necessary. Such investments will not be forthcoming unless timber is priced at its full scarcity value. The full scarcity value of timber is defined as a price that fully covers logging, transportation, and user costs" (62).

The loss of rent for natural forest harvesting is generally related to non-competitive bidding processes and often to political, rather than economic, criteria for allocation of resources by governments (this behaviour is not peculiar to tropical countries). As well, governments may estimate rents on the basis of volume cut rather than volume of standing timber, thus creating an incentive to waste or trash wood that does not attract high market prices (Repetto, 1988). It is rare for governments to demand that companies bear or even share the cost of infrastructure for forestry operations (surveys, roads, grading, repairs of environmental damage, and townsites and other amenities for the labour force). Incentives for reforestation and for follow-up to ensure high success rates from planting are virtually non-existent. The anticipated public benefits from resource allocations to timber companies include employment, multiplier effects, technology transfers, infrastructure that can be utilized for other purposes, and foreign-exchange earnings. As noted in the case studies reported in the next several chapters, anticipations and actual returns have rarely coincided.

Because the forests are undervalued, timber prices depend more on felling, extraction, labour, and transportation costs than on the cost of raw material. These are lower in underdeveloped countries than in the industrialized world. Guppy, among others, argues: "Because they are relatively so cheap – whatever the consumer countries say at the United Nations Conference on Trade and Development (UNCTAD), or timber trade representatives say in the press – world demand for tropical timbers has grown faster than that for any other major raw material" (1984:955).

The World Bank anticipates continued demand. In its view, the demand (seemingly a magical condition, not to be modified by human intervention) will be so great that supplies will have to be obtained from the boreal forests and lower-quality North American hardwoods and from plantations in the tropics. The bank expects to provide up to 30 per cent of the costs of reforestation (plantation) programs, toward a total of between $1.5 and $2.4 billion per annum (Spears, 1980). FAO estimates of demand are even greater. Data for 1950–80 show consumption of tropical hardwoods to far exceed the rise in living standards for the importing industrial countries (FAO, Yearbook, 1980). At the same rate of growth in consumption of all woods as that traced between 1950 and 1980, FAO estimates that by the year 2000, world timber stocks per capita will have declined from

Table 7.2
Africa, Asia, Central America, South America, and Oceania: sawnwood production, imports, exports, and consumption in selected countries, 1991 (1,000 m³)

	Production	Imports	Exports	Consumption
WORLD	457,477	86,172	87,529	456,120
AFRICA	8,435	3,225	1,409	10,251
Nigeria	2,706	0	29	2,677
South Africa	1,792	80	107	1,765
ASIA	100,849	16,156	7,879	109,126
China	20,521	1,034	131	21,425
Hong Kong	421	312	295	439
India	17,460	19	35	17,443
Indonesia	9,145	0	789	8,356
Iran	173	133	0	306
Japan	28,264	9,400	11	37,653
South Korea	4,041	946	142	4,845
Malaysia	8,929	28	4,982	3,974
Myanmar	466	–	39	427
Pakistan	1,520	87	–	1,607
Philippines	723	0	57	666
Singapore	206	986	705	488
Thailand	939	1,534	56	2,417
Turkey	4,928	26	43	4,911
Vietnam	885	26	506	405
CENTRAL AMERICA	1,383	1,115	104	2,394
SOUTH AMERICA	25,764	374	2,081	24,057
Argentina	1,446	46	41	1,451
Brazil	17,179	265	479	16,965
Chile	3,218	0	1,246	1,973
Ecuador	1,641	–	18	1,622
OCEANIA	5,208	1,344	893	5,659
Australia	2,751	1,238	20	3,969
New Zealand	2,198	45	776	1,467
Papua New Guinea	117	0	0	112

Source: FAO, 1993.

142 CM to 114 CM in developed countries and from 57 CM to 21 CM in less-developed countries (Guppy, 1984: 951). Shortages will be endemic; thus plantations become essential correctives. One is advised to treat such market forecasts with caution. The trade journals invariably predict whatever suits their readers. The World Bank, FAO, and other international agencies tend to support forecasts of growth in plantations and their markets, providing the technical

expertise and much of the funding to ensure that the predictions are fulfilled.

Even at these high rates, exports represented only a fraction of logged timber, and that not because the exporting countries consumed the remainder. Their own consumption had risen, but exports account for a much higher proportion of the usable logs. The problems were rather that too often the forests were high-graded – only the most profitable species were taken out – and the logging was not done with care to ensure that the momentarily less valuable trees were left in good shape. Norman Myers noted that sometimes as few as 6 or 8 out of 150 per acre might be taken, yet many others were destroyed: "Commercial trees are often limited to those with the widest-spreading crowns, which may be as much as 50 feet across; when one of these giants is felled, it is likely to cause several others to be broken or pulled down" (1981:6). Other writers concluded from surveys in Southeast Asia that average logging practices left up to two-thirds of residual trees broken and the soil so impacted that most of the forest was damaged. Following logging, impoverished peasants or detribalized indigenees, often moved into the clearances and established subsistence crops; victims themselves, they further degraded the remaining land.

Long-term planning is not notable in any of the forested developing countries. Short-term plans with limited objectives involve the granting of concessions, eviction of encroachers, and the daily politics of "fast buck" logging. In consequence, the returns on forestry for the regions where forests once stood or where bare remains are still being logged do not change the general level of poverty.

REGIONAL CASE STUDIES

Malaysia

In the late 1980s, Malaysia was the world's leading exporter of tropical hardwood, accounting for 58 per cent of tropical log exports and 81 per cent of Asian exports. Of the country's total harvest, 65 per cent of logs were exported. Timber and wood product exports accounted for over 11 per cent of the country's total export earnings, only fractionally less than the earnings from oil and greater than those from palm oil or rubber through most of the 1980s. The logging industry accounted for some 13 per cent of the gross national product in 1990.

Malaysian lowlands have been virtually deforested: first by colonial oil palm and rubber plantations, and more recently by commercial logging. In 1985 the government banned the export of sawlogs and in 1989, of all logs from peninsular Malaysia. Log exports are now entirely from Sabah and Sarawak on the Island of Borneo. An "official" explanation is that peninsular Malaysia has developed wood-processing industries while the islands have not, and it is the case that 669 sawmills were located on the peninsula by 1989, while only 152 were located in Sabah and 126 in Sarawak; similarly, plywood and veneer mills were more numerous on the peninsula. (Sirin, 1990: 249). This explanation, of course, overlooks the paucity of remaining timber on the peninsula and the relative abundance on Borneo when the bans were put in place. Ironically, the condition imposed by Britain for giving up its colony to Malaysia in 1963 was that Sarawak have the power to establish its own land policies.

By virtue of isolation, but even more because of the land-preserving practices of indigenous people, Sabah and Sarawak had somehow saved their rainforests until the early 1960s. Japanese buyers discovered here a resource that appeared limitless at the very moment when other hardwood sources were becoming less plentiful. As well, settlement – some of it connected with government transmigration policies – was claiming land for agriculture. Within little more than a decade, the vast forest was criss-crossed with logging roads and dotted with logging camps; its soils were impacted by huge tractors and trucks. The mighty dipterocarps were falling at a rate of two acres a minute in the late 1970s by the count of one researcher, Gurmit Singh (1981:181). *World Wood* in 1987 estimated that the harvesting rate in Sabah was four times the regeneration rate (Oct. 1987:45–6). The environmentalist group Sahabat Alam Malaysia claimed that at the 1986 rate, Sarawak would lose 30 per cent of remaining state forests by the mid-1990s. That estimate appears to be on the low side, and many observers argue that the native forest is already nearly exhausted.

The International Tropical Timber Organization (ITTO) estimated in 1990 that primary forests would be logged out by the turn of the century. Stan Sesser, in a *New Yorker* report, put this estimate to the coordinator of the World Rainforest Movement in Penang. He responded: "Sarawak was cutting 12 MCM per year in 1988. The ITTO report, which is very conservative, was based on that figure. But the next year the cut rose to 18 MCM. I would say there are from five to ten years left" (Sesser, 1991:47).

Most of the forest is logged by short-term contractors. Since they have no processing mills of their own, the contractors have no incentive to maximize utilization. The tax revenue system encourages the export of logs. A National Forest Policy was promulgated in 1978, but forestry is managed by separate state governments. Most tax revenue is collected by the federal government, but state governments are allowed to collect land-related revenue, including timber export duties. Higher export duty is collected on sawlogs than on sawn timber. Naturally, state governments seek to maximize these revenues, which means supporting the export of logs rather than wood products. *World Wood* estimated that the harvesting rate in these regions should account for half the government income (Oct. 1987:45–6). Even at that, the resource rents are so low it is cheaper for Japan to import wood for essentially wasteful applications than to use its own sources where these would be appropriate and available.

The taxation system is one issue; general corruption is another. In a 1987 election campaign in Sarawak, two powerful (and family-related) timber barons each told tales intended to demonstrate that the other's group controlled excessive timber concessions. Between them, it turned out, they controlled about 30 per cent of Sarawak's forest land (Jomo, 1992a:iv–v). Indeed, it appears that the quickest way to become very rich in Malaysia is to become a politician. As understated by Gurmit Singh, "officials charged with enforcing regulations at the ground level and in collecting levies are faced with daily temptations to look the other way from the powerful and rich loggers. It is not surprising then that employees in such a department would not be too enthusiastic to work in the research and reforestation sections, when they are aware that the revenue and enforcement sectors provide opportunities for supplementing their income" (1981:186). Singh also says that the Japanese traders and shippers make more money than state coffers in Malaysia and that it is difficult to buy construction timber within Malaysia. Such charges are commonplace. Survival International has repeatedly documented political corruption, and lists of landowners in one region after another are effectively those of political office-holders and their relatives (see, for example, Sarawak Study Group, 1992:3–28).

However, the chief beneficiaries of the logging industry in Malaysia are Japanese importing and construction companies. As late as 1989, 63 per cent of the logs were destined for Japan, 15 per cent to South Korea, and 12 per cent to Taiwan (Jomo, 1992a:ii). In Japan they are used for plywood moulds in the construction industry. Moulds of this kind could equally well be made of plantation wood; pine is frequently used for this purpose elsewhere. After one or two uses, the moulds join everything else from the construction site as garbage in landfill. The explanation for this bizarre and wasteful application of magnificent and increasingly rare dipterocarp wood is that it costs less to log a tree in Sarawak, import it to Japan, mill it, and make plywood than to use Japanese-grown trees (Sesser, 1991:66). The cause of the price differential lies partly in the cheap labour of Sarawak and partly in the value of land and the yen in Japan, but largely in the deficient resource-rent practices and corrupt system of concessions in Malyasia. The problem here is not that there is an oversupply of competing tropical hardwoods on the Japanese market – the contrary is the case; it is, rather, that prices reflect decisions by government bureaucrats or politicians with little reference to market forces or the country's long-term interests.

Pulp production is limited to one mill, Sabah Forest Industries, on the island of Borneo. This company is wholly owned by the government, and the mill, with capacity for 125,000 tons of paper, draws on the mixed tropical hardwood forest in a 271,200-hectare concession area. In 1992 a new group – Genting Sanyen Paper, backed by Genting International, a gambling operation in Malyasia which also has a corrugating mill in Singapore – opened the country's first large brown-grade paper machine at a completely new mill in Selangor. It anticipates having a world-class papermill with a capacity of 1 million tons per year, and plans are under way for development of a newsprint mill. One of the incentives for the Selangor mill construction, noted by its chief excutive officer, is that Singapore has become a somewhat less-than-friendly environment for paper production (*PPI*, Nov. 1992: 66). Waste paper will provide the major fibre source, available in Malaysia because the country imports 3 million tons of paper and board annually, but the company is contemplating the establishment of a pulpmill on Borneo. Further pulp production would depend on land acquisitions for plantations. Imported pulp for a paper industry might be more economical, but plantations are being installed.

Paper producers in Malaysia that make printing and/or writing grades include Sabah Forest Industries and North Malaya Paper Mills, which supply domestic markets and export small quantities. There are also some ten tissue mills, whose combined output exceeds domestic consumption, with export markets in Australia, New Zealand, and Singapore. Since domestic consumption of all paper products is increasing, exports are expected to decline (*PPI*, Nov. 1992:55,59).

Plantations were started in 1973 with *Pinus caribaea, P. merkusii*, and *Araucaria* on a pilot-scheme basis in peninsular Malaysia and in 1982 on Sabah under a policy somewhat ironically labelled the Compensatory Forest Plantation Project (CFPP). The Sabah operations are intended for the pulp and paper mill, which will need "compensatory" wood in light of dwindling supplies of tropical timber. The project plans to establish some 188,000 hectares of plantations growing *Gmelina arborea, Acacia mangium, Albizzia falcataria*, and *Paraserianthes falcataria* based on fifteen-year rotations; the Sabah Forest Development Authority (SAFODA) plans to add about 100,000 hectares, and the Sabah Softwood Company will top that with another 61,000 hectares during the 1990s, for a projected total of 350,000 hectares on Sabah (Sirin, 1990:248). Another plantation started in the 1970s is a joint venture between a government body, Sabah Foundation, and North Borneo Timbers, a British firm, at a 60–40 equity ratio, in a logged-over area at Tawau in the southeastern part of Sabah (Udarbe, 1990:219).

The director of the forestry economics unit in Malaysia's Forestry Department, Lockman Sirin, notes that "the wisdom of cutting down natural tropical forests and reafforesting with fast-growing species ... was totally lost on most people. The urgency which prompted the CFPP, that is the need to produce timber quickly within 15 to 20 years, was not seen as strategic to these critics" (1989:261). Rarely is an official so naively blunt about the reasons for deforestation, and nowhere else are plantations frankly labelled "compensatory."

Beneficiaries of the papermill development include the Japanese producers of the Kobayashi trimming machine, Sulzer Escher Wyss, which supplied the waste-paper treatment plant, Dorries and Voith-Dorries, which supplied the papermill machinery and rebuilt other machinery purchased second-hand from Australia, and other companies in Europe and Japan whose wares were needed (*PPI*, Nov. 1992:69).

Not numbered among the beneficiaries of forestry are the tribal peoples of Malaysia, particularly those of Sabah and Sarawak. As is the case for other subsistence economies, many of the native communities of Sarawak have survived on shifting agriculture, fish, and wildlife for their livelihoods. The Penan were nomadic hunters in the forests. Both land and habitat have been eroded as forestry concessions destroyed the forest. The native people disputed the right of the government to assign their lands to the state and then to grant logging concessions without respect for customary land rights. The Kayan, Kenyah Dayak, and Penan peoples gained international attention by blocking timber operations. A seven-month-long blocade to stop the logging of their ancestral lands culminated in November 1987 in arrests of forty-two native people, charged with "obstructing" government policy and "unlawful occupation of State lands." After the 1987 demonstrations, the Sarawak Legislative Assembly made it an offence to obstruct roads constructed by timber licensees or permit holders, and the fines and jail terms were severe (Sahabat Alam Malaysia, 1990:10–14). In the following months, logging resumed. Rivers became silted and polluted, the food supply for native peoples dwindled, and the government provided no negotiating options. Again the Penan and Kelabit peoples set up a blocade in May 1988, and more have been erected in several areas since then. The history is replete with arrests, detentions, fines, and abuse. The government's position on shifting cultivators on the outer island (they no longer survive on the peninsula) is that they deforest some 50,000 hectares per year but that "Government efforts to settle the shifting cultivators have been effective in reducing the extent of deforestation attributable to this cause" (Sirin, 1989:253).

By 1988 the world beyond Malaysia was beginning to gain knowledge of the situation. Survival International and the World Rainforest Movement were active in disseminating information, and in July of that year the European Parliament unanimously adopted a resolution calling on member states to suspend imports of timber from Sarawak (European Parliament, 1990a). In 1989 an investigating team from the U.S. congressional staff and public-interest groups reported that logging procedures had degraded the environment and adversely impacted the health and well-being of the native populations of Sarawak. A unanimous resolution in the European Parliament in 1989 appealed to the Malaysian authorities to stop arresting and detaining native peoples (European Parliamant, 1990b). The resolution also called on the International Tropical Timber Organization to include native land rights in its terms of reference for a mission to Sarawak.

Of the native people, now numbering 9,200 altogether, the Malaysian prime minister, Mahathir Mohamad, said in 1990, "There is nothing romantic about these helpless, half-starved and disease-ridden people." His government has decided to relocate them, by way of "endeavouring to uplift their conditions." He was responding to criticisms by Britain's Prince Charles, who had accused Malaysia of waging collective genocide. "If we do not correct this," said the prime minister, "the world will tend to believe the statement made by someone of such status and influence" (Youngblood, 1990). In response to the international attention, the Malaysian government announced in 1989 that it would "soon" impose a total ban on log exports from the islands, but this policy turned out to mean a gradual phasing out by 1995. The rate of deforestation increased the following year.

Myanmar

About 70 per cent of the world's remaining teak forests are situated in the highland regions of what is now called Myanmar (formerly

Burma). These and other hardwood trees cover about 42 per cent of the country's total land area. About 3.3 million hectares are classified as degraded, and 8.8 million hectares as "taungya" or agro-forestry reserves; 16.1 million hectares are closed forests. In addition, there are 227,710 hectares of industrial plantations and 84,948 hectares of fuelwood plantations (FAO, 1991a:6).

Though there is a history of cutting in the south, the north was preserved by the colonial government. Earlier, the last Burmese dynasty had monopolized the cutting and sale of teak, and export duties on this and other hardwoods were important sources of revenue for governments in pre-colonial and colonial periods. British timber merchants and Burmese contract loggers took over government monopolies in the nineteenth century. The Tenasserim deciduous monsoon forests were ravaged in the mid-nineteenth century, and British timber merchants moved on to the Sittang and Salween river valleys and then into Lower Burma, before the British annexed the remaining portion of the Burmese kingdom in the final Anglo-Burmese War of the late nineteenth century (Adas, 1983:95–6). Despoilation of hardwood forests notwithstanding, the British administration did attempt to conserve the remaining highland and monsoon forests. A memorandum issued by the governor-general of India in 1855 declared teak trees to be government property, and a new Forest Department in Burma established forest reserves. Shifting cultivators were singled out for tough supervision, but commercial logging was eventually brought under control.

Conservation policies did not extend to the evergreen monsoon forests of the Irrawaddy delta lowlands, which succumbed to the fate of lowlands elsewhere: they became rice paddies. Michael Adas (1983) has provided a detailed account of the destruction of coastal mangrove swamps and lowland kanazo forests within a space of merely two or three decades. Among the acts of the British imperial government was one of the first transmigration projects undertaken in Asia. These lowlands were lightly populated in the 1850s. The British administration provided assistance to settlers from more populated India and upper Burma; it encouraged the development of rice exports, improved port facilities, and built roads, railways, and canals to transport both rice and settlers. By the turn of the century, lower Burma was deforested.

Post-independence policies in Burma, now called Myanmar by its military leaders, have provided inconsistently for some conservation and minimal reforestation. The World Bank contracted with the Sandwell Company to conduct a feasibility study of forest resources in 1977–78, after which the bank established two major teak plantation projects. The first, in Pegu (or Bago Yoma) and called the East Pegu Yoma Project, was financed to the extent of US$3.9 million. The second was financed by the American Development Bank with a loan of US$2.3 million. Foreign aid coincided with a peak period for plantation forestry and a steadily rising price for teak logs. Foreign-exchange earnings from forestry exceeded those from agriculture for the first time during the 1980s.

The United Nations Development Plan and the FAO provided assistance in 1981 to establish "nature conservation and national parks" which would develop the tourism business in safaris and maintain game sanctuaries. But logging has been more in evidence than conservation or replanting; in fact, the plantations have generally failed because they have not been accompanied by the labour-intensive management they require. The 1991 FAO report gave extraction rates for the previous year as 400 thousand cubic tons of teak, 1,630 thousand cubic tons of hardwood, 1,768 thousand cubic tons of fuelwood, and 88,210 thousand numbers of cane (FAO, 1991a:3, based on data from the Myanmar Ministry of Planning and Finance, 1990).

A century-old teak tree can fetch up to $40,000 on world markets of the 1990s. Unfortunately, its market value is too great, and the military rulers of Myanmar have provided timber concessions of such extent that some observers estimated the disappearance of teak by 1995. The FAO somewhat understated the problem in noting that the annual cut of teak in the late 1980s exceeded the annual allowable cut by as much as 26 per cent in some regions. An Associated Press journalist claimed that UN preliminary analysis of satellite photographs indicated 485,000 hectares of tree cover disappeared annually since 1985 (Grey, reprinted in *Vancouver Sun*, 4 Aug. 1990). The Associated Press report says that while Thailand concessions work along the frontier, joint-venture firms from Taiwan, Singapore, and Hong Kong log interior areas. In fact, the Myanmar government has created policies that permit foreign companies to log and to buy the production from wood-based industries in return for investment. Although the growth rate in forestry increased by 34.2 per cent between 1987 and 1988–89, its financial returns to the country have been disappointing.

The chief beneficiaries of more open access to the teak forests are Thai logging concessions. The ban on logging in Thailand (described in chapter 9) was accompanied by agreements between the Thai government, represented by army commander General Cahovalit Yongchaiyut, and the military dictatorship in Myanmar to allow Thai logging companies to cut teak and other tropical species in the neighbouring country. Thai-owned truck-loggers, often linked to the Thai army, are rapidly deforesting the world's remaining teak (*Economist*, 7 April 1990). The December 1988 agreement specified border-crossing points for the transfer of logs into Thailand. Burmese customs officers were stationed on the Thai side of the border to assess and stamp them. By February 1989 the Burmese government said that twenty concession areas were already under contract, with authorizations for total exports of 160,000 (metric) tons of teak logs and 500,000 (metric) tons of other hardwood logs. Hamish McDonald, reporting to the *Far Eastern Economic Review*, estimated the 1980 value of revenues to Myanmar at US$112 million (22 Feb. 1990:17). Another twenty contracts were negotiated within the next few months, all of them recommended by the Thai government; companies not so recommended were shut out. McDonald identified several of the concessionaires as senior military men or companies with shareholders affiliated with politicians of the then-ruling Chart Thai party or its coalition partner.

Logging has provided the cover for the Myanmar military to wage war against the Kachin, Karen, and Mon insurgents who once inhabited these forests. Several thousand refugees from Myanmar's southwestern state of Mon have fled into Thailand while escaping heavy fighting between government troops and Mon insurgents. Border crossings are occasionally bombed near the eastern state of Karen (*Jakarta Post*, 7 May 1990). World attention to these struggles and environmentalist lobbies against the decimation of this last intact teak forest succeeded in July 1989 in getting a government announcement banning new logging licences. Yet McDonald relates in February 1990, "Recent visitors to border areas report new roads carved through the jungle, machinery and elephants massed, and a heavy stream of log-carrying trucks heading into Thailand. So much timber was coming across by last October [1989] that the Thai concessionaries were lobbying for permission to re-export logs and timber from Burma to ward off a fall in prices" (*Far Eastern Economic Review*, 1990:18). Among the ironies of this situation are that many of the watersheds for the now ostensibly protected forests on the Thai side of the border are located in the Burmese concessions.

The FAO study of 1991 notes that the border trade with Thailand accounted for about 30 per cent of the total export value of forestry products in 1990 (FAO, 1991a:4). The Myanmar government (and the FAO reporter) boasts that this is a positive step toward privatizing the economy and liberalizing trade.

In addition to logging, forests are diminished by irrigation and hydroelectric power dams. Watersheds are poorly managed, and the Ayeyarwady River is reported to be carrying an annual sediment load

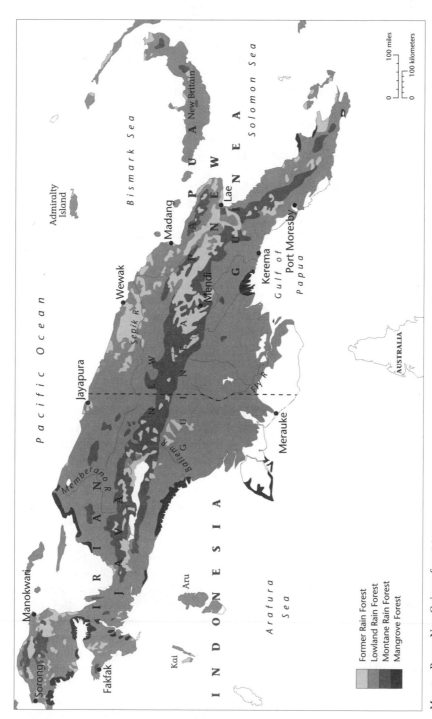

Map 7 Papua New Guinea: forest types

of 299 million tons per year. This ranks in the category of the Yellow, Ganges, Amazon, and Mississippi rivers; the results include flooding and droughts in the lowlands. As well, some 500,000 hectares of forest land were converted to non-forest uses in the 1975–80 period, and this deforestation continues. It is often blamed on hill tribes here as elsewhere. However, most of the taungya takes place in remote, hilly regions and in forests classified as degraded (FAO, 1991a:32). Illicit trade in animals, particularly in elephant tusks, has not been controlled effectively, and this activity is yet another, and often damaging, use of forests.

In April 1990 the United States Senate passed a bill to impose a ban on all Burmese imports, including teak and other hardwoods used by Thai furniture makers. U.S. markets for teak furniture are the most important ones for Thai manufacturers, accounting for over one-third of exports in 1989. The reasons for this bill are not particularly associated with concern for forests; all products from Burma were included in the ban (*Nation* [Bangkok], 27 April 1990). The stated reason was to deny a source of hard currency to the military regime. The Thai government is reported to have established contracts with the governments of Laos and Cambodia as alternative sources (Smucker, 1990).

Papua New Guinea

Colonial forestry policy in Papua New Guinea (PNG) was quite unlike that in other regions. Approximately 97 per cent of PNG land was unregistered customary land when Papua was colonized by Britain and New Guinea by Germany. Neither colonial power annexed this land for the crown, though both imposed government controls on its use and the sale of its produce. An external trade in logs was thus inhibited under colonial rule and began only under Australian tutelage in 1971, when for the first time, customary owners were permitted to sell timber directly to outsiders.

When PNG became independent from Australia in 1975, a new government sought means of improving its revenues. There followed a generally confusing period while levels of government and ministries tried to learn new roles; in the interstices of responsibilities, some ministers and others found ways of improving their revenues from log exports. In 1979 a white paper proposed an increase in the volume of log exports. This was intended to assist in raising revenues for public works. The paper also proposed the establishment of a state marketing authority, so that the government would have a window on the industry, and a series of guidelines governing transfer pricing, permits, and sales. Foreign ownership was to be permitted, but strictly controlled.

The white paper happened to coincide with the imposition of bans on logging in Southeast Asia; at first these bans were relatively weak, but they became progressively stronger, and importers began seeking alternative sources of roundwood. The buyers were from Japan, Korea, Malaysia, and Singapore. Thomas Barnett, the commissioner appointed to inquire into Papua New Guinea forestry for the prime minister in 1987–89, provided details on this process; what follows is from his account, as reported to a conference in Australia in 1990. The 1990 Forestry Bill was still in draft form at that time. As documented by Barnett, the decentralized and disorganized forestry service was incapable of controlling foreign and domestic operations in the following years.

Operations commenced illegally, Forest Working Plans, if submitted at all, were widely ignored, logging tracks were pushed through at the whim of the bulldozer drivers, hillsides and river banks were logged and the immature future forest resource was being bashed and trampled in the reckless haste to get the logs down to the waiting log ships. The dazed and disillusioned forest owners watched in disbelief as (usually) Asian operators removed their trees before moving on to the next area, leaving environmentally disastrous logged-over hillsides, temporary mud roads and rotting log bridges to erode and cave in to clog the watercourses. (1990:11)

Barnett found only one company, the large Japanese-owned Stettin Bay Lumber Company, to have established extensive plantations and good roads. It was, he said, "clearly looking to a long term presence in its timber area."

The volume of log exports increased, and the volume of local processing dropped dramatically after 1979. The commission's final report showed log exports to have increased from 472,500 in 1979 to 1.4 million in 1987 and sawn timber to have declined from 62,000 board feet to 2,700 board feet in the same period. The National Forestry Development Programme of 1987–91 required the volume of log exports to double by 1991. Barnett documents illegal and barely legal operations throughout the PNG. In 1989 he reported that the national forestry policy was either non-existent or so violated that it was useless; that control of forestry operations was ineffective, and logs were being taken out with neither advance government agreement nor supervised care for the remaining timber; that local forest owners were receiving few benefits; and that the governments were being regularly cheated out of revenue through transfer pricing, tax evasion, and fraud. Corruption and fraud were widespread and included national leaders. The commission's recommendations in 1989 covered the various issues. One year later, when Barnett gave his paper at the Australian National University, the recommendations remained on the shelf; only one government leader had been punished, companies in breach of the law continued to operate, and overcutting was still occurring. In October 1989 the World Bank reported on PNG. It noted environmental damage caused by the expanded logging program, inadequate legal arrangements, various technical and economic defects in log marketing, and generally incoherent and wasteful management of the forests.

The Philippines

The Philippines hosted a forest industry much earlier than neighbouring countries in Southeast Asia, and while forestry was not the only cause of deforestation, it was a significant one. As was true of other Asian territories, the invading colonial powers of the mid-sixteenth century had variable success in subduing tribal people in the Philippines. The Muslim sultanates in Mindanao and the hill tribes in northern Luzon resisted Spanish authority until well into the nineteenth century. Where the colonial rulers succeeded with the lowland groups, they co-opted Filipino chiefs as tax collectors and other officials. The spread of wet rice agriculture, promoted by priests and officials alike, encouraged the peasants to take up more sedentary lives and gradually converted the forests that had covered most of the islands. Estate agriculture followed, with plantations of sugar, abaca (also known as Manila hemp), coffee, indigo, and tobacco. The highlands remained forested until the twentieth century, but the plains of Luzon were mostly cleared by about 1920 (McLennan, 1972; Roth, 1983:32).

Deforestation of central Luzon caused both floods and the drying up of lakes and swamps, as the same process later did in Thailand. The loss of forest cover, combined with the creation of hard pan in the rice fields, damaged the soil and reduced the water-table. The soils could not retain rain water, and so floods became a regular cause of devastation. What remained in much of the central plain was grassland. American conquistadores of the twentieth century blamed the hill tribes for the degraded grasslands. Slash-and-burn agriculture offended their sensibilities, and their tendency to seek culprits in the least sophisticated sectors of the population was manifested here as elsewhere. Their blame was not entirely misplaced; earlier Spanish attempts to force Filipinos into labour on tobacco and sugar estates had encouraged many peasants to head for the hills. Their numbers swelled as potential agricultural land in the lowlands was monopolized by their overlords. Overpopulation in the hills had the usual deleterious impact on forests, and swidden agriculture, though still alive in name, was no longer the conservationist strategy it had been when tribal people were sparse and in control of their own territory.

Forestry began in earnest in 1904 with the granting of a concession to log some 300 square kilometres of dipterocarp forest in northern Negros to an American lumber company. The Insular Lumber Company was headed by a manufacturer of sawmill equipment in the United States, and the operation was technically well served (Roth, 1983:46). Selective logging soon gave way to clearcutting. Barrington Moore Sr, an American forester of the period, expressed strong reservations about the *modus operandi* in two articles published in the United States in 1910, which had some persuasive power in Manila. Subsequently a government-sponsored investigation of the reproductive capacities of logged-over areas reported that cutting trees at 40 or even 50 centimetres in diameter effectively killed off the dipterocarp forest. The delicate ecosystem was fatally disturbed when the canopy of larger trees disappeared; moreover, on higher elevations, dipterocarps needed nearly four centuries to grow to a diameter of 40 centimetres, so sustained yield logging was simply impossible in the hills (Roth, 1983:47–8).

According to Nelflor Atienza, the senior forester in the Filipino Department of Environment and Natural Resources in 1989, forest land lost during the previous two decades amounted to 7.7 million hectares; a total of 8.5 million hectares remained, of which only 1 million hectares had commercial potential. Log-export bans were imposed in 1986, in anticipation of increases in the exports of processed wood products (lumber, plywood, and veneer), but these sales did not materialize. Meanwhile, destructive logging practices continued, and in Atienza's view, these were a major cause of deforestation. Selective logging was prescribed, but timber concessionaires damaged residual growing stock when they cut selectively and also clearcut large portions of their areas, eroding soil with heavy machinery. As well, illegal cutting of timber has destroyed forest land. Projections about growing stock from old-growth and logged-over areas are pointless, as the old-growth forest is virtually non-existent (Atienza, 1990: 296–302). Plantations might become important in the future, but little land is currently seeded.

Vietnam

In the early 1940s, 46 per cent of Vietnam's 33 million hectares was forest. That proportion fell to 29 per cent in 1975 and has dropped further since then; it is estimated in 1990 at about 20 per cent. Two-fifths of the country is barren wasteland according to Professor Vo

Qiuy, a Vietnamese ornithologist who is trying to get international aid to reforest the country (Palmer, 1990). Much of the destruction was caused by heavy bombing and the U.S. use of the defoliant Agent Orange during the war. The forest was cleared along roads and canals, in total 325,000 hectares, to minimize the danger of guerrilla attacks. There are 25 million vast bomb craters, some turned into fish ponds or vegetable patches, but most are unusable for agriculture (Kemf, 1990).

India

Between 1947 and 1977, India's forests decreased from 40 per cent to 20 per cent of total land cover. Severe ecological damage is endemic, and both tribal and village populations have lost access to wood resources essential to their survival (Shiva et al., 1982). Social forestry, a government plan announced as the means of increasing fuel sources for the poor, has gradually become a way of creating private eucalyptus stands (Shiva and Bandyopadhyay, 1983). While the losses have been great in the post-independence period and the de-communalizing of common lands has deprived peasants of essential resources, losses were even greater in the nineteenth and pre-independence twentieth centuries, and the privatization of land was a leitmotif long before independence.

Lowland forest depletion is attributable in India, as throughout Asia more generally, to the expansion of agriculture. The initial waves of deforestation may be traced for each region to the influx of British administrators. As agriculture expanded under British supervision, so too did the population, and the paradoxical increase in numbers of landless and impoverished villagers intensified pressures on forest lands at higher elevations, as similar processes did elsewhere in Asia and Africa.

The deforestation of the western Himalayas and the Assam region has been examined by Richard Tucker (1983, 1988). He traces small forest depletions to the Pahari people, who had traditionally sown a mixture of grain crops and maintained common pasture lands in the hill country. Their populations were small. However, timber became a commercial commodity before the area was occupied by the British. Private Indian timber contractors had already created a brisk trade in construction timber, primarily the hardwood sal (*Shorea robusta*) from the foothills for the expanding cities on the deltas of Indus and Ganges in the early nineteenth century (Tucker, 1983: 156–7). The growth of urban centres drawing on rich agricultural settlements further depleted forests.

Military fortifications and the construction of the most remarkable railway system in colonial territories effectively finished the job in northern India. Under the Raj and under princely states adjoining Raj-controlled territories in the mid-nineteenth century, timber operations gradually moved upward into the coniferous region, where a slow-growing tree known as deodar provided durable wood ideal for construction and railway ties. Pine, spruce, and fir species were not commercially exploited until after independence, when roads were constructed at higher elevations. Oak was also used commercially and by villagers for fodder and mulch, and soil erosion in some areas was due to the overexploitation of oak stands (Tucker, 1983:147–9). But before 1947, the deodar forests were exploited by the British.

Exploitation began in accessible regions, largely for commercial profit; it then moved to more remote stands in the Punjab and eventually in Kashmir for military defences. With the construction of railways across the subcontinent from the 1850s through the 1880s, the trees of the Himalayas became railway ties, and what remained entered the larger Indian and international markets. Deodar forests of the Punjab hills were requisitioned for two major rail lines, one of them designed to carry wheat exports from the Punjab to ports for European markets (Tucker, 1983:158–62.) The railways took priority in allocations of timber-cutting rights, but as they were built, so too were new markets created along their path for construction timber and urban fuelwood supplies. The Himalayan and submontane forests were nearing depletion by the 1880s, and the deodar forests of Kashmir and sal timbers of Nepal were eyed as the next great woods to be buried beneath trains. Throughout this history, the hill tribes were gradually transformed into wards of the state. As elsewhere in India, they lost control of the land and of its many products.

Colonial history in the Assam region, on the easternmost fringe of the Himalayan range, involved less damage to mountain areas. Sal is the major species on the lower hills; subtropical broadleaf evergreens, including hollong (*Dipterocarpus marocarpus*), cover the steeper ranges near the borders of Burma and China. These mountainous terrains were remote and inaccessible, and for this reason resisted depredations. Tribal shifting agriculturalists maintained their subsistence plots, their numbers being too few to seriously threaten the forests.

The lowland forests of Assam, however, were displaced by rice farming and tea plantations. The Brahmaputra lowlands were populated by a rapidly growing number of Assam-speaking Hindu rice farmers; and further downriver in Bengal, by Muslim peasants. British tea planters moved into the region in the mid-nineteenth century and imported plantation labour from outside. These diverse groups all had claims to the remaining forests as India and Pakistan moved toward independence. The sal forests of Sylhet in the south became part of Pakistan, and the reserves remaining in India were substantially reduced as Hindu migrants fleeing Pakistan imposed new pressures on the land base, roads were constructed into the mountains, and political turmoil reduced the capacities of the administration to sustain forest reserves (Tucker, 1988).

British colonial policy had not been indifferent to the fate of forests, even as colonial interests persistently undermined loftier intentions. The first Indian Forest Law, enacted in 1878, established reserves and protected forests, and in the early nineteenth century, state forestry services created sal plantations. By 1900, protected forests covered 8,300 square miles. A plantation in the Punjab, the Changa Managa forest, had some 4,047 hectares under management; in this same region an annual harvest of 80,000 trees had been cut for railway ties during the 1880s. Deodar was much less easily regenerated.

Ramachandra Guha (1983) notes that these reserves put forests firmly under government control. Access rights, now more often regarded as privileges, were flexibly interpreted by local authorities. Tribals and villagers in the western Himalayas had free access, but tribals in several other areas gained access by buying concessions or were denied access. Local policies reflected local political conditions; where tribal and village cooperation was essential to British security, as near the Tibetan border; or where strong village communities were restive, access was more liberally granted. While the reserves were important oases in a largely deforested region, they were not intensively managed, and they were variously protected during the First World War as military demand for timber affected the region. The demand for timber became even greater in the Second World War and after independence (Tucker, 1988).

The post-colonial government established an immense bureaucracy in forestry as in other areas, but deforestation continued. Vandana Shiva and colleagues at the Bangalore Institute of Management estimated that 3.4 million hectares of forest were destroyed from 1951 to 1973: 71.5 per cent by agricultural expansion and 11.8 per cent by irrigation or other river valley projects. Projects initiated with the optimism of the green revolution resulted in soil erosion, floods, and droughts, and it was the poor villagers and the remaining tribals who typically suffered from the intended reforms (1982:158).

Under the banner of environmental concerns, a new Forest Act of 1980 reinforced the policies initiated under the Raj. Sanctions against encroachments and the discretionary powers of the bureaucracy were increased, and access to communal forests – indeed, the existence of communal forests – was further delimited. Guha argues that "the bill is based on the (correct) premise that the tribals compete with the mercantile and industrial bourgeoisie for the produce of the forests ... [It is] designed to further exclude the forest dweller from utilisation of the dwindling forest stock" (1983:1943).

China

China has a total forest area of about 119 million hectares, most of this in the northeastern provinces, but with some growth in southwestern provinces. It has started a major program of reforestation in areas that were cleared during war and civil war in the 1940s or exploited for fuel sources since then. Though its standing forest is limited, it has some 5,000 papermills and 9,000 pulpmills, and it ranks fourth among the world's paper and board producing countries and sixth among pulp-producing countries. Most of its pulp is made from non-wood plant fibres, especially rice and wheat straw, sugar bagasse, and bamboo (*PPI*, Oct. 1992:46). As economic development takes place, China's domestic demand for paper products increases, and it may be at the leading edge for technical advances in non-wood fibre production and utilization in small mills appropriate to both fibre sources and community stability. Meanwhile, its import of paper products is increasing.

China is importing foreign equipment from the Canadian Industrial Consortium for a new, large kraft pulpmill, based on bamboo and masson pine fibres, to be built in Sichuan province, and much of its existing machinery is produced by European manufacturers, including Voith, Beloit, Dalian, and Valmet. Thus is China, though differing from its neighbours in many other ways, already integrated into a global economy. On the other side of the equation, however, China is beginning to invest in pulpmill capacities in other regions, including a mill in British Columbia. Through this measure, similar to the model introduced by Japanese papermakers, it can obtain wood-based pulp for its own paper manufacture.

Brazil

The Asian forests were earlier, and are still more, integrated into the world's market system than the tropical forests of Latin America. However, some forests in Latin America have been destroyed for commercial purposes or through settlement and other activities, with incidental commercial gains. The Brazilian araucaria forest is a particularly visible case. It once covered between 20 and 25 million hectares along the Atlantic coast in the three southernmost states of Brazil: Paraná, Santa Catarina, and Rio Grande do Sul. It had been reduced to about 445,000 hectares by 1980. John McNeill (1988) examined its history and informs us that deforestation was gradual before 1930, more rapid up to 1945, and then very rapid until about 1970. Nearly 3 per cent of the total cover was cut annually in the 1950s in the state of Paraná; that is, about 250,000 hectares. The rate slowed after 1970 because so little was left. Between 1973 and 1977, McNeill reports, "only about 80,000 hectares of araucaria disappeared, but that amounted to 8 to 10 percent of existing stands in 1973" (1988:17). Deforestation of araucaria stands continues in eastern Paraguay and southern Chile.

The timber trade developed with the opening of a railway in the state of Paraná in 1883. Over one hundred sawmills were operating by 1908, and these were a major source of employment. The araucaria was useful for construction, railroad ties, furniture, and other purposes, and its usage became more widespread through the 1950s and 1960s. In the 1960s, 40 per cent of the cut went to exports and the remainder to the Brazilian market. But the surviving stands were already remote from markets, and the international market was providing alternatives. Argentina, the major importer, had started its own araucaria stands and could also obtain softwoods from Chile. By 1970 the Brazilian export trade was almost finished.

Altogether, official lumber production accounted for about 2 million hectares, "perhaps 10 to 15 percent of the original araucaria zone." McNeill points out that even with a generous allowance for poaching and illegal exports, one is forced to conclude that the rest of the forest – burned intentionally or accidentally in the rush to settle the land – was simply wasted. Among the identified ecological impacts is a reduction in the navigability of the Iguaçu River, periodic flooding from disrupted drainage, and drought apparently connected to local climatic changes. The state of Paraná and the central government of Brazil have been slow to enact forest conservation laws or to enforce existing legislation. Reforestation legislation dates from 1965, but actual rates of planting are small, even by official statistics.

The araucaria grows slowly, reaching a height at maturity of between 25 and 30 metres; undisturbed, it survives to an age of three hundred years in subtropical regions. The much faster-growing species loblolly pine and eucalyptus are the chosen replacements. Araucaria regenerates poorly on its own and is easily dominated by other species in second-growth forests; thus McNeill concludes that the disappearance of the araucaria forest will be permanent (1988:29). The only optimistic note here is that plantation owners have recognized the need for patches of species other than eucalyptus and loblolly pine in order to reduce the risks of monoculture forests. Some araucaria stands have been planted for this purpose.

A second Brazilian example is the mahogany trade of the Amazon region during the 1960s and 1970s. Once roads were built, a speculative boom on land values drove the logging of territory thought to be useful for settlements or ranching. The roads supported a sawmill industry, which turned out to be more lucrative than cattle. The Amazon's contribution to Brazilian roundwood production rose from 14.3 per cent in 1975 to 43.5 per cent a decade later. The state of Rondônia was a major point of origin from which this raw material was forwarded to lumber and pulp mills in Mato Grosso, Minas Gerais, and Bahia.

John Browder (1989) examined the impact of roundwood and lumber production, especially from natural stands of mahogany, in the mid-1980s. He concluded that the industry provided substantial off-farm employment in "burgeoning boom towns." As forests in the older regions were depleted, the sawmill industry moved further into the interior, where there were still tracts of undisturbed forest land. In the region of Rolim de Moura in Rondônia, Browder found that 40.6 per cent of the urban labour force in 1985 was employed in the industry, with half the output going to the export market. Licensed lumber mills numbered 89 in 1975 and 1,639 in 1984, capital investment likewise rose, from under US$5 million to about US$307 million in the same period. In some regions, lumber production was the only industrial activity, and for four of the Amazon's six states and territories, industrial wood production accounted for over 25 per cent of the total value of industrial output. Export subsidies provided by the government fuelled the expansion between 1981 and 1984. In 1985, federal export policies were revised, and the mahogany export trade collapsed (Browder, 1989).

Browder's study points to several effects of the boom periods in natural forest products such as mahogany. One is heightened employment for woodworkers and also for workers in transportation and related service industries. Paradoxically in this case as often in others, the export trade came at the expense of the workers' living conditions. In Rolim de Moura a shortage of housing because of a lack of locally produced building materials left workers with poor accommodations at high prices. The most obvious negative effect, however, was to the forest. Harvesting operations were extremely wasteful and destructive. According to Buschbacher and colleagues:

The end result is thousands of square kilometers of cut-up forest criss-crossed with soil exposed by machinery, and laden with dead slash. The extensive canopy openings and the dry slash on the forest floor turn a normally fire-resistant ecosystem into a fire-susceptible ecosystem. Fires that are set in nearby pastures to control weeds easily spread into the logged forests. The logging roads provide a route for rapid spread of fire into the forest interior. Few forests in the Paragominas area have escaped fire damage. (Buschbacher et al., 1987)

SUMMARY

All the countries of Southeast and South Asia have experienced massive deforestation from whatever cause. Industrial forestry has become a more destructive agent in tropical forests since about 1960. Decorative timbers, in particular teak, have long had world markets. These and other tropical woods were cut and sold until the mid-1970s or early 1980s in most forested regions, after which time, log export bans were imposed. The bans were partially in response to depletion of reserves, but more often to opportunities for more value-added industry. As bans went into effect, loggers moved on to adjacent countries: Thai loggers to Myanmar, Philippine loggers to Indonesia, Malay loggers to Sabah and Sarawak, in each case increasing the total Southeast Asian deforested area. In the regions where bans have been imposed, plywood and sawnwood industries have been established; thus logging continues, but the logs are not exported. Japan is the chief importing nation for tropical hardwoods. China, unlike other developing Asian nations, appears to have established a substantial pulp industry on the basis on non-wood fibres and to have succeeded in planting or replanting some forest land. It is increasingly engaged in the global market economy, and it has begun to invest in pulpmills elsewhere to feed its own paper industry.

Indigenous people have suffered loss of hunting or agricultural land wherever large-scale logging has taken place. Formerly independent and travelling in small family bands, they have been dispersed or else forced to live in confined areas where they cannot sustain themselves through traditional activities. Forced evictions, physical abuse, confinement, hunger, poverty, and dispersal of indigenous peoples are widespread throughout Southeast Asia and are documented by local groups such as Sahabat Alam Malaysia and international journals such as the *New Yorker*. As long as customers purchase wood from regions where these actions are known to take place, and as long as logging companies enrich their owners by denuding forests, indigenous people are victims of tropical forestry. Referring to the conflict in Sarawak, Raphael Pura of the *Asian Wall Street Times* argues that "it isn't just forest ecology that is at stake ... a more basic set of issues underlies

this feud. Environmental concerns take a back seat locally to a fundamental political tussle over who gains from the exploitation of the
state's forests" (Pura, 1990).

The export trade has had lesser historical impact in Latin America.
However, since the 1940s the araucaria forest of the Brazilian Atlantic
coast and large quantities of mahogany and other decorative woods
have either gone to world markets or been destroyed by careless
logging operations. Brazilian tax incentives sustained the mahogany
trade in the early 1980s; these were later changed and the trade has
diminished.

8 The Tropical Forestry Action Plan and Plantation Forestry

Government planners, international bankers, and aid officials attended an international conference on tropical deforestation in Bellagio, Italy, in 1985. They termed the danger to tropical forests one of the most serious environmental threats of our time, and while they spoke, an estimated 500 hectares per day disappeared. Deforestation is an ecological disaster. It is also a social disaster, and for many indigenous forest dwellers or poor people forced to encroach on forest lands, it is an act of genocide. The proposed solution, according to the international planners, is to invest in improved forestry and plantations.

Buyers, investors, consultants, and research scientists are concerned about the fate of the forests, though for varying reasons. Thus there was no surprise that the opening conference on the Tropical Forestry Action Plan in 1985 was enthusiastically greeted and its resolutions endorsed throughout the commercial world. A serious environmental threat, now acknowledged, would be addressed by a US$4.5 billion package to rehabilitate, manage, replant, and preserve forests around the world. Half of this amount was proposed for Asian countries, the remainder to be divided between Africa and Latin America. Included was a $1 billion package for industrial forestry, mainly plantations, viewed here as a way of reducing the pressure on remaining natural tropical forests while providing developing countries with a source of income. There were actually two plans published in 1985, one created by scientists in the World Resources Institute, together with representatives of the World Bank and the UN Development Program; the other by the FAO Committee on Forest Development in the Tropics.

They were subsequently united in a common action plan, which established priorities for investment: rehabilitation of upland watersheds and semi-arid lowlands; forest management for industrial use; fuelwood, agro-forestry, and energy; conservation of forest ecosystems; and institutions for administering the plan.

INTERNATIONAL CONCERNS AND CRITICAL PERSPECTIVES

Deforestation caused by agents other than forestry was the target. It was with reference to these other agents, which almost incidentally deforest the tropics, that the assistant director-general of the Forestry Department of the FAO opened one of the conferences on the need for a Tropical Forestry Action Plan: "One of the main causes of deforestation in the world, and in the destruction of the broadleafed forests of the tropics in particular, is due to the fact that these forests are not really used. If there is no use for the forests, the forests become what tropical forests are now: areas which just support other economic activities or sectors" (M.A. Flores Rodas in FAO, 1987b).

Flores Rodas has been criticized for failing to recognize the direct uses of forests by indigenous peoples and shifting cultivators. In fact, their use of forests is recognized but lamented in international conventions. The introductory section of the Tropical Forestry Action Plan states: "It is now generally recognized that the main cause of the destruction and degradation of the tropical forests is the poverty of the people who live in and around them and their dependence on the forest lands for their basic needs." The plan addresses ways of alleviating their poverty, including plantations for fuel and commercial forestry purposes.

Advocates of the FAO plan argue that it will redistribute income, provide employment for the poor, enrich soils on marginal and degraded land, eliminate the incentive to log natural forests, and give a source of foreign-exchange earnings to underdeveloped countries. Even apart from the commercial potential, plantations will likely be the essential source of fuelwood, and the World Bank argues that more than 50 million hectares will be required for this purpose by the year 2000 (Panayotou and Ashton, 1992:31).

Critics are not persuaded. They note that the proposals are consistent with the interests of investors in the forest industry and the governments of many southern-hemisphere countries where tropical forests are located, not those of forest dwellers or the poor who use tropical forests for subsistence agriculture. They point out that the same multilateral institutions which now want to save forests were the ones that financed their destruction. The critics argue that plantation forestry deprives peasants of food sources by turning land over to export crops; it evicts the poor, untitled farmers from agricultural land and at least indirectly causes the burgeoning of shanty towns on the outskirts of Third World cities. Plantations, they say, deplete the water-table, endanger watersheds, soak up nutrients, and displace rather than substitute for tropical forests; in short, they are very poor means for economic development.

There is a self-fulfilling aspect to the discovery of tree plantations as an alternative to logging natural forests, and it is this characteristic that makes them suspect. The forest diminishes, settlement and agri-culture move into "empty spaces," or the land dries up and fails to respond to what small silvicultural treatment is offered. Natural regen-eration, if it is possible, takes a long time; developing countries are in as much of a hurry as the developed countries, and the latter urge them on through their international associations, funding practices, and market incentives. The chief forester of Malaysia, quoted in the previous chapter, exhibited a mind-set all too prevalent in the devel-oping world and enthusiastically cheered on by the developed coun-tries: clearcut the natural tangle and get on with planting quick-return plantations.

In the view of Charles Secrett, forest coordinator for Friends of the Earth International, "Plantations in tropical forest areas generally fail. They usually involve clearing existing forests to establish them. The plan does not warn against clearing existing forests or stress that plantations should only be established on already degraded land. And far more effort must be put into defining what sustainable, renewable harvesting of tropical forests requires" (*World Wood*, Oct. 1987:39–40).

Two scientists, M.S. Ross and D.G. Donovan, reviewed the action plans and raised some serious reservations:

On close scrutiny the activities proposed under the "New" Plans do not appear strikingly different from those forestry activities implemented in the past by various aid agencies. Reviewing the experience of the last forty years, the conclusion must be that by and large the old formulas have not been successful. Certainly it is by no means clear that these very same activities pursued with greater vigour or monetary might will reduce to any significant extent the wasteful destruction of the tropical forest occurring at present. (1986:122–3)

Ross and Donovan argue that foreign-aid funding alone cannot address deforestation problems. Governments and peoples of the affected countries control forest uses and abuses; at present they either have positive incentives to allow deforestation or no positive disincentives.

Funds for rehabilitation and plantations will not overcome that basic situation or lead to conservation of the remaining forests. If ultimately poverty, the need for fuelwood or foreign-exchange earnings and employment, or vested interests in logging override concerns about deforestation (as seems most likely), foreign aid will not solve the problem. Since countries differ in history, governments, cohesiveness, population distribution, and land ownership, very specific actions would be needed for each country. Further, as long as developed countries import tropical wood or otherwise encourage logging of tropical forests, the aid packages are wasted. That critique notwithstanding – and in some measure it captures the reflective voice of the less enthusiastic audience to the Tropical Forestry Action Plan – the FAO and other agencies have proceeded to fund rehabilitation projects in those countries they deem most critically in need and to invest in industrial forestry and plantations in many tropical and subtropical regions.

Debates cannot be concluded until there is clear evidence about the impacts of plantation forests. The evidence in the mid-1980s was far from clear. The ecological impacts on soil, water levels, air, and temperature in plantation areas and on gene pools contiguous to them were still under study and as controversial as the social impacts. What works in a subtropical region such as northern New Zealand may be quite inappropriate in the moist tropical rainforests of Asia. Moreover, pine plantations on rotations of fifteen to thirty years may have very different impacts on soil and water levels than the fast-growing eucalypts or gmelina varieties now being planted in the tropics. The unknowns did not prevent the conference from endorsing plans for both more intensive management of existing forests and additional plantations throughout Africa, Asia, and Latin America. These proposals included an increased rate of industrial plantation from 4,000 to 30,000 hectares per year over a five-year period beginning in 1987 in Malaysia, an increase from 78,000 to 100,000 hectares per year in Indonesia, an increase from 120,000 to 240,000 hectares per year in India, a doubling of the rate of plantations from 160,000 to 320,000 hectares per year in Brazil, and similar growth in other countries.

ANTECEDENTS AND EXTENT OF PLANTATIONS IN THE 1990S

The idea of plantations was not new in the 1980s. During colonial periods in both South America and Indonesia, tree crops such as bananas, citrus, cacao, oil palm, and rubber or shrub crops such as coffee and tea or beet crops such as sugar were planted. There was a lengthy history of plantations of decorative woods and wood for railroad ties, mine props, and fuel in both tropical and temperate countries.

The Nordic countries – Norway, Sweden, and Finland – reforested land at least a century ago for industrial purposes and longer ago than that for other uses. The United Kingdom is currently afforesting land at a rate of some 30,000 hectares per year, especially in northern England and Scotland. The Iberian Peninsula's pine and eucalyptus plantations are well established. North America has more recent plantations, most of them in the southern United States. Canada has relatively little investment in plantations, though there are reforestation programs in parts of the country. The hinoki and sugi forests of Japan are plantations. In very recent times, China has reforested and afforested terraces adjoining rivers, where land had been severely eroded as a consequence of earlier deforestation; these terraces are plantations, frequently containing very few species. Plantations intended for industrial forestry date from the 1920s and 1930s in New Zealand and Australia.

The FAO study of 1985 set the rate of plantation establishment in tropical countries at 1.1 million hectares per year, with a total area of about 12 million hectares having been planted by 1980. A subsequent FAO study commissioned by the Swedish Agency for Research Cooperation within Developing Countries, published in 1992, provided an estimate for 1980 of 35 million hectares in plantations in developing countries, 10 million of these in tropical countries (Lanly, 1982, FAO, *Yearbook*, 1988, 1992). About 100 million hectares were in plantations worldwide, and of these, about 40 per cent were for fuelwood production. In the first half of the 1960s, the Asia-Pacific region planted some 373 thousand hectares; Latin America, some 103 thousand hectares. By the last half of the 1970s, Asia was planting 1.5 million, and Latin America, 1.9 million hectares. Africa, although far behind Asia and Latin America, added some 123 thousand hectares by 1980 (Panayotou and Ashton, 1992:172).

SPECIES

Plantation crops include intermediate trees in the tropical forest sequence, such as gmelina, or varieties exotic to the plantation regions, of which various pines and eucalypts are common. Once established, they may include genetically engineered species, usually involving some hybrid combination of intermediate and exotic varieties. Plantations are normally monocultures, since among their anticipated benefits are standard-sized, homogeneous crops easily planted, easily managed, and easily harvested. They introduce new ecological systems in place of whatever existed naturally. The new systems may displace marginal forest lands and may improve soil conditions; or they may displace tropical, subtropical, or temperate forests. Since by their nature they are much less complex biomass regions than tropical forests, they provide a less textured forest with fewer life-forms.

The objective of industrial plantations is to produce short-rotation trees for low-density sawlogs, wood chips, and paper pulp. These trees include *Eucalyptus, Acacia, Paraserianthes* (or *Albizzia) falcataria, Anthocephalus, Araucaria, Gmelina, Pinus,* and *Leucaena* species, the choice to be determined according to soil, climate, temperature, water conditions, and intended function. Most of the tropical plantations in Africa and Latin America are monocultures, with eucalyptus and pines comprising between 50 and 80 per cent of the areas; in Asia a larger number of species, including teak and acacias, are planted (FAO, 1992).

Companies in subtropical and temperate climates of New Zealand, Australia, Chile, and the southern United States began major plantations of radiata pine (known as Monterey pine in the United States) during the 1950s and 1960s. This tree can be grown in relatively short cycles (between eighteen and thirty years) and has high productivity per hectare. During the 1980s, 1 million hectares per year were planted in the southern United States (Sedjo, 1989:2). Caribbean pine is becoming a new fibre source in Central and South America. Elliotii pine is grown in South Africa and Brazil. Several varieties of pine are grown in the Brazilian state of Paraná. Lodgepole or jack pine is indigenous to Canada, and first growths are still harvested; lodgepole has also been planted in Sweden.

The most popular plantation tree is the eucalyptus. The genus *Eucalyptus* contains over five hundred species with widely varying characteristics. These are indigenous to Australia, Papua New Guinea, Indonesia, and the Philippines, where they grow in semi-arid and subtropical environments ranging from coastal to subalpine altitudes. Australian eucalyptus trees have been planted on other continents since the late eighteenth century, usually for decorative purposes. At the end of the last century, plantations were seeded in the Mediterranean basin to produce construction wood. Eucalyptus plantations in Brazil and elsewhere became important sources of fuel in this century.

Though used as a fibre source for pulp since the 1930s, eucalypts were not a preferred source until the late 1970s. By then, research and experimentation had begun to achieve homogeneous monoculture plantations, and appropriate pulping technologies were developed to exploit the characteristics of the species fully. Eucalypts are extremely sensitive to differences in soil conditions, nutrients, and other growing conditions, and they require considerable fertilization to reach high growth rates in short rotations (ideally 100 cubic metres per hectare in five to seven years). They are also sensitive to temperature changes and easily damaged by competing vegetation. None the less, in good growing conditions and with intensive management, they mature in short rotation cycles of between three and twenty years. Most can be harvested in seven-year cycles, and they can be coppiced and regrown through three complete cycles so that three crops from one seedling are now commonplace. In plantations they require less land and thus less physical infrastructure than softwoods. In some soils they produce over twice as much pulp wood as pine, and they are everywhere easier to cut and transport. Although still exotic in most regions, eucalypts are resistant to insects and fungi; that may be a temporary advantage. The benefits for pulp production include high bulk, porosity, strength, good fiber bonding, and good drainage and sheet formation. They produce paper with a smooth finish. In short, they have excellent properties for pulp, paper, and paperboard manufacture and are now the preferred species for most warm-climate plantations.

Eucalyptus globulus is now promoted by industry experts as the best fibre source for computer papers. Developed for commercial purposes, the species produces, in addition to computer papers, a range of tissues and printing-grade papers. It is highly valued because it creates a white pulp, either not requiring bleaches or easily bleached (best if harvested before trees reach their tenth birthday), and its natural variability can be controlled through biotechnology, so that a plantation can produce a uniform wood with a desired density. It shares the general disadvantages of hardwoods in that it lacks what is called "tearing strength" for papers requiring traction in the manufacturing process, specifically newsprint, but this defect can be overcome by mixing some long fibres (such as pine) and clay into the pulp. Eucalypts generally have high opacity and bulk, both advantages for some fine paper grades.

ECOLOGICAL CONCERNS

If natural forests are destroyed and replaced by plantation crops, the gene pool is dramatically reduced. Monoculture crops of eucalyptus or pine cannot regenerate the pools of seeds for mangoes, rambutan, durian, rattan, or resin gum, for example. However, there is insufficient evidence to determine whether monoculture plantations have long-term irreversible negative impacts on the total gene pool in contiguous regions.

The effect on soils is another concern for which insufficient evidence is available. The *Eucalyptus globulus* plantations of Portugal have been in existence for three decades and have been the subject of considerable controversy. So far there is no hard data to support the argument that these plantations are turning the land into a desert, but there are sober observers who worry that such a change is occurring. A Portuguese-Swedish research project conducted in 1981–85 found the plantation trees to have strong root systems capable of controlling soil erosion, though there is surface erosion in the first generation. The long-term effects will not be evident for many more decades (Kardell et al., 1986).

J. Davidson, in an FAO-commissioned report on Bangladesh, argued that the standards against which soil loss should be judged had to be clarified before evaluations made any sense: "If eucalypts were to be used to replace a natural, undisturbed moist deciduous forest, such as those found in the south-east of Bangladesh, water runoff and soil loss probably would be increased. If, however, eucalypts were planted on degraded land with no trees, sparse tree cover or poor degraded forest … then runoff and soil loss can be expected to decrease" (Davidson, 1985:2). Several studies in India and Israel suggested that water yields were higher from catchments under eucalyptus than from other genera. As well, these studies indicated that nitrogen and phosphorus content in the floor litter for some eucalyptus species were either the same as or higher than is normally found in coniferous tree regions (Ghosh et al., 1978). On the basis of these research reports, Davidson concluded that eucalyptus usually does not make the soil more acidic or in other ways harm it and that the tree usually enriches soil on degraded sites (1985:7–8).

There is some evidence that in the humid tropics, rapidly growing eucalyptus plantations consume more water and regulate water flow less well than natural forests, but the evidence is not conclusive. Several writers note that erosion control is not a function well served by eucalypts, because their strong surface roots compete with ground vegetation. However, they appear to control nutrient run-off as well as natural forests. Again, there is conflicting evidence on the overall effect of eucalypts on the water-table. They apparently adapt to different water situations, using groundwater where it is available or relying on surface water where it is not and conserving soil water when the soil is dry (for discussions, see Crane and Raison, 1980; Wise and Pitman, 1981; Raison et al., 1982; Poore and Fries, 1985; Davidson, 1985; Turner and Lambert, 1986).

If large areas once covered by tropical forests are transformed into monoculture plantations, there will be microclimate change in the process. The canopy layering of the tropical forest protects the ground and controls the temperature and humidity of the forest. Plantations by contrast consist of same-age trees, thus there is no layering and reduced ground-level protection. Although observations indicate that temperatures are higher in plantation regions, systematic measurements and comparisons are not yet available.

Gary Hartshorn of the Tropical Science Center, San José, Costa Rica, examined many of the ecological issues in 1979 and concluded that the establishment of plantations might be a means of saving more sensitive and vital tropical forest areas. In his view, the abandoned agricultural lands and unproductive pastures prevalent in the tropics would be biologically suitable for plantation forestry. Indeed, tree plantations would enrich soil in such regions, lessen erosion, change microclimates in beneficial ways, and improve local habitats. Hartshorn warns that "tree plantations are not suitable for every site or deforested area. Soil capability studies are an essential prerequisite for determining general suitability for plantation forestry. Detailed knowledge of site capability is just as important as information on appropriate plantation species" (1983:86). The most sensitive sites are in watersheds. Frequently these regions have been deforested by slash-and-burn agriculture or by populations in desperate need of fuelwood. Hydroelectric and irrigation reservoirs are damaged by higher-level watershed erosion and deforestation. Reforestation of these areas is essential, but plantation forestry is not usually the appropriate solution since harvesting activities generally damage the watershed.

The establishment of plantations on degraded soil, while preferable to plantations in such sensitive sites as watersheds, may also produce problems. Monoculture crops may deplete soils of specific nutrients, and successive plantations may cause imbalance in the small nutrient pool in the soil. Complete utilization of wood or fibre resources may also deprive the area of wood-stored nutrients and abruptly expose the soil to tropical rains, causing further erosion. Logging itself severely damages the soil. Bulldozing removes much of the fertile topsoil (one of the problems with the Jari plantation in Brazil). Heavy machinery compacts the soil and reduces the infiltration capacity, leading to greater surface run-off and erosion. These various concerns suggest that while plantations may succeed and may be beneficial in degraded soils, they need to be developed with much care for the soil, and harvesting should be done by labour, rather than with heavy machinery. Hartshorn emphasizes that the various problems have not received adequate study, and documentation is lacking on significant aspects of plantation forestry.

One of the problems with situating plantations on grasslands and degraded soils is the expense of controlling the growth of imperata (alang-alang) grass until the trees mature sufficiently to provide their own protective cover. Those who want quick returns on their investments are rarely prepared to undertake the costs of rejuvenating the soil, though eventually these areas can be afforested and will produce reasonable wood yields.

Other aspects under investigation include pests and diseases. Among the advantages of pine and eucalyptus plantations where these are not native species is that they are free of the diseases that normally affect them in their native settings. The question is, How long can they sustain resistance to the indigenous pests and diseases of the new location? There is a substantial literature on this issue indicating that pests and diseases not infrequently switch from native hosts to introduced tree species; further, that they are particularly attracted to monoculture stands (Hartshorn, 1983:88–90). One thesis is that pest and disease problems will increase as the plantation area increases, regardless of the crop's geographic origin (Strong, 1974:1064–6). On balance, Hartshorn is of the opinion that "rampant deforestation of the tropics requires major expansion of tree plantations to meet the projected demands for forest products." He does, however, warn that there is still much research to be done and that there are ecological risks.

The published literature on monocultures, soils, pests, and other aspects of plantation forestry is inevitably behind new developments when innovations occur at rapid pace, but intensive research is well under way. Monocultures are recognized as dangerous; a typical strategy now is to intersperse monoculture crops with natural forests (many of them planted anew) or to mix the trees in the monoculture area with vegetable crops. Another approach is to use different hybrids and clones throughout the planted areas. With respect to pests and diseases, companies are developing biological-control methods, often introducing natural predators into natural forests interspersed with the plantation crops, and this strategy is proving to be quite successful. Animals and birds that were nearly extinct in some tropical and subtropical areas are now being reintroduced into regions next to the plantations as a means of controlling pests. Cloning, rather than hybrid development, is another technique for eliminating trees that are susceptible to particular diseases or fungi. Soil enrichment is the subject of extensive research and experimentation, sometimes carried on via chemical fertilizers but now more frequently undertaken by leaving tree cuttings on the forest floor after harvesting, varying crops, planting non-commercial crops in particularly degraded areas, and using similar imitations of nature's means to improve soils. In short, while the objections to plantation forestry have substance, there is also evidence to the contrary. Much more research is in order before more plantations are established, but economic pressures in some regions and world-market and investment conditions will not encourage countries to wait for evidence.

ECONOMIC CONSIDERATIONS

An economically viable plantation capable of exporting substantial quantities of wood (or providing the fibre for domestic processing for lumber, pulp, and paper markets) needs vast acreage. The countries with such "empty" spaces are those with tropical and subtropical forests – that is, provided the forests are clearcut and the forest dwellers dispersed. Such requirements raise economic and social, as well as ecological, issues.

Roger Sedjo (1983) used a simulation model to examine the economic viability of plantations in selected regions, given the price and cost conditions that existed in 1979. The results, though limited to conditions in that year, are worthy of examination. The data of 1979 were important to investors in determining where they would put their capital and corporate energies over the critical decade of the 1980s. Sedjo began with a 1978 FAO study of all plantation forests, totalling 90 million hectares in the mid-1970s (at that time, 2.8 billion hectares of land worldwide were classified as forest). These included the second-growth stands of Europe, North America, and the USSR, as well as tropical and subtropical regions. He modelled thirty-two conditions on sixteen sites for regions where industrial plantations actually existed at the time. The sites included two in the U.S. South, two in the Pacific Northwest, five in South America, two in Oceania (Australia and New Zealand), three in Africa, one in Europe (Nordic), and one in Asia (Borneo). For each site he included plantations that provided only pulpwood operations and others that were integrated with either standard-quality or with lower-quality saw timber (in some cases, both). The objective was to determine which provided the highest net value.

Initial investment costs were separately considered. Processing and harvesting costs were assumed to be the same in all regions, so that the comparisons would reflect only the economics of growing timber and not differential labour or other costs. Another assumption – valid in 1979, but dubious a decade later – was that tropical and temperate southern-hemisphere softwood lumber is of lower quality than its northern-hemisphere counterpart. Comparisons are made for the "base case: present net value" under two assumptions about price structures on world markets. Sedjo's findings, in summary form, are these: (1) all but one of the cases (Nordic) passed the feasibility criteria based on plantation management cost, stumpage price estimates, and biological considerations; (2) slow-growing Nordic forests were the poorest performers (the Pacific Northwest was more profitable, but not outstanding and well below southern plantation results); and (3) the *Eucalyptus, Gmelina,* and *Pinus caribaea* plantations of central and Amazonian Brazil, the *Pinus taeda* plantations of southern Brazil, and the *Pinus radiata* plantations of Chile were outstanding performers. African plantations, especially those in South Africa, generated high returns and slightly exceeded the returns for the U.S. South. New Zealand performed very well and Australia slightly less well, but neither quite as well as the U.S. South and Borneo. Overall, then, the ranking, with slight variations according to the assumptions, put South American plantations at the top, followed by Africa, the U.S. South, Borneo, New Zealand, Australia, the Pacific Northwest, and the Nordic region. Finally, (4) the non-integrated operations that produced only pulpwood were somewhat stronger performers than the integrated operations that produced pulp, paper, and possibly other wood products. However, the variance depended a good deal on slight variations in world-price assumptions. The results were also affected by the assumptions about lower sawlog prices in southern plantations.

In later work on the same subject, Sedjo (1989:5) and Sedjo and Lyon (1989) examined timber sources and projections for the next half-century, and concluded that the U.S. South and the non- traditional "emerging" plantation regions (New Zealand, Chile, Spain, Portugal, South Africa, Brazil, and others) would become responsible for most of the increases. Lanly and Clement (1979) estimated plantation growth to the year 2000 would include enormous increases in all of the tropical and subtropical regions; as we approach that date, these predictions are proving to be correct.

There is some dispute about the long-term economic future for eucalyptus pulp. Bruce Zobel (1988) argues that the enormous expansion of pulp eucalyptus plantations, especially in Brazil but also throughout Asia, Africa, India, Indonesia, and Europe, will overtake demand by the turn of the century. Further, he contends that other hardwoods, in particular gmelina, will become competitive with eucalyptus and that these other hardwoods can be grown on less-favoured lands where wet, heavy clay soils with high pH do not support eucalyptus plantations. Temperate-zone hardwoods such as aspen may also become competitive within a decade, as various costs even out and experimentation leads to more homogeneous plantation trees.

There is also a possibility of further technological, and especially biotechnological, change that could erode the southern-climate advantage. Eucalyptus trees die in sub-freezing temperatures, but current experiments in France are reported to be resulting in sturdier varieties that can withstand the cold (Williams, 1988:24). Experiments along these lines are also being undertaken in the south, especially at the Klabin nurseries in mountainous Monte Alegre in Paraná, Brazil. Genetic selection of species capable of withstanding mild and short frosts has produced a sturdy tree suitable for higher elevations and more rigorous climates. This development has potential applications in the northern hemisphere and could halt the advances of the lower-altitude growth regions in southern countries.

It is possible that Zobel is right, but at this stage all prognoses are speculative and depend on a variety of dubious assumptions. We do not know, for example, whether electronic media will supplant newspapers and other printed material or whether the general growth in literacy throughout the developing world, and especially in Asia, will offset such technological changes. We also do not know whether current research on eucalyptus will result, finally, in the development of genetically engineered trees quite different in characteristics from the present varieties. We cannot predict how quickly the northern countries will respond to the changing market conditions or how rapidly global investors will switch their capital locations.

My own interpretation is that there will indeed be overcapacity for some time, but that its effects will be far more devastating to northern producers than to those in the south simply because of the differential in costs. The eucalyptus plantations are far enough along the development path and the companies are so far advanced now in experiments and research that we can be sure we are not witnessing a short-term wonder. The plantation economies are in for a good long run, and they have the competitive edge to stay in the business.

SOCIAL IMPACTS

It is one thing to say that plantations are at least ecologically preferable to deforestation or that they are economically viable, but neither statement leads to the conclusion that they are therefore socially beneficial. The problems are numerous. Though southern plantations may be cheaper for corporations than reforestation of northern mountain regions, they require an upfront investment that poor farmers cannot mount. And though they need seven years to reach maturity, rather than fifty, even seven years is too long for the poor to wait. People have to eat, and eucalyptus plantations do not provide food.

Moreover, plantations are economically viable only if they are large, and they are most profitable when they are attached to a modern pulpmill. Therefore those who can afford the investment – merchants, bankers, corporations, élites of various kinds – are strongly tempted to claim ownership to large areas that might otherwise provide subsistence crops. Plantations, then, evoke battles over land rights, and in these battles it is usually the poor who lose. As well, since degraded lands tend to occur in dispersed patches, there is less incentive for large companies to invest in clearing these than in claiming territory endowed with good forests or with forests that could be otherwise regenerated. The following chapters describe numerous examples of that preference.

Markets for export crops are notoriously volatile, and forestry markets have always been boom-and-bust affairs. In boom periods, the market seems to be insatiable, and investors build more mills to catch up with their optimistic forecasts; the downturn typically coincides with overcapacity. International corporations can withstand these fluctuations. They may reduce employment or close down operations in one place, stay in the market with produce from another, and put expansion plans on hold. Smaller operations simply go under, because all their funds are tied up in unprofitable mills or plantations. The exceptions may be those mills whose owners are manufacturers (papermakers, for example) and whose product (pulp and/or chips) is purchased by the owner on a long-term inflexible contract. For most producers, the market is less firm, and with new technologies and new fibres constantly appearing, there are many risks. Countries dependent on such structural conditions are in jeopardy.

Where the new forestry has been most successful, in Brazil and in Chile, old towns have been rejuvenated by the employment opportunities in the mills. The most modern mills provide not only wages, but hot lunches, medical benefits, clothing, family housing, schooling, and playgrounds. As well, they have furnished more than mere jobs to numerous recent university graduates who form their research teams. These employees, production workers and researchers alike, not to mention a cadre of office, technical, and administrative workers who are sometimes engaged through contract arrangements, are beneficiaries of the new forest industry. The mill workers are not, however, the whole of the population affected by the new plantations, and the burgeoning army of landless peasants and indigenous peoples is increased whenever land is taken over for plantations.

Multiple or mixed-crop plantations have been explored in several regions. There is an ecological rationale for planting mixed crops over large areas simply to avoid all the potential defects of monoculture crops (FAO, 1992; Panayotou and Ashton, 1992:171–89); some studies indicate that crop productivity is enhanced in certain cases, so there may also be economic advantages. There is another, social advantage in regions where poor peasants compete for the land. Non-timber products such as sugar cane, fruit, spices, and fuelwood, grown together with timber, would enable local people to obtain earnings and food while growing a cash crop.

The plantation form known as "social forestry" or "community forestry" may include allowance for plantation workers to plant subsistence crops underneath or interspersed with tree crops, advancement of funds for seeds and subsistence to farmers who grow tree crops for nearby pulpmills; or aid in establishing plantations for fuel. Some of these projects are funded by FAO and the World Bank as precursors to or extensions of the Tropical Forestry Action Plan. They have not generally received a good press. To give one example, twenty pre-1985 projects, all financed by international aid agencies in India, were surveyed by Bharat Dogra (1985). In his judgment, every one destroyed tropical forests, and their major social impact was to deprive villagers of fuelwood. The chief beneficiaries were pulp companies, which obtained low-cost wood supplies. Similar accounts are given by reporters elsewhere. Three foresters who examined a range of examples concluded: "Social forestry, as it is being practised, is an external intervention in villages and also in nations ... It is just an arcadian version of industrial forestry's trickle down theory: it has not gained credence by having its scale diminished. It fails, as industrial forestry failed, to see poverty as an issue daily produced and reproduced by the structures of society" (Dargavel, Hobley, and Kengen, 1985:22).

This critical perspective notwithstanding, many of the advocates of social forestry, as evidenced in the FAO plan, contend that small-scale agro-forestry plantations empower local villagers and provide them with both a cash income and fuelwood. Critical to success is a shared understanding between villagers and either donors or technical experts of what a project is for and how to judge it. By way of examining this issue, Jeff Romm (1986) creates a hypothetical plantation project where local foresters attempt to provide a fuelwood or cash-crop source to villagers both because it will help them and because it will enable them to avoid using wood from a larger pulp plantation nearby. The officials earnestly provide the expertise; the villagers provide the labour. Is the project a success? Much depends on several conditions not taken into account by the technicians, including the importance of cattle, alternative grazing grounds, and opportunity costs for labour invested in plantations.

A critical issue is land ownership and control. If the state, manifested in the form of forestry officers, assumes ownership of land that villagers still regard as their common property, social forestry projects devised by officials are unlikely to succeed once the villagers recognize the conflict. Where governments are concerned with owning all land, policing it against encroachment overrides other social objectives. Romm suggests an alternative where the government does not claim ownership, but rather allows for de facto control of land by villagers engaged in social forestry projects. He points out that results under this model may include more productive use of forests and better conservation practices; obviously, the cost of policing is reduced as well.

If forestry is intended to provide profits to a privileged few, and if external aid agencies, buyers, suppliers, and investors accept the legitimacy of that objective, then it follows that reform of destructive forest policies will not come easily. The alternative is long-term planning of resource development for the purpose of enhancing the lives of inhabitants and sustaining the forest. Such planning would necessarily begin with the needs of the poor since, until those needs are met, as even the FAO action plan observes, all else fails. It is difficult, however, to find an example of any long-term planning other than the Japanese mountain villages over the 1950–80 period. Social forestry projects are not examples of planned development: they are scraps thrown out to the poor in the course of a wild drive to exploit forests right now through plantations and industrial forestry.

There is, finally, the one issue that refuses to die, despite the claim by most governments that plantations are established on otherwise useless land. I know of no objective study that quantifies the areas where that claim applies, as compared with the plantations situated on land made available by logging productive forests. Since companies invariably assert that the forest was no longer intact (usually blaming shifting cultivators for this shameful situation), they prevent any objective examination of how much land is being deforested in order to plant commercial crops. Hartshorn's arguments and those of the FAO and many other well-intentioned helpers are all based on the notion that the marginal land can and will be regenerated and the forested land left to nature, if only international funding is made available.

PLANTATIONS IN NEW ZEALAND AND AUSTRALIA

New plantations have been starting up in tropical regions in the 1980s and 1990s (examined in the next several chapters), but New Zealand was establishing industrial pine plantations in subtropical conditions by the 1950s. A brief history of New Zealand's globe-trotting company, Fletcher Challenge, reveals the nature of the expanding global forestry economy as the company moved to North and South America before being recycled under pressure of changing markets and financial conditions. Australia, with a mild temperate climate in the south and a subtropical to tropical climate in the north, has more recently planted pine. While eucalypts are indigenous to Australia, they have not been primary sources of wood fibre, but plantations are now being established. Both New Zealand and Australia are aiming for a greater share of the Japanese chip and pulp markets. Profiles of these two developed regions – both in the southern hemisphere, yet by virtue of settlement patterns not among the developing countries – provide some insight on well-advanced plantation systems.

New Zealand

The plantation economy of New Zealand began and continues in controversy not unlike that now occurring elsewhere. Birch (*Nothofagus*) forests were felled and burned to make way for plantations. Earlier, 2,000-to-5,000-year-old North Island kauri, members of the family Araucariaceae, had fallen to the axes of settlers, boat builders, and whaling stations, a decimation finally stopped in 1974. The kauri forest usually included tall softwood trees of the Podocarpaceae family, which became important to the New Zealand timber industry. Further south was a mixed forest of podocarp and hardwoods of the *Metrosideris* species; and still further south, mixed podocarp and beech. After many of these trees were felled, governments began to consider the possibility of clearcutting what remained and planting commercial pine. Radiata pine was planted in the 1950s on state and private land on the North Island, and plantations were expanded steadily through the 1970s and 1980s. The planters argue that the natural cover was already destroyed; their critics, that they deliberately destroyed, rather than renewed, what was left (Searle, 1975; Bertram and O'Brien, 1979; McClintock and Taylor, 1983; Roche, 1991; see also numerous publications of the New Zealand Forest Service, other New Zealand ministries and public agencies (various dates), and the New Zealand Forestry Council, 1980, 1981, 1982).

The export of logs to Japan began in 1958 and of sawn thick squares in 1967. By 1972, sales represented a quarter of total roundwood removals. However, they dropped thereafter and were down to 10 per cent of total cut by 1982, as Douglas fir and hemlock from the United States and old-growth softwoods from the USSR vigorously competed for the market (Fenton, 1984). The alternative to log exports and sawn timber was pulp. Much of the forest policy through the late 1970s and early 1980s was dominated by the interests of an emerging pulp-mill industry. Exotic pine plantations, totalling 336,000 hectares in 1950, covered 939,000 hectares by 1982. The state pioneered many of the new developments and worked closely with – some would argue was captured by – the pulp industry. As with all such large-scale developments, the growth transformed small rural communities and pitted conservationists against developers.

Linkages to Japan and Korea are of some duration in several pulp and paper mills and panel board plants. Carter Holt Harvey (CHH) is linked to Oji Kukusaka Pan Pacific Limited in a joint venture at Whirinaki in Hawkes Bay established in 1972. Winstone Pulp Indus-tries has a joint-venture pulpmill with Korean companies Samsung Industries and Chonju Paper Manufactures. C. Itoh of Japan, whom we meet again in Latin America, is the *sogo shosha* linked to several wood-chip exporting businesses in New Zealand. Sumitomo Forestry Company of Japan has an interest in a fibreboard mill. Even with substantial levels of foreign investment, however, the New Zealand state maintained a strong presence in the forestry sector until the late 1980s (Roche, 1991:2–6).

In the midst of battles over forests in 1988, the government under-took a major restructuring of its forestry management arrangements, joining in rhetoric, as well as in action, with the global trend toward privatization. It announced that it would sell state forests amounting to 550,300 hectares. Widespread and intense public opposition to the sales did not alter the plan, and eventually forty-three of the ninety available forests were sold. Successful bidders included two of New Zealand's largest companies, Tasman Forestry and Carter Holt Harvey, and a number of joint ventures and consortia with Japanese, Hong Kong–Singaporean, and Chinese–Hong Kong investors (Roche, 1991:8–14). Foreign investment increased, as Southeast Asian invest-ment companies gradually joined a group that included Shell (in Baigent Forests Ltd), Resources Elders (soon to be renamed New Zealand Forest Products), and C. Itoh. The largest of the new entrants was London Pacific, owned by a Singaporean-Malaysian group of inves-tors, which ended up in receivership by 1990 (Roche, 1991:3–6). Michael Roche assessed the privatization program, first in terms of the stated reasons for it given by the government; on these grounds, it failed because it did not generate the anticipated funds. In fact, half the land remained under government ownership because bids were too low. In his opinion, "The forest sales more fully incorporated New Zealand's plantation forests into the Pacific rim timber economy (and internationalized economy generally). From a theoretical perspective the sale is interesting because it represents an instance where a nation state's government has actively encouraged the internationalization process by seeking to withdraw from forest processing and ownership" (1991:17).

By 1990 there were five pulp and paper companies in New Zealand exporting to Australia and Japan. In order of size they are Carter Holt Harvey and Winstone, both producing thermomechanical pulp, and Fletcher Challenge (Tasman), Caxton, and New Zealand Forest Products (previously Elders Resources, bought out by CHH in 1990), each manufacturing kraft pulp and a small range of tissues and paperboard. The two groundwood pulp companies rely on state forests. Tasman also uses state-forest supplies but has, in addition, its own plantations. Caxton buys kraft pulp from others and has established plantations. New Zealand Forest Products began with 80 per cent of its land in a private forest estate, but this proportion has fallen in recent years (Fletcher, 1988:61).

The president of New Zealand's largest pulp producer, Hugh Fletcher, says that it would take the *Los Angeles Times* only a month to consume as much paper as New Zealand uses in a year. Even so, Fletcher Challenge (FC) travelled far on funds originally generated in New Zealand, and while it was cut short by financial shortfalls in the 1990s, its voyage revealed a good deal about the globalization of this industry.

According to Bruce Jesson's 1980 history of the company, the original Fletcher family firm became public and diversified within New Zealand in the immediate postwar period, holding to its founder's policy of owning its own fibre sources. In the 1940s the second generation owned a substantial and diversified empire of over one hundred companies, including interests in Australian Newsprint Mills, which itself held a 28 per cent interest in Tasman Pulp and Paper. Tasman won the bid to build a new papermill when the government of New Zealand called for tenders to purchase 23 million cubic feet annually from the maturing Kaingaroa state pine forest on the North Island. While Fletcher (called Fletcher's at that stage) built the mill and town of Kawerau, the New Zealand government provided total backing and was the main shareholder. When another paper company, New Zealand Paper Mills, suffered spasms following an unfortunate merger, Fletcher began buying its shares and took over the company for the first time in 1969. Over the next decade Fletcher expanded into the Australian market. Even though in the late 1970s, it was experiencing a debt crisis, it managed to strike an agreement with the New Zealand government whereby the government underwrote the

Fletcher takeover of Tasman. Shortly afterward, Reed International sold its shares in Tasman to Fletcher. Bowater had already sold its shares to the government and Fletcher, and it ended its marketing contract with Fletcher in 1978. Two years later the New Zealand government sold all its shares to Fletcher for $37 million, and Fletcher in turn sold a third of its shares to Challenge. Thus was born the modern company, Fletcher Challenge, so incorporated in 1981. Thus began, as well, the Fletcher Challenge odyssey abroad with substantial liquid funds.

In Hugh Fletcher's own account of the company's history (1988), its diverse operations in non-forestry sectors – everything from home appliance outlets to sheep farming – allowed it to finance the acquisition of Crown Forest Industries (Crown Zellerbach) in British Columbia at a time when Canadian and other forest companies were reeling from a major recession. Other take-overs were similarly possible because of the company's ability to shift funds between sectors as their different cycles affected finances at different points. Fletcher observed that paper products have steady growth prospects, while market pulp prospects are always unstable. In his view, new plantations in Latin America and the Iberian Peninsula would reduce the advantages of traditional producers and lead to potential oversupplies of market pulp. Thus he concluded that new greenfield capacity should be avoided, and it was on this rationale that the company moved to obtain existing plants in British Columbia, New Zealand, and Minnesota, rather than construct new mills.

Having heard and read claims that Fletcher Challenge was able to expand by virtue of not paying taxes in New Zealand, I asked, during an interview with vice-president Geoff Witcher in September 1984 in Wellington, whether this claim was valid. He agreed that the company had not paid corporate taxes for the previous two years. The reason was that "we live in a democratic society. The government sets the overall game plan." In his view, FC, by producing a quality product for export, was offsetting the balance-of-payments problems which had caused an economic downturn in New Zealand in the early 1980s. In addition, the company received tax credits for employing Maori workers in replanting.

While situating itself with a softwood supply in British Columbia, Fletcher Challenge purchased (through debt swap financing) radiata pine plantations and half of the Papeles Bio-Bio newsprint mill in Chile in 1986 (later increasing its holdings to 100 per cent). The Chilean plantations were scheduled to reach maturity at the same time as pine plantations in New Zealand, and the purchase forestalled competition on Asian markets. FC then moved into Brazil in 1988, obtaining half the shares of the Papel de Imprensa SA (Pisa) mill through debt swap financing (*PPI*, Aug. 1988:7; *Appita*, Sept. 1988:349–50). The intention was to build a 240,000-ton-per-year newsprint mill, drawing on 50,000 hectares of pine plantations in Paraná state, together with locally purchased wood. FC also purchased a sawmill at Jaguariaiva. In the same year the company became the majority shareholder, with 50 per cent of the stock, in Australian Newsprint Mills.

The downturn of economic fortunes in the early 1990s finally affected FC, but its restructuring also reflects changing management views on the need to own plantations. Intensive management, genetic improvement of stock, fertilizing, thinning, and pruning forests are all necessary for plantation owners; the question for such companies as Fletcher Challenge is whether they would have better returns on their investments in manufacturing operations if they relied on the purchase of market pulp, as many of the Japanese companies do. Both influences may be involved in FC's decision to sell off a number of sawmill operations in British Columbia and restructure operations in New Zealand, entering into joint ownership with American institutional investors for 60,600 hectares of plantations. The plantations, though now 39 per cent owned by the new partners, remain in FC's national inventory. In 1993 the company sold 49 per cent of its Chilean newsprint and forestry operations to a group of U.S. institutional investors, including the radiata pine plantations and the Papeles Bio-Bio mill at Concepción. The overall intention of the various sales was to reduce debt, but it also signals a reversal of the founder's guiding principle of owning the company's supply of wood.

Australia

Australian native forests and woodlands cover some 1,205 million hectares (13.7 per cent of the total land area), but only 41.3 million hectares are classified as forest. In addition, pine plantations cover 802,000 hectares, and eucalyptus and other hardwoods, about 42,000 hectares. Increasingly, these forests provide raw material for the woodchip export trade, while Australia imports higher-value lumber and paper products (Penna, 1987:11, 32–3). The felling of the native forests followed much the same pattern as in New Zealand, and the supply of timber was short by the 1940s (Watson, 1990:5–9). Radiata pine plantations began to take their place in the 1950s, and small sawmillers were gradually ousted by larger companies aided by government afforestation policies and financial support. By the 1990s, the industry as a whole was highly concentrated, particularly in pulp and paper production. Australian Newsprint Mills, 50 per cent owned by Fletcher Challenge, dominates the newsprint market through two mills. In other papers, the AMCOR group has large shares of several of the other sector companies, including Bowater-Scott and Kimberly-Clark in tissues and Australian Paper Manufacturers in packaging and industrial papers. Carter Holt Harvey of New Zealand exports softwood chips from Australia, as does Tasmanian Pulp and Forest Holdings, owned by Associated Pulp and Paper Mills (APPM), which in turn is majority owned by Fletcher Challenge; FC dominates Tasmania's forest economy (Dargavel, 1982; Penna, 1987:36).

John Dargavel (1983) argued that although the biological and technical opportunities for pine plantation forestry in Tasmania were excellent, their economic returns were poor. State forests were extensive, but private investors were few and far between. APPM took 58 per cent of the pulpwood and 30 per cent of the sawlogs, and had harvesting rights over 54 per cent of the state-owned commercial forests. Estimates of the returns to growers were in the neighbourhood of 2.3 per cent, where 6 per cent would be a marginally profitable undertaking. Penna noted that Tasmania's public native forests, with limited exceptions, were covered by five pulpwood supply agreements to four companies, and much of the product consisted of wood chips exported to Japan (1987).

New proposals, based on both eucalyptus and pine plantations, were mounted for expansion of the forestry sector in the late 1980s. The impetus lay in anticipation of ever-growing markets in the Pacific-basin countries. North Broken Hill–Peko (Australia), together with Noranda of Canada, proposed a eucalyptus plantation and pulpmill operation for Wesley Vale in northern Tasmania. After heated debate, the government endorsed the project subject to effluent requirements, which the joint partners refused. The president of Noranda is reported to have noted that he could get a better deal in Alberta. Reasons beyond environmental concerns might have prompted the reaction, however: six Japanese companies that were originally scheduled to participate in the consortium withdrew. Dargavel argues that the anticipated sale of pulp to Japan was unrealistic and that the decision of the two partners to go as far as they did was based largely on state offers of subsidies (1989b).

Australians are engaged in tough debates about the future of forestry. The Australian Conservation Foundation has its view (Cameron and Penna, 1988); industry and governments, theirs. As usual, the trade-offs and compromises are difficult to reach where industry is needed for jobs, but internationalization introduces complex demands that change the country. Ecological issues continue to be unresolvable because interpretation and speculation are more prevalent than evidence. Finally, the economic calculations are problematic in a global market, where the major buyers have a multiple-sourcing strategy and competition between sellers is intense.

SUMMARY

Plantations are more efficient producers of industrial roundwood than tropical forests. There are ecological issues attached to plantations of pine and eucalytus, including long-term impacts on water-tables, atmosphere, temperatures, and soils and the problems associated with monocultures. There are also economic and social issues. The social issues are similar to those in the exploitation of tropical forests, specifically marginalization of landless peoples. Fuel and food needs of the poor are not met by plantations that use land areas earlier capable of sustaining them. Investment in plantations nevertheless is growing, with support from the UN, the FAO, the World Bank, the IMF, and numerous private or national aid programs and banks. In favour of plantations are the arguments that they reduce the likelihood of deforesting the tropics and provide a cash crop for developing countries. It is not self-evident that these defences are valid. Many forests are cut in order to plant commercial crops, and the cash of such crops is not necessarily infused into the social fabric of developing countries. Subtexts of the Tropical Forestry Action Plan are widely criticized. Cynics read the text as: "plantations could be very profitable, and the poor have to be dissuaded from camping on land claimed by wealthier citizens." In considering the pros and cons of this contentious issue, one asks whether the subtext so discredits planning that *laissez-faire* is preferable. The persistent underbelly of poverty is the measure of the status quo. Planning may be suspect, but at least it brings the issue into the public realm.

New Zealand and Australia established plantations several decades before they became prevalent in other parts of the southern hemisphere. Both countries aim for the Japanese market, competing with producers of eucalyptus and pine wood and pulp. The odyssey of the New Zealand firm Fletcher Challenge provides a glimpse into the globalization process: in this case, a boom-and-bust adventure.

9 Thailand, the Land
No One Should Use:
A Case Study

The commercial world at its doorstep, Thailand appears to be a rapidly developing country, one of Asia's tigers, with a growth rate in double digits. Its capital city, Bangkok, is literally sinking under the weight of this development, and moving along its congested streets is an exercise in futility. Thailand's population has ballooned since the beginning of this phase of rapid development, and the country can no longer support its people. Prostitution and venereal diseases are widespread. The water is severely polluted. Many of the villages have suffered from floods because of denuded hillsides and silted rivers. Natural forests, which were once lush and covered the land, have almost disappeared. In their place are dried patches with alang-alang grass cover. Hill tribes complicate the country's politics and are blamed for deforestation, but it is not they who have brought about the tragedy that is modern Thailand.

Thailand is not now a major forest economy, though in the early days of the teak trade it was prominent. It is a targeted country, however, with a number of investors, both Thai and foreign, seeking opportunities to establish pulp plantations and mills. It hesitates at the starting gate. Government and companies are eager for investment, but the people, hostile to further deforestation and alienated from their traditional land, oppose it.

NATURAL FORESTS

The Thailand Forest Act of 1941 defines "forest" as the land area which no one has authority to occupy or use (Arbhabhirama et al., 1988:142). The definition notwithstanding, less than a third of the

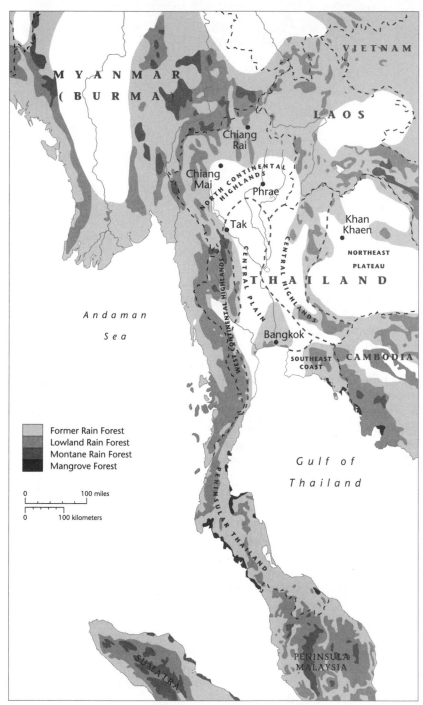

Legend:
Former Rain Forest
Lowland Rain Forest
Montane Rain Forest
Mangrove Forest

0   100 miles
0   100 kilometers

VIETNAM

MYANMAR
(BURMA)

LAOS

Chiang
Rai

Chiang
Mai

NORTH CONTINENTAL
HIGHLANDS

Phrae

Khan
Khaen

Tak

NORTHEAST
PLATEAU

CENTRAL
HIGHLANDS

THAILAND

Andaman

Sea

WEST CONTINENTAL HIGHLANDS

CENTRAL
PLAIN

Bangkok

SOUTHEAST
COAST

CAMBODIA

Gulf of

Thailand

PENINSULAR THAILAND

SUMATRA

PENINSULA
MALAYSIA

Map 8   Thailand, Laos, Vietnam, Myanmar, and Cambodia: forest types

forest that covered Thailand in 1913 remained in the late 1980s (Feeny, 1988:120). Norman Myers (1980) estimated a loss between 1961 and 1978 of over 50 per cent of these forests. Official data record a reduction from 20.92 million to 16.09 million hectares – that is, nearly 5 million hectares – in the five-year interval between 1975 and 1981 (Sahunalou and Hoamuangkaew, 1986a: table 1). Arbhabhirama et al. (1988) estimated that between 1961 to 1985, forest cover declined from 53 to 29 per cent of total land area, excluding para rubber plantations (see also Gro-Thong, 1988:225).*

Of the remaining forest cover, about 60 per cent is in wet-belt evergreens; the remainder is deciduous. Evergreen forests include tropical rainforests in southeastern and peninsular regions, where contact with the monsoons is direct and precipitation is high. Dry or semi-evergreen forests are scattered along valleys and low hill ranges; the principal trees are *Anisoptera*, *Dipterocarpus*, *Hopea*, *Tetrameles*, and *Lagerstroemia* species. On the Korat Plateau, where poor acid soils predominate, a coniferous forest includes primarily *Pinus merkusii* and *P. kesiya*. In addition, evergreen forests include both fresh-water and mangrove swamplands, and beach forests near sea coasts. Deciduous forests occupy the dry belts (Wacharakitti, 1988:123).

In the northwestern highlands, a dense evergreen forest once provided the natural cover for watershed catchment areas that conserve and regulate water flows to the lowlands. But cultivation of crops or logging on steep slopes has caused soil erosion and sedimentation of downstream rivers. Watershed areas have lost their capacity to absorb rainfall and slowly release it; instead, downstream areas are frequently flooded during the rainy season.

Logging or farming in the lowlands, or the undulating ground between plains and foothills, has displaced much of the dry deciduous dipterocarp forests. Most of these areas are now being farmed in the northeastern regions, but agriculture is marginally productive and is not an economical use of the land. In the foothills themselves, commercial forestry might be the most economically rewarding land use, but these areas are already largely deforested.

HISTORICAL CONTEXT

The Teak Trade

Exports of teak, sappanwood (which no longer exists), and other hardwoods to Hong Kong, Singapore, China, Java, Japan, Europe, and

---

* The Thai measure for area is the rai, which equals 1,600 square metres; 6.25 rai equal a hectare (or 10,000 square metres).

America were common in the 1870s. Teak forests were under the control of Lao chiefs until 1899. The chiefs issued leases to Burmese, Shan, and Chinese traders; these, in turn, employed hill tribes people and Burmese workers to extract teak, using elephants and river rafts in the process. The trade expanded after the 1870s, as British interests gained control over the resource. British foresters shaped and staffed the Royal Forestry Department, created in 1896, and authored a Forest Protection Act in 1897. British-built railways facilitated the flow of teak from the north to Bangkok, and steamers transported it to British colonies, Britain, and Africa. Gradually the British East Asia Company and British civil servants succeeded in transferring owner-ship and control of the forests from the feudal chiefs to the royal government in Bangkok, ostensibly to control the concessionaires, encourage replanting, and reduce excessive logging. Teak plantations were created as early as 1906, and an autonomous government cor-poration, the Forest Industry Organization (FIO), was established to undertake reforestation at the village level. The decline in teak resources continued, however (De'Ath, 1992). The British remained in control until the onset of the Second World War.

One of the provinces most devastated by this activity, Phrae, lost much of its natural forest to the teak trade. Local villagers and hill tribes became enmeshed in the hunt for big trees; elephants were the source of hauling power then as now. When the East Asia Company withdrew, other, mainly Thai companies moved in. Foreign businesses were issued no new timber leases after 1952, and by the 1960s the continued depletion of teak stocks was due entirely to indigenous firms. The government imposed export duties and finally, in 1977, a ban on exports of teak. This did not stop the trade. The Royal Forestry Department became more active in imposing restrictions, but the villagers became yet more adept at poaching. According to a local police chief in 1990, "ninety per cent of the people we arrest are poor villagers," their alternative sources of subsistence having already been destroyed by the log trade. The financiers of the activity cannot be identified. Both police and forestry officers are accused, often by each other, of accepting bribes and being active participants in illegal logging. The action is not infrequently violent as well as illegal, and death threats, gun battles, and other unsavoury behaviour are inter-spersed with elephants, villagers, and truck loads of finely crafted teak furniture. David Feeny estimates a ratio of three to one for illegal to legal logging (1988:120). Thai truck-loggers, having virtually deci-mated the country's remaining trees, have now moved to Myanmar (Burma) to destroy the last remaining intact teak forest of Asia (as has been described in chapter 7).

Other Timber Trades

Timbers other than teak were also exploited for the export trade and later for the manufacture of plywood in the state-owned Thai Plywood Company, established in 1951. Small-scale plantations of teak, eucalyptus, leucaena, and acacia were established by the government. Thailand was self-sufficient in timber before the mid-1960s. As logging continued and consumption of wood products increased with growth in population, Thailand began importing timber in 1967.

New legislation in 1983 extended logging rights in privately owned and national forest reserves. It resulted in a 61 per cent increase in the total value of forest product exports in 1985. Logging concessions were governed by legislation that created incentives for reforestation by private companies. The argument was that if these companies were to have a long-term stake in forests, they would replant. This argument, common in every forested country, was not supported by any evidence; natural forests were cut, not harvested, and the companies moved on to other forests. Land-use policies, moreover, were inconsistent, and the Royal Forestry Department never had adequate funding or staff. Government agencies defended their territories against one another, and public and private companies competed; the laws and regulations, even where appropriate, were unenforceable. Reforestation lagged far behind logging, and logging of watersheds was piling up sedimentation in rivers and causing floods.

Rubber Plantations

In the Surat Thani and Nakhon Si Thammarat provinces of Thailand, "rice was cultivated in the lowlands, fruit trees in the foothills and coconut trees near the coast. Plantation crops such as rubber were cultivated in the better drained lowland and gentle slopes in the foothills" (Panoyotou, 1989:8). Higher slopes were forested, and mangrove forests at the coast protected fisheries. These traditional land-use patterns that had sustained the regions for many centuries were fundamentally altered, beginning about the early 1960s, for the expansion of rubber plantations on the highlands and shrimp ponds at the coast. Most of the natural forest had been cleared for the plantations before the end of the 1980s. Soil erosion and persistent flooding failed to stop the growth of the plantations.

To deal with these problems, aerial seeding of fast-growing trees was undertaken on mountain slopes wherever some topsoil remained. However, rubber trees are not deep rooted and are not able to provide the cohesive agent for eroded soils. After the massive flooding of southern provinces in 1988, the role of rubber plantations finally came under renewed scrutiny. Many of the plantations are in areas classified as protected forests or are in watersheds; they are located on steep slopes near headwaters. An international team of experts argued that the protective cover which had always provided the necessary brake on soil erosion and silting had been removed, primarily for rubber plantations. They also noted that a 1969 survey had indicated 45 per cent of the area in the forest reserves was considered suitable for rubber, a proportion their own study found to be excessive (Zinke, 1989).

Logging Bans in 1985–89

Thailand's deforestation was rapid and devastating; villagers paid the price in the ravages of extensive flooding. Support for logging bans was widespread, though merchant companies and some government officials objected. Limited bans were imposed in 1985, and total bans in 1989. Illegal logging continued, some of it carried out by the loggers, some by the poor in search of fuelwood, and some by villagers tempted by cash from outsiders; but the legal, massive logging of natural forests came to an end. A victory for the anti-logging forces was achieved; it was short-lived, however. The cut-over lands were declared suitable for plantation forestry. Subsequent legislation provided numerous tax concessions and property rights to private individuals and firms for the planting of eucalyptus on public lands in the northeastern provinces. No provisions were made for the livelihoods of up to 10 million people who lived on the land. Labelled encroachers because they had no land title, they came into persistent conflict with private and state companies.

INDIGENOUS PEOPLES AND ENCROACHERS

The population of Thailand increased from 18 million in 1953 to 52 million people in 1986. Deforestation is often blamed on overpopulation, with special vitriol reserved for the poor and the remnants of tribal groups. Obviously, this small country cannot sustain over 50 million people, so there is a basis for the more general argument. However, missing from it is the context for the growth in population. This includes the history of logging concessions, wars and revolutions in neighbouring countries, and the impacts of commercial and industrial forestry. When we examine this context, it is clear that overpopulation or hill-tribe incursions are not sufficient explanations for the devastation or for the millions of people who have no means of subsistence.

The Hill Tribes of the Northeast

Nomadic hunter-gatherers still exist in the northeastern hills of Thailand and the adjoining regions of Laos and Myanmar. The Thai government has recognized nine groups of hill dwellers as "minority groups" for whom public welfare may be provided. These are the Karen, H'mong (Meo), Lahu, Yao, Akha, Lisu, Lua' (Lawa), Khamu, and H'Tin, most of whom now engage in some form of agriculture (Thaitawat, 1990). They are variously descended from, or have apparent linguistic or other cultural affinities with, ancient Chinese, Tibetan, and Laotian peoples. The Thai, who are now dominant, arrived later, and only in the twentieth century have they taken control of the highland regions in the north.

The hunting-gathering Phi Tong Luang, or "spirits of the yellow leaves" as translated from Thai, are exceptions: neither are they recognized officially nor do they practise agriculture. Numbering about 150 individuals, who travel in family bands throughout the remote Nan and Phrae provinces of Thailand, they have been forced farther into the mountains and frequently inhabit watershed areas within forest reserves, entirely excluded from Thai society and speaking languages unknown beyond their small bands. Their habitat has been largely destroyed since 1970, and the few individuals who have survived often work for other hill tribes, particularly the H'mong, in exchange for food. Members of the Phi Tong Luang have been used as tourist attractions. Tourist agencies charge fees for picture-taking sessions, both in their natural habitat and in circus-like settings in Bangkok; some have been shown in sentimental movies. Those who survive both servitude and mockery are impoverished and malnourished; the majority have died, victims of abuse, malaria, forced relocations, and the death of their culture. The fate of the Phi Tong Luang repeats a history experienced earlier by the Sakai tribes in southern Thailand and by many other tribal peoples throughout Asia.

The hill tribes that have developed settlements or who still practise swiddening are frequently targeted as the major culprits in northern Thailand's deforestation. Many speak no Thai, do not hold Thai citizenship rights, and pay no Thai taxes, and they escape control by government authorities. Augmented by the influx of tribal people from neighbouring countries, where political and military turmoil obliged these people to move, and also by landless lowland Thais moving into mountain areas in search of farmland, their population has grown. They have been pulled into market relations with others and protected, willy-nilly, by modern medicine; as a result, they have lost the spiritual bases and the physical limitations that once kept population densities within the range of the land's support system. The Asian Development Bank estimates hill-tribe cultivators at about 1.4 million people and claims that they clear some 500,000 hectares each year (Arbhabhirama et al., 1988:173; Feeny, 1988:143).

Swiddening in one northern Thailand region inhabited by Lua' villagers, researched by Peter Kunstadter in the early 1960s, involved one-year rotations over ten years for plots in the nearby forests. Swidden land was communal, the property of the village and subject to reallocation rules determined by village religious leaders. It was used to grow rice; Kunstadter observed that there was sufficient soil moisture for vigorous regrowth after a crop was reaped. The surrounding forest was preserved as a spiritual home. The village had already undergone dramatic changes since the Lua' and other tribal groups created original settlements in the distant past. In the twentieth century, the Bangkok government took control of the northern region, imposed head taxes, and forbad swiddening. These policies, together with competition over land between Lua', H'mong, and Karen village groups, reduced their independence and their control of their own farming practices. The cash economy and market, government services, roads, and schools, the influx of outsiders, and the conversion of some villagers to Christianity all contributed to the dilution of their culture. Kunstadter documented their gradual shift in land-use system from swiddening to irrigated fields similar to those of the lowlanders. In his words, "Instead of maintaining the diversity of their environment with their swidden-rotation system, they are increasing its homogeneity by leveling the land into terraces, building larger irrigation systems, and permanently changing the land from forest to farm" (1988:104).

Landless Peasants

The Reserved Forest Act was passed in 1964; before that, 80 per cent of the land was without title. Once land was designated as "reserved," people already living on it were classed as "squatters." When the Royal Forestry Department began granting concessions for logging and later for the planting of eucalyptus, they were evicted. International attention and local demonstrations finally persuaded the government to modify the eviction terms. In 1988 it announced that communities existing prior to 1967 would receive land title; those established after

1967 but before 1975 would receive deeds stipulating rights to use the land. These rights could be inherited but not alienated. Communities established after 1975 might use the land temporarily. But these modifications were not designed solely for poor farmers: among the communities established after 1975 have been planters with long leases at very low rates (Puntasen et al., 1990:28). Encroachment is a way of life for the rich as much as the poor, and commercial logging is as much the reason for it as the need to grow food and find fuelwood.

Hill tribes and poor peasants contribute to deforestation, and as their numbers increase, their impact on the land becomes ever more destructive. Their impact, however, occurs in a historical context of the commercial exploitation of teak and other natural forests, in the course of which their subsistence capacities were steadily eroded by merchant traders and property laws that favoured the rich and more recently by the development of industrial plantations, hydroelectric dams, highways, and industry. Logging of natural forests was stopped at the end of the 1980s. At about the same time, voices became louder in the defence of both forests and the poor. These voices argued that forests could not be saved if millions of people lacked fuel and food; plantation eucalyptus or pine would not provide livelihoods. Their arguments have had some effect; the Thai government has become more concerned with the social impact of forestry policies and less obsessed with the potential benefits from plantations and pulpmills.

PULPWOOD PLANTATIONS AND MILLS

Pine forests are indigenous to the hill regions of the north. Regarded as a nuisance by a forest service used to teak and hardwoods, the pine forests were still more or less intact in the late twentieth century. Jaakko Pöyry, the international consulting firm called in to virtually every country examined in this book, advanced a familiar plan. Called sustainable forestry, it involves the harvesting of contiguous areas over a span of time equivalent to the growth period required for commercial purposes (about a quarter-century); the labour was to be supplied by local villagers. Plantation pine would replace the natural forest. This proposal is still under consideration, though opposition to it from villagers in the pine forest region has been strong and loud.*

* I am indebted to Ann Danaiya Usher for permitting me to read her study of Thailand's forestry dilemmas in draft form. Since the proposed book is not available for public reading and quotation, it has not been listed in the bibliography. Her many articles in the Bangkok Nation form the basis of the work.

Eucalyptus Plantations

Eucalyptus is the more likely source of pulpwood. Australian eucalypts were introduced to Thailand in 1941, but they drew little attention before the late 1970s. The species *Eucalyptus camaldulensis,* also known as the red river gum, was promoted by the Royal Forestry Department and the National Energy Authority for community reforestation and as a source of fuelwood (Gro-Thong, 1988:224–6). Though officially introduced as a good source of firewood and charcoal, this tree has other possibilities. It is not ideal for lumber, but is used for making window and door frames; it is also a secondary material for plywood. A source of nectar and pollen for bees, it might support a honey industry. Its leaves contain oil, another possible commercial use. If planted on a substantial scale, its seedlings might sustain a commercially profitable trading business. But its most remunerative use is in pulp manufacture. Successful eucalyptus pulp plantations elsewhere became the model for Thai business interests and the government.

The 1985 National Forest Policy created the legal framework for the expansion of pulp plantations. It established objectives for development: at least 25 per cent of the forest land was to be used for commercial activity and 15 per cent for conservation. The state would support industries that linked harvesting rights with pulpmills. Incentives for the promotion of silviculture by private enterprises would be provided. In the associated Sixth Plan for forest land, the Royal Forestry Department would plant trees, especially at higher elevations. Tourist and similar businesses were proposed for national parks, but were subsequently abandoned in response to public protest (Arhabirama et al., 1988; Puntasen et al., 1990).

The government Forest Industry Organization and the private sector both planted eucalyptus, and companies obtained "promotional privileges" for plantations, chip mills, and pulpmills. Puntasen and colleagues estimated that a total of about 8,000 hectares of eucalyptus plantations, half of them privately owned and most of them less than four years old, were in place by 1990. About 3,200 hectares were located in the northeast. Five provinces throughout Thailand were identified as tree-growing centres, and incentives were provided for planting the eucalypts.

One of the international companies that announced its intention to establish a plantation was Shell Thailand. Shell, the giant oil company that has entered the pulp business, was establishing plantations throughout the southern regions in South Africa, Brazil, Chile, the Congo, and New Zealand. In 1987 the company announced that it would establish a eucalyptus plantation in the Khun Song Forest Reserve of Chanthaburi province, from which it would export wood chips to Japan, Korea, and Taiwan (Usher, 1990). The original land area was about 20,000 hectares, but the government later halved it when government foresters determined that the rest of the area was still covered with tropical forest in good condition. Five years later, Shell was still awaiting permission to go ahead despite various public-relations exercises and the hiring of a London-based firm to assess the ecological feasibility of the project. The apparent reason for the failure was that villagers had mounted strong opposition and the international community had expressed hostility on ecological grounds.

While Shell was awaiting government approval of its plans and threatening to move its investments to Indonesia or Malaysia, a local agro-forestry magnate, Senator Kitti Damnerncharnvanich, quietly purchased land rights from villagers and applied for small leaseholds from the government until he had amassed a good-sized plantation in Chachoengsao and other eastern provinces. He proposed a 1,000-tons-per-day pulpmill for Thailand's eastern shore to produce eucalyptus pulp for markets in Japan, South Korea, Taiwan, China, and Australia. This mill, which would have been the world's second largest, was to be financed in a joint venture with Nisshu-Iwai Sumitomo of Japan (*PPI*, Nov. 1988).

The senator criticized Shell for trying to get its land in a single location all at once. But his family and employees were caught cutting tropical forest in order to plant eucalyptus, and what became known as "the Suan Kitti incident" was not helpful to Shell. Ann Danaiya Usher, who covered the incident for the *Nation* of Bangkok, speculated that Shell had also cut tropical forest: the game, while not over, had become considerably more controversial (Usher, 1990: "The Tree Farm That Never Was"). The Canadian International Development Agency (CIDA) was caught up in the scandal over Suan Kitti. It had agreed to partly finance the Canadian consultant H.A. Simons Ltd, which was bidding to conduct a feasibility study in competition with Jaakko Pöyry, similarly subsidized by the Finnish government. Of CIDA's participation, the head of the team appointed to review Thailand's commercial forestry policy observed, "It is good that CIDA has laid down environmental criteria for their projects so that their support for industry does not have negative impact on the rural sector. But are they willing to follow this through to its logical conclusions? ... If you think about it, CIDA is actually helping subsidize the Suan Kitti company ... [The agency] seems to have contradictory goals in Thailand" (Saneh Chamrik; quoted by Usher in 1990e).

Wood Chips and Pulpmills

Pulpmills were less rapidly established than plantations, and meanwhile a flourishing wood-chip trade developed. Companies vied for the "promotional privileges" to produce wood chips for export. The V.P. Eucalypt Chipwood Co. Ltd, owned jointly by Thai-Taiwanese interests and located in Chachoengsao, was on stream by the end of the 1980s, with a capacity of 240,000 tons per year. Taiwanese interests in the output of plantations was strong from the beginning. The major investor in this company, Sahaviriya Group, was a distributor of computers and office machinery.

Japanese investment has taken the form of a consortium of fifteen pulp and paper companies, including Mitsubishi Paper, Jujo, and Oji, known as the Thai-Japan Reforestation and Wood Industry (TJR). The Japanese External Trade Organization (JETRO) anticipates an eventual import of up to 2.6 million metric tons of wood chips annually from Thailand (*JETRO Magazine*, Sept. 1988). TJR, however, is attempting to obtain its chips directly from farmers. Usher reported in the *Nation* that in 1987, some 200,000 families in nine provinces within 300 kilometres of eastern seaboard ports would plant about 184,600 hectares of eucalyptus. The Japanese consortium introduced the idea of forest cooperatives and contracted with village cooperatives to buy all trees at a minimum guaranteed price. There seemed to be differing information about who would provide the loans to farmers for the purchase of saplings. Among the candidates, who were reluctant to agree that this was a role they had accepted, were the major Thailand agriculture bank and the Japanese government's Overseas Economic Cooperation Fund. Also ambiguous was the accuracy of yield estimates, though there is no ambiguity about who would bear the burden of risk. Price might be guaranteed; income was not. And eucalyptus plantations, like any crop, might fail. The villagers would still have to pay off the loans, wherever the funding came from (Usher, 1990g: "Pulp Links, from Oz to Siam").

In the mid-1980s, Thailand imported pulp and paper in growing quantities: 200,000 tons of paper and 100,000 tons of pulp. Though the country's first papermill was founded in 1923 and a few more mills were established over the next half-century, their raw materials were waste paper, bamboo, rice straw, grass weed, kenaf, and sugar bagasse; their production capacities were small. The non-wood pulp is a suitable feedstock for the manufacture of writing paper, corrugated board, lower-quality wrapping papers, and some other general-use papers (Gro-Thong, 1988:228–9). It was hoped that with its own source of short-fibred pulp, the country could reduce imports. It could export wood chips and, once pulpmills were established, pulp and possibly paper.

In fact, growth in capacity and output for the non-wood pulpmills was rapid and substantial in the 1980s. A measure of these developments may be obtained from comparative annual growth rates. For the industrial sector as a whole between 1974 and 1985, a period marked by rapid acceleration in foreign investment and the establishment of garment and electronics factories, the growth rate was 7.8; for paper and pulp industries, 8.1. Consumption of all types of paper increased from 260,000 tons per year to 630,000 tons per year. Production of industrial paper, which constituted 71 per cent of total demand for paper in 1985, increased from 97,000 tons per year to 294,100, an annual average growth rate of 10.6 per cent (Dow et al., 1987; cited in Puntasen et al., 1990:3). The three major producers of pulp in 1990 were Phoenix Pulp and Paper (capacity 94,000 tons/year), Siam Pulp and Paper (two mills, with capacities of 43,000 and 73,000 tons/year), and Bang Pa-in Paper Mill (capacity 13,000 tons/year) (Bank of Thailand, *Monthly Report*, Aug. 1990; cited in Puntasen et al., 1990:6, table 3). With rising prices for pulp and paper in the late 1980s, more applicants sought planting and pulp-manufacturing privileges.

Apichai Puntasen and colleagues describe scandals connected to projects undertaken jointly by the public and private sectors involving Thai and Japanese or Taiwanese companies (1990:21–5). Officials in the Royal Forestry Department, according to this account, were involved in eucalyptus plantation auctions that may have unfairly benefited one of these joint ventures. Japan and Taiwan provide potential large markets for Thailand's plantation pulp. There is also a growing market for paper products in Thailand itself. Despite the rapid expansion of pulpmill capacities, production is not meeting domestic demand. Imports of over 43,000 tons in 1989 were added to domestic production of nearly 147,000 tons. The country was gradually becoming more self-sufficient, but the market was still expanding (Puntasen et al., 1990).

By 1993, pulp production had grown to 200,000 tons, though 160,000 tons of this was non-wood. Paper and board production had expanded to 1,418,000 tons. Imports of pulp were slightly greater than production, but paper imports were gradually declining and stood at 325,000 tons in 1993. Two new pulpmills, with a combined capacity of 300,000 tons per year, were in the planning stages. Phoenix Pulp and Paper expanded its production of eucalyptus pulp to 100,000 tons per year and announced plans to build a new wood-free paper machine in a joint venture with Siam Paper. At the same time, it switched to bamboo pulp in its first mill. The Thai Paper Banbong mill in Ratchaburi province also increased its capacity for coated printing and writing papers to 170,000 tons per year, enabled in this venture by technical assistance from the Japanese papermaker Hokuetsu (*PPI*, July 1994:68). Thailand's first newsprint mill began production in the spring of 1993, using waste paper imported from the United States. The project is a joint venture between Thailand's largest newspaper publisher, Thai Rath, and the Korean papermaker Shinho, and the International Finance Corporation has a 15 per cent share. Overall, then, by 1993 Thailand was becoming more self-sufficient in paper products, was still reliant in considerable measure on non-wood pulp, and was not yet (nor likely to become) an exporter of either pulp or paper.

External and Internal Linkages for Commercial Forestry

As noted in other chapters, there are many external, as well as internal, business interests in plantation forestry. Apart from timber merchants (of whom the Taiwanese and Japanese consortia and companies are primary in Thailand) and forestry corporations, there are also companies selling machinery services in mining, dam, and highway construction. They increase the pressure on governments in developing countries to "open up" tropical forest areas, promising an increase in wealth from such liberal behaviour. It is not only the pulp companies that benefit from new plantations; it is, perhaps even more, the manufacturers of pulping machinery and chemicals, of harvesting equipment, and sawmill blades.

There are also the service industries: the multitude of experts, financiers, and consultants. A complete industry is now associated with conducting feasibility studies and "master plans" for the clearcutting of forests and their replacement with plantations. Jaakko Pöyry, as but one of that company's numerous ventures, provided a plan for Thailand. Aid programs from industrial countries sponsor research stations in underdeveloped countries aimed at discovering the appropriate species for plantation forestry, and although the industrial nations pay for the research (and incidentally cause problems by providing industrial-country salaries for their own researchers on the sites), in the end it is the industrial countries, their corporations, and their consumers who benefit most from the results. The World Bank has been a major force in the establishment of a forest industry in developing countries. It has provided many consumer forecasts and much of the funding on its usual tough terms, which leave little room for sentiment.

PROTESTS AND AGRO-FORESTRY ALTERNATIVES

Reforestation becomes a controversial subject in these circumstances. Are eucalyptus plantations really forests? Should the acreage given over to them be included in government statistics on reforestation? Many Thai villagers have protested against such definitions. Many have also pointed out that incentives for establishing plantations are so great, and land is so cheap under pulp leases, that planters gain most from clearcutting natural forests under the concessions. The government responds that its Reserved Forest Act requires conservation of all forests except those "overly exploited." While there are clear bureaucratic definitions of what this term means, the incentives are so great that companies take the risk of disobeying the law. The Suan Kitti affair was a demonstration of that fact.

Villagers at Baan Namkham in the Ubon Ratchathani province of Thailand tried a novel method of preventing the clearing of common forests and their replacement with eucalyptus plantations. In March 1990, Buddhist monks ordained the trees. Villagers contended that cutting ordained trees would be equivalent to killing a monk, a grave sin. "We peasants have no power to fight the authorities. We have to turn to the power of goodness to help us," one of the local farmers explained to a reporter. The villagers have lost land that used to provide them with food, firewood, herbal medicine, and supplementary income from sale of forest products. Baan Namkham was the first village in Tambon Khampia, Trakan Phutphon district, to be "reforested" as the Forest Industry Organization calls its commercial eucalyptus-plantation scheme. Villagers claim that what actually happens is that officials chop down natural forests to make way for the plantations. They also claim that the planted eucalyptus trees (which they call "demons") suck up groundwater, harden the land, and kill other trees nearby with their strangling roots. "If they don't respect the sanctity of the yellow robe, then we don't have any hope left," say the farmers, who argue that not only the forests, but also the wood islands left in rice fields for cattle fodder are essential to their survival (Ekachai, 1990a).

Professional foresters have argued that the large stands of *Eucalyptus camaldulensis* which they are planting are similar to natural forests in their environmental impact. Their claim has occasionally been accompanied by questions about their opponents' expertise to make judgments. The former rector of the major forestry university took his colleagues to task at a conference in Bangkok in 1990, saying: "I admit that I am stupid. I feel even more confused and foolish after having heard foresters intimidate those who are interested in the eucalyptus issue by telling them not to comment because they don't know anything about it not having studied forestry at university" (quoted by Usher, 1990g).

Among the numerous objections mounted by villagers and tribals in Thailand (as elsewhere) is that the plantations deprive them of their livelihoods. The Thailand Development Research Institute (TDRI) conducted a study in 1990 and discovered that, in fact, these large plantations are very profitable for the rich, but devastating for the poor (TDRI, 1990). These findings are contrary to government, company, and foresters' claims that the plantations will especially benefit poor farmers. Among the issues raised in the report are the differential impacts of credit arrangements: the poor farmers can engage in plantation agriculture only by borrowing on non-institutional markets at high rates. They are frequently unable to survive through several years while awaiting a harvestable crop. Thus most poor people cannot share in the benefits. If, moreover, they have insecure land tenure, they are often pushed off the land to make way for the crops of rich farmers, merchants, or corporations. In addition to the evidence regarding inequitable income returns, the report noted that the international literature provided evidence that eucalyptus plantations reduce the water-table and affect neighbouring crops where moisture and nutrients are in short supply. Other reports note that soil preparation for eucalyptus plantations destroys the natural balance of the region. Flora and fauna are totally wiped out. Even dry sticks have to be burnt lest they later ignite and burn down the eucalyptus saplings (*Nation* (Bangkok), 1990a).

As long as there are millions of poor families roaming the country in search of fuelwood and agricultural land, the decimation of the forest will continue irrespective of attempts to save it. Support for reforms that would enable the poor to share in the fruits of the land is widespread. It even includes industrial planters, who note that labour is required for planting and silviculture; the landless poor could be harnessed to the plantations via community organizations. Thai government policies so far reflect a mixture of these sentiments: opportunities have been established for poor people to gain a homestead and plant crops interspersed among plantation trees, but at the same time, the trees are not the original forest cover, and the plantations are controversial because of both their ecological effects and the ambiguity of motives in their establishment. Government policies take form within two general formats; one is licences to farm, the other an experimental redevelopment of the taungya system.

Right-to-farm usufructuary licences (called STK) have been issued to large numbers of encroachers, with the expectation that these licences would give them a sense of ownership and an incentive to invest in the land. Such individuals are employed by the Royal Forestry Department to reforest and to maintain standing forests. About a million families are involved in this program, and optimists anticipate that it will serve its two primary purposes: to reforest the land and to provide for positive rural development.

Reforestation programs emphasize fast-growing trees, including pine, Persian lilac, eucalyptus, leucaena, casuarina, acacia, duabanga, acrocarpus, and mahogany. These programs, like the taungya system described below, emphasize community development as well as reforestation.

The taungya system traditionally involved shifting cultivators and villagers in the ongoing management of forests. The Forest Industry Organization (FIO) was established to encourage their participation in the management and harvesting of teak forests. The traditional taungya organization had been adapted to shorter-rotation crops and more settled villagers. The FIO maintains forest tree nurseries that supply plant materials for reforestation to organized villages. Villagers are permitted to plant subsistence crops interspersed with tree crops in exchange for tending the tree crops. Wages are paid, and the farmers are allocated 0.6 hectares per family for housing and home gardens. They are permitted to utilize up to 1.6 hectares per year for intercropping of tree plantations. Their homestead land cannot be sold, but it can be inherited, and some service facilities – schools, health centres, temple or church, electricity, water supply, and roads – are provided within the area. The density of plantations is determined by each unit, though a normal system involves 625 trees per hectare with spacing of 2 by 8 metres or 4 by 4 metres, thus allowing for interspersed crops. Various agricultural crops are grown in combination with different tree crops according to the region's soil and water conditions and the market demand. Experiments have been conducted with such cash crops as bananas, pineapples, coffee, and cotton, in addition to maize, nuts, beans, onions, and gourds. Farmers who maintain tree crops with high survival rates are rewarded. A family that utilizes 1.6 hectares per year for three consecutive years, plus one new planting plot and two previous planting plots totalling 4.8 hectares, will receive bonuses in addition to regular wages and subsidies.

The first village was established in 1968. As of the mid-1980s, there were forty-two forest villages with a total population of just under 10,000, or nearly 2,000 households. Half of these households were in northern Thailand. The now-aborted Thai Plywood Company was also organized as a forest village system, but it was not successful (Arbhab-hirama et al., 1988:162–3; Sahanulu and Hoamuangkaew, 1986a:72–5).

Although it is too early to assess adequately the costs and benefits for villagers and the government or to determine how well reforestation has been advanced by the modified taungya and other agro-forestry systems, some case studies have been carried out. One project in northeast Thailand initiated in 1979 as a UNDP/FAO pilot program was intended to engage the participation of nearly 1,300 households and 8,000 people. Deforestation of the region had been caused by logging, maize plantations designed for export, and road construction. While none of the resident farmers had property deeds, ownership claims were recognized by the local population, and the land-tax office had accepted payment and issued receipts for taxes, an implicit recognition that the government was pressed to acknowledge. An FAO report on the project notes that it was designed to provide these villagers with economic benefits through the planting of eucalyptus on degraded natural forest land. The original target was the creation of six agro-forestry villages more concentrated in location than the then scattered distribution of population and the establishment of village cooperatives, health centres, schools, and other infrastructure. Much time was spent in persuading farmers to relocate and accept land entitlement limitations. Fruit-trees, cotton, soybeans, upland rice, and kenaf were to be interspersed with the eucalyptus, but at the end of the project, maize was still the single most important crop. By 1985 about half of the sampled farmers had started to plant forest trees, though full participation in the project, particularly in the management of village woodlots and in silvicultural practices, was not widespread. Even so, 1,163 hectares of degraded forest land had been planted by 1988, and some cooperation between forestry officers and villagers had been established (Amyot, 1988).

The province of Chachoengsao, northeast of Bangkok, encourages its own people, however organized, to jointly plant forests within the compounds of temples, schools, and foundation offices, at roadsides, near water resources, and on other public lands. Eucalyptus is generally used in these plots, and some have other plants interspersed. Some success was evident to me during a field trip in 1990. Where public or private plantations had not been installed, deforested land had become hard and sun-baked. In the same and neighbouring forest districts, eucalyptus, teak, rubber, and other plantation species had been successfully planted. The forestry officer who kindly guided me through this area provided information to local people regarding the benefits of eucalyptus plantations (Umprai, 1990). They are not, and never will be, substitutes for natural forests; they are, however, more beneficial than alang-alang grass or bare, sun-baked land.

A foreigner's perspective on Thai forestry practices was provided by a professor of economics at the University of Washington in Seattle, who was a visiting scholar at Thanimasat University in 1990. Robert Halvorsen argued that the problem was simply one of "externalities." If companies were obliged to pay for the full costs of relocating villagers and ensuring that the soil was properly managed for long-term use, then forestry would be entirely beneficial in places such as Thailand. He suggested that the government grant encroachers legal property rights in land, so that they would have an incentive to make it productive and in order to control erosion. At the same time, the land should be removed from the category of forests, so that agriculture and agro-forestry could be carried on. The logging ban should be removed, because illegal logging ends up doing more harm than legal logging; the government could better control the legal concessions (Halvorsen, 1990).

The economics of plantation forestry would surely be very different if the full costs of relocating villagers were included, just as, in North America, including the full costs of extraction, reforestation, and guaranteed employment of workers would alter the equations of forestry. But in Thailand, and in most other countries where tribals and villagers still have cultural identities, relocation is not simply an economic function, even if it were genuinely underwritten by those who stand to profit from logging. There is no way of paying for the loss of territorial homes; relocation is a major sociological issue, not merely an economic externality. As to government controls on legal concessions, one has to imagine a government removed from concessions, legal or illegal, to suppose that these would work. Such arm's-length management is not characteristic of forest economies in Asia (or, for that matter, though the forms differ, in the industrialized countries).

Much of the forest is already destroyed; the risk of planting eucalypts on the remaining land may be less great than the danger of leaving it to alang-alang grass. If indeed it is possible to plant crops under the eucalypts, or to intersperse them and thereby reduce the risks of monocultures, then a partial solution for the farmers may be found. It would depend on the farmers having cooperative rights, a great deal more legitimate authority to determine the outcomes, and subsidies to provide for their survival while the trees are growing. Such a solution has little appeal for large pulp companies, but if it were the essential means of obtaining fibre, then they might accept it. The economics of the pulp trade do not require that pulp companies own the plantations; they could still be profitable (as Japanese consortia demonstrate) if they had to purchase the trees on an open market.

SUMMARY

Thailand has a deforested land, millions of landless poor who continue the deforestation process because they lack fuel and food, and a history of severe flooding following deforestation of watershed areas. It has a forestry department that is understaffed and not very effective as a manager of forest lands; it has a history of illegal logging and corruption; has villagers who are angry and who protest the further destruction of their land by industrial forestry. It has a growing sector of eucalyptus plantations, but much of the output consists of wood chips sold to Taiwan and Japan; it has pulpmills using vegetable fibres and promises of investments in wood-based pulpmills. The country is not an industrial power in forestry; it is exploited by its own leaders, used by merchants elsewhere, and a long way from discovering solutions to its social and economic problems.

10 Indonesia – Peddlars, Princes, and Loggers: A Case Study

The Wai Seputih river in Sumatra is banked by tropical forests scheduled for destruction. Monkeys still play at the edges, but other wildlife is not visible – perhaps an occasional small wild boar, rarely a butterfly. It is a hot day, and the motorized longboat in which we travel toward a plantation area contributes to the general cacophony of the busy waterway. This is a major route to the channel between Sumatra and Java, a route ideally situated for the flotation of logs as soon as the concession companies get their act together. They have mills ready for wood, and they have potential buyers; they need investment capital to establish the plantations on what has been classified as degraded land.

Since the publication in 1963 of *Peddlers and Princes*, Clifford Geertz's influential study of Indonesian rural life, this scattered country has changed enormously, though the traditional village structure remains. More of the peddlers of Java are now touched by, if not actively engaged in, the global economy, and none more so than the entrepreneurs of the wood and pulp industries.

Of Asian countries with forests, Indonesia was one of the first to impose logging bans and begin the development of indigenous wood-products industries. Its economy has been closely tied to Japanese markets and investors, though its plywood has a wider, global market. It has a nascent pulp industry using waste paper and non-wood fibres but now moving quickly into wood pulps. If the country chooses to continue destroying what remains of its natural forest, it could substantially expand its pulp plantation economy. Government plans to increase annual pulp production from 1.7 million tons to 14 million tons by the year 2000 are regarded by outsiders as unrealistic, but a

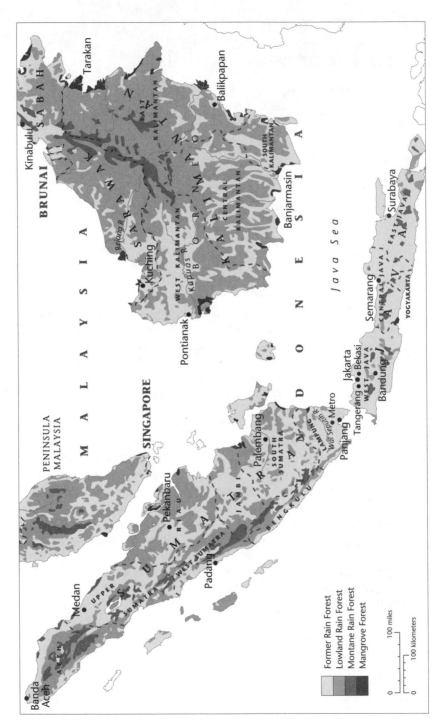

Map 9   Sumatra and Kalimantan (Indonesia) and Sarawak and Sabah (Malaysia): forest types

target of about 7 million tons, using plantations in Sumatra and Kalimantan, might be achievable (*World Paper,* June 1994). Anticipated benefits are – as usual in such developments – expansion of exports and general economic growth, increased affluence and employment, industrial expansion, and upgrading of technological capacities and labour skills. The chief architects are local and foreign investors, many tied to military or other influential élites or government, and it is they who anticipate the benefits. The costs include deforestation and other environmental destruction, obliteration of tribal cultures, and dispersal of tribal peoples and landless peasants. The victims do not anticipate benefits.

NATURAL FORESTS

Indonesia is a collection of islands that comprise twice the size of Western Europe in total area and contain a population of 180 million. The country is estimated to have 65 per cent of the total forest area in Southeast Asia and about 10 per cent of the world's tropical rainforest. However, estimates of forest area and of the extent of deforestation vary considerably. In 1983 the Indonesian measure known as the Consensus Land Use Plan (TGHK) estimated forest land at 144 million hectares, or 65 per cent of total land area. Estimates by an independent Indonesian study group are closer to 108 million hectares, and even that figure may be exaggerated. The study group shows the average annual deforestation rate over the decade of the 1980s to be 1.2 per cent, ranging from 0.3 per cent on Bali (already largely deforested) to 1.6 per cent on Kalimantan, 1.7 per cent on Java, and 1.8 per cent on Sumatra. The forest area was calculated by this group to have been about 119 million hectares in 1982, reduced to 108 million hectares in 1990 (Wahana Lingkungan Hidup Indonesia, 1992:12–15). N. Guppy estimated that by the mid-1980s, over 15 million hectares were covered with alang-alang grass and that another 40 million hectares in Java and Bali were severely eroded (1984:937).

The TGHK classification allows local stakeholders to develop regional classifications by consensus. Four land-use categories are identified: production (government retains ownership but may grant concessions to timber entrepreneurs), conversion (for farming and miscellaneous plantations), protection (in regions of critical watersheds), and conservation (for the protection of biological diversity). As noted by P.F. Burgess for an International Timber Trade Organization survey in 1989, "it hardly seems likely that the system will ensure for the permanent forests the long-term protection from alienation which is essential for sustained-yield management" (1989:118). By the 1990s, the conservation category included 19 million hectares, much of this in remote or unloggable terrain. The conversion category amounted to 30 million hectares. Productive forest land accounted for 64 million hectares, 90 per cent of which was already under concession. In Pontianak, West Kalimantan, where log booms fill the harbour and busy entrepreneurs broker sales to merchants, the local museum provides a graphic depiction of how the classifications work in practice. The mountainous northern region, politically sensitive due to border disputes with Malaysia and hill-tribe incursions, accounts for much of the 37 per cent of total land in protected forest. Conversion forest amounts to 10 per cent of the total, and production forest to 29 per cent, these portions in the more accessible central region.

In the natural forests of Java, teak is the major species, but on the outer islands where the native dipterocarp forests are most abundant some 120 other native commercial species are prevalent. These include the meranti (both red and white varieties), keruing, kapur, bandkirai, and damar laut, found on most of the islands but in greater density on Sumatra and Kalimantan. On the eastern islands, merbau, lara, sonokembang, and kaju arang hardwoods are common. There are some pure stands of agathic softwood and of eucalpyts and casaurina. The hardwoods may be manufactured into veneers and plywood, various construction materials, and furniture.

Deforestation prior to 1970 was caused in Indonesia as elsewhere by numerous agents: slash-and-burn agriculture, clearances for transmigration projects and roads, uncontrolled logging before export bans on roundwood were introduced, and a thriving plywood industry that superseded the log trade. The pulpmill industry is scarcely two decades old, though there were already forty-seven mills operating in 1994 and several more were under construction. Most mills were located on the island of Java, but the wood comes mainly from other islands. Government policies, concessions, and subsidies have encouraged the industry's expansion; plantations are hailed as the solution to deforestation and degraded land. Where plantations have been developed, the major species are pine (*P. merkusii*), acacia (*A. mangium*), albizzia (*A. falcataria*), and eucalyptus (*E. deglupta*). Albizzia is possibly the fastest-growing species in the world: from seedling to full size takes only four to five years. Acacia requires closer to eight years, but is particularly well suited to the soils on outer islands. Eucalypts adapt less well to conditions on the islands.

HISTORICAL CONTEXT

The Netherlands controlled the region now called Indonesia from the early seventeenth century through to the onset of the Second World War, when Japan occupied the country. With no national bourgeoisie when it declared independence in 1945 and achieved it in 1949, the postwar Indonesian state dealt with a foreign-dominated economy, though all political parties were committed initially to policies of economic nationalism. Domestic capital, both Chinese and Malay in origin, created alliances with political parties. Through these alliances, licences, credit, and other concessions were distributed, and they constituted the potential basis for an emergent national bourgeoisie linked to foreign capital. In the same period, the political power of the military increased. By the late 1950s, the military had gained control of most of the new state corporations expropriated from Dutch enterprises, and it provided the support base as presidential rule overtook parliamentary cabinet government in 1965 after a bloody civil war (Robison, 1986:46–65).

Forestry was a relatively small business conducted mainly by local entrepreneurs for the domestic market prior to the creation of a national forest policy in 1967. To obtain logs, foreign investors were obliged to enter into production-sharing agreements. The Indonesian state forestry enterprise, Perhutani (established in 1961 and responsible to the minister of agriculture), secured monopoly control of much of South and Central Kalimantan, together with three million hectares in East Kalimantan during the early 1960s (Manning, 1971:36). Originally, it was capitalized by the government to develop production, processing, and marketing of teak on Java. As the Kalimantan Forestry Development Corporation, it entered into production-sharing projects with Japanese firms, with a view to the development of some 2.5 million hectares in various locations of Kalimantan. Japanese participation was secured because alternative sources of wood supply in Malaysia and the Philippines were declining, and Indonesian policy inhibited straight purchase of logs from local (primarily Chinese) contractors. The interests of the Japanese consortium and those of the Indonesian government, however, were not entirely coincidental – the buyers wanted logs; the sellers, long-term investments – and the various projects included in the agreement were largely unsuccessful.

The New Order after 1967

Military control of the state continued after the overthrow of Sukarno in 1965, but foreign capital was treated with more hospitality. Two laws enacted in 1967 as part of the "New Order" were crucial to the development of forestry in this period: the Basic Forestry Law and the Foreign Capital Law. The Basic Forestry Law established concessions, and while it acknowledged customary ownership rights, it warned that these would be subordinate to superior public interests. About six hundred concessions were established, many of these obtained by the Indonesian military (Robison, 1986:254). The Foreign Capital Law invited foreign investors to establish joint ventures with Indonesian firms. Companies were exempted for two to five years from paying income taxes. They were required to pay only an annual fee of five cents (US) an acre plus production royalties per cubic metre of timber of between fifty cents (US) and seven dollars (Grossman and Siegel, 1977:2). They enjoyed import exemptions on machinery and parts, obtained investment allowances, and had permission to employ foreign managerial and technical personnel (Robison, 1986:170).

In a 1967 article in the prestigious *Bulletin of Indonesian Economic Studies*, a contributor enthused about the "vast and as yet almost untapped resources." Dahlin Thalib went on to chart these resources: only 7 per cent had been explored; about 13 million hectares were immediately suitable for commercial exploitation; the Department of Forestry had listed forty-three locations on various islands ready for exploitation (1967:91–5). At that time, native hardwoods of the rainforest were the focus of attention, with an estimated 413 billion cubic feet available (about thirteen times the volume in the Philippines).

Enamoured of the potential, Indonesia thus entered a boom period based almost entirely on log exports. In 1959 timber accounted for only 1 per cent of exports; by 1974, it accounted for nearly 10 per cent, and of all non-oil exports, about 32 per cent (I. Palmer, 1978:123). By the end of 1969, 5 per cent of the total identified timberland (5.8 million hectares) had been granted in concessions worth $165 million. In the first Five Year Plan, the targeted harvest for 1969–70 was 1.2 MCM; 4.6 MCM were produced. The 1973–74 target of 5.1 MCM was quadrupled (I. Palmer, 1978:120). In a study of West Sumatra, Hendra Esmara noted a doubling of price for sawn woods and logs between 1967 and 1969, because production could not match demand. At that time 3 per cent of forest land was exploited. Within the next two years, at least six joint ventures with foreign companies were arranged (1971:46–7). In 1971, Riau (central Sumatra), the rest of Sumatra, Maluku, and Sulawesi combined produced some 1.8 MCM of logs, valued at US$19.4 million (Manning, 1971:31). The timber boom in East Kalimantan, however, exceeded that of Sumatra and all other regions. Timber exports from that region grew from under 1 per cent in 1967 to nearly 8 per cent of total exports by 1971, and by then timber export values surpassed all traditional export crops except rubber. Kalimantan as a whole produced over 5.5 MCM valued at US$68.3 million in 1971, of which more than 4 MCM and US$55 million were attributable to East Kalimantan (Manning, 1971:30–1). The vast bulk of this timber was shipped as logs to Japan.

Most foreign investors in forest industries were Japanese, and, as before, many Japanese companies were consortia of forestry firms whose interest was in the acquisition of raw material for construction or pulp industries. Under the old order, the Japanese Forest Development Corporation and Perhutani had jointly planned a $35.5 million investment involving twenty-three production units and a targeted production of 800,000 CM. Only five of these units were completed by 1970, with combined production of 130,000 CM, and the joint venture was formally disbanded. Perhutani was left with a $10 million debt (Manning, 1971:36n13). The partners could not agree on repayment of debt claims, and Perhutani was unable to make new agreements with the Japanese firms. Despite this setback, Indonesia welcomed other Japanese investments, but Japanese investors were more interested in procuring raw material than in constructing mills.

Excessive reliance on the Japanese market was a central problem for Indonesian contract companies and for the country more generally. Any change in the Japanese economy or any development of alternative wood sources elsewhere for Japanese buyers would immediately affect Indonesian exporters. A 1971 incident highlighted the vulnerability of East Kalimantan producers. They had on hand some 400,000 CM of hand-cut logs that failed to meet the standards for mechanized production introduced by the plywood producers of Japan during a minor recessionary period. The stock deteriorates quickly and must be sold within four months of cutting, but the Japanese refused to purchase it and no other Southeast Asian buyers could absorb this quantity. Transportation costs were too high to court more distant markets. Prices fell and much of the timber was wasted (Manning, 1971:50).

Though Japanese firms were most numerous, Malaysian, Filipino, Korean, and American firms also participated. Sorianoy (Philippines) and Korea Development Corporation accounted for over two-thirds of total foreign, and over half of all, forestry investment commitments (Manning, 1971:38–40). Weyerhaeuser, which then owned more timber than any other private corporation in the world, obtained concessions in East Kalimantan in 1971 (as International Timber Corporation Indonesia). Grossman and Siegel conducted a study of its operations in 1977, by which time it was exporting $66 million worth of logs annually. It had already established a strong export market in Japan, with sales of softwood from the United States and hardwood from the Philippines and the east Malaysian state of Sabah. Indonesian timber was needed because these other sources of hardwoods were becoming exhausted. On paper, 35 per cent of Weyerhaeuser's holdings were held by Indonesian interests; in fact, the partners were seventy-three top military leaders under the corporate title of the Regional Development Agency. Weyerhaeuser withdrew from Indonesia in the early 1980s when the government began pressing the company to establish processing plants.

The state retained a presence in forestry through Inhtani. This, like other state corporations in the resources sector, provided the nexus through which foreign companies obtained timber concessions and were regulated. In forestry, as in other resource and strategic sectors of the economy, scandals have been numerous; both capital accumulation essential to further growth and personal accumulation used for luxuries derived from dubious relationships and mutual favours between business groups, military personnel, government agents, and foreign firms.

Among the temporary beneficiaries of the New Order were provincial governments. Initially, they retained autonomy over the granting of concessions up to 10,000 hectares, and they received 70 per cent of the licence fee and royalty receipts. In East Kalimantan these royalties provided 57 per cent of total provincial revenue in 1970, resulting in the proliferation of small concessions, many of them logged by Malaysian contractors. Poor planning, overcutting, insufficient forest management and harmful silviculture practices, and poor labour practices were all evident. Small firms did not have the capital to develop long-term plans or processing facilities. After 1971, however, the central government reduced the provinces' granting power to only 100 hectares, insufficient for timber operations. This change decreased the benefits for the provincial government, but it also reduced the multitude of ill effects from small operations (Manning, 1971:37n16 and table 8, 44, 58).

The new investment laws required Indonesian participation in management for all foreign firms whose proposed investment was under $2.5 million. Many of the Indonesian partners, however, were only nominally involved in forestry or other enterprises. The military was particularly prominent in new logging concessions. On shares of under 3 million rupees, military groups and individuals entered joint ventures with foreign or Chinese firms with total investments closer to 100 million rupees. The imbalance would strongly suggest that the military's contribution was the concession, traded for a share of profits. A 1971 study by Chris Manning showed that five paramilitary organizations had obtained over 1 million hectares of concession area in East Kalimantan, representing 10 per cent of the total forest land set aside for production and 20 per cent of that in production at the time. These organizations were generally linked to foreign companies in joint ventures, or they employed Malay contractors. One operated on the Malaysian side of the border and refused to pay royalties to the provincial government. A second, one of the largest and a joint venture between a military organization and an American company, caused numerous problems for the provincial government because it relied on contractors, failed to produce despite having control of some of Kalimantan's most valuable stands, was badly organized, refused to comply with provincial regulations or adequately report production levels, and clashed frequently with immigration and customs officials (Manning, 1971:57).

Of twelve companies surveyed by L. Wells in another 1971 study, three had active Indonesian partners; the others were providing services such as interpreting government regulations and supplying contacts, and two were involved at only the most nominal level, receiving a contract fee (Manning, 1971:49; citing SEADAG Report, 25–26 Feb. 1971:9). In this group of twelve, three joint partners were paramilitary, and one was a group of retired foresters. Such partners are unlikely to be involved for the purpose of acquiring business and technical skills in order to develop their own enterprises, as would be consistent with the intentions of the investment laws on joint ventures.

According to Richard Robison, "large numbers of these concessions fell into the hands of the military and their Chinese clients, who in a large number of cases entered joint ventures with foreign companies which financed the ventures but often subcontracted the actual operations out to Philippine or Malaysian loggers. This, however, was one area where domestic investment grew, and by 1975 it exceeded foreign investment" (1986:186). Planned foreign investment vastly exceeded domestic sources in 1970; there was clearly slippage between investment approvals and realized projects. The Foreign Investment Board and Domestic Investment Board data cited by Manning indicates an anticipated investment value for foreign firms in 1970 of US$384 million, compared to US$53 million for domestic sources (1971:39, table 6).

The domestic investors identified by Robison all had close relationships with either the military or particular political factions of the government. According to Robison's information, Yap Swie Kie (Sutopo Jananto), whose group was built on forestry concessions, enjoyed close links with the military in the late 1960s. He entered into joint ventures with Japanese partners in the 1980s. Kang King Tat (Yos Soetomo) had one of the largest timber organizations in the country, with external ties to investors in Hong Kong, Singapore, and Japan. In the mid-1980s, Soetomo and other members of the Sumber Mas group were arrested for tax fraud. Mirawan HS had a large concession for Sulawesi "black timber."

Employment

Employment in forestry expanded throughout the late 1960s. Total direct job creation attributed to foreign investment as of August 1974 was over 400,000, about half of the jobs in East Kalimantan, which had at that time only 4.6 percent of the total population. The remainder were in Riau and Jambu on Sumatra and in Sulawesi (I. Palmer, 1978:116, table 5.9). This distribution was determined by the location of natural tropical forests. Unlike the textile and tourism industries, forestry provided some development opportunities on the outer islands rather than on overpopulated Java. The outer islands, however, had insufficient local labour supplies, and these operations therefore provided employment potential to migrant workers from Java.

Employment of Indonesian workers was required under the contracts, except where nationals lacked the necessary skills. Training programs were required where there were skill deficiencies. However, these rules were vague, and many companies failed to observe them. Malaysian and Filipino firms were the least likely to employ Indonesians, having easy access to low-paid contract teams of skilled workers from their own countries. Indonesian companies were likewise inclined to hire Malaysian and Filipino contract teams because they already had the skills to work with mechanized logging and milling equipment. These teams were available in Indonesia because logging bans in their own countries had left them without work. Manning reported that all the Indonesian firms included in his study of twenty-four East Kalimantan firms employed such teams, and one paramilitary firm employed over five hundred foreigners; some of the American firms likewise employed contract workers. Indonesian workers were more likely to be employed in West Kalimantan, where the trees were smaller and hand logging had not yet been superseded by machinery (Manning, 1971:46–7).

While it is true that many of the employment benefits were foregone through the export of logs and inadequate investment in processing, local capital did not have sufficient reserves to develop processing plants – certainly not at that time, pulping facilities. As well, the workforce did not have the skills to operate mechanized facilities during this early phase of development, and domestic consumption levels were then and still are in the 1990s lower than for any other of the Southeast Asian countries. Workers were generally ill housed in barracks, with poor sanitation and rough living conditions; early concession agreements did not specify working conditions, and welfare levels were generally low. Manning notes an exception to the general practices in the Sorianoy agreement completed late in 1970. It required the company to provide schools, houses, hospitals, and electricity for its own employees and for the local community. The company also agreed to assist shifting cultivators in resettlement (Manning, 1971:48).

Log Export Bans in 1975–85

Foreign-investment laws and subsequent pronouncements by the Indonesian government were clear about the long-range objectives: to create secondary industry, upgrade the labour force, reduce reliance on exports of raw materials, and generally improve the standard of living. The export of logs was always controversial, with numerous spokespersons arguing that such exports pre-empted future employment and returned low revenue compared to the potential revenue of more value-added forest products.

The intention was to shift to higher value-added production, but capital costs for mechanized sawmills, plywood plants, and paper plants were beyond the capacities of domestic capital in the 1960s and early 1970s. After the oil price increases of 1973–74, the state had substantial sums available for state-led industrialization, and domestic capital was strengthened. The government's priorities shifted from the provision of basic needs and agriculture-supportive manufactures to value-added processing industries related to raw materials, including timber and oil. But even with added strength and state backing, domestic firms were not able to move forward alone into wood-based industries and paper production.

To cope with this problem, the government pressured foreign investors to advance into processing industries. New foreign investment in logging was prohibited in May 1975. Tax incentives were offered to foreign capital in pulp, paper, and plywood manufacture. In 1978 the government doubled the export tax on logs from 10 per cent to 20 per cent, likewise to induce development of domestic timber processing facilities (Lindsay, 1989). Log exports declined, but investments in new plants were not immediately forthcoming. Domestic contractors could still obtain quick returns with relatively low investment in logging and log exports, and Japan imposed a 20 per cent tax on imports of processed logs.

Despite a temporary decline in timber markets in 1980, the government limited log exports to 32 per cent of total output, and in May 1981 it prohibited log exports altogether for companies that had not invested in processing. Companies with mills could export logs on a ratio of 1:4, and those in the course of establishing mills could export at an interim ratio of 2:1; the remainder of the logs had to be locally processed. By 1980 forest products contributed 7 per cent of export revenue, and the proportion continued to decline. The Indonesian share of Japan's hardwood log import volume decreased from 44 per cent in 1979 to 28 per cent in 1982; the latter figure was 56 per cent of Indonesian log exports altogether. By the end of 1982, log exports comprised only 38.4 per cent of the total value of all wood exports. In 1985 the government prohibited all log exports (Hunter, 1984). This move resulted in the merger of various companies and the development of new joint ventures. Japanese investors moved into plywood milling. Other international firms, including Weyerhaeuser, opted out, but new investors from Taiwan, Japan, South Korea, and Hong Kong moved in to take their place (Robison, 1986:187–8).

Domestic processing of veneer and plywood, while less extensive than the government had planned, expanded rapidly in the 1982–85 period. Major importers were the United States, the European Economic Community, the Middle East, and the Far East. The competition for these markets consisted of other ASEAN and Pacific rim suppliers: Singapore, Korea, Taiwan, the Philippines, Malaysia, and Brazil. Japan, though taking more than half of Indonesia's logs in 1982, took only 17.4 per cent of its veneer and 1.7 per cent of its plywood. None the less, Indonesia became the world's largest exporter of plywood in 1982; ironically, this success depressed the world price of plywood, while the reduction in hardwood logs on the market caused their world price to increase (Lindsay, 1989:121–2; Hunter, 1984:101).

Sawmills are generally less costly to build than pulpmills, and investments in them steadily increased throughout the 1980s. Sawn timber exports of meranti increased in the early 1980s, with Italy, Singapore, and other countries providing new markets. By 1990 there were over four hundred sawmills in operation. These two new value-added exports did not provide equivalent revenue as log exports during the same period would have, but by 1986 the revenues from actual exports were finally becoming roughly equal to the assumed revenue from logs had the bans not been in place. In addition, there were more jobs, and more of the jobs involved the acquisition of new skills.

It is not clear, however, that the log ban has had any beneficial environmental effect. As much volume of timber was cut. Moreover, according to Holly Lindsay's analysis, more energy was used in the processing, and the fuel was subsidized by the government. It would appear that plywood and sawmill producers had not developed the technology to burn their own waste products efficiently. As well, many of the new mills were inefficient in other respects, and the incentives led to overinvestment in a short period, with the usual result of numerous bankruptcies.

There were many problems with the new concessions in Indonesia. Not the least of these were vague regulations and ill-defined forest boundaries. Environmental concerns were frequently raised, but also frequently ignored, because the potential profits were vast. The victims of industrial logging and transmigration were not only trees; they were, most particularly, the tribal peoples. Native people hunted, fished, and gathered their subsistence; or they engaged in shifting agriculture until the timber boom reduced their free-roaming areas. Timber ceased to be a free good, as it had been in traditional forest cultures. It was no longer available to local inhabitants for traditional long-houses and canoes or as fuel.

Rice farmers suffered displacement as a result of the timber economy. Farming was stopped in some areas, and fields were flooded in others. Companies occasionally dammed rivers in the dry season and then let unbound logs tumble down in the wet season. When a river was dammed, the shortage of water affected everyone along its route. When the logs rushed downriver, they smashed into houses and destroyed latrines and rice fields (I. Palmer, 1978:126). In some areas, the timber concessions all but forced the farmers into wage-labour employment.

These problems notwithstanding, over the long run the Indonesian government succeeded in substituting secondary industry for raw materials exports. Before examining the pulp and paper sector, let us consider the dimensions of the social problems connected to displacement of native people and poor farmers.

INDIGENOUS PEOPLES, ENCROACHERS, AND TRANSMIGRANTS

Catherine Caufield alerted the world to the fate of tribal peoples in Indonesia in her book *In the Rainforest.* She described the native peoples of the Mentawai Islands (Siberut, Sipora, and North and South Pagai) as hunters, fishers, and shifting cultivators whose extended families cultivated sago palms, bananas, and taro on small clearings in the forest before 1960. They collected forest products, but cut only the sal tree for the purpose of making canoes. Their religious taboos ensured the protection of wildlife stocks, and cultural practices helped them maintain a fairly stable population level. Yet over the past sixty years, and particularly since 1960, missionaries and the government have persistently discouraged traditional housing patterns, virtually destroyed the culture, and obliged the Mentawaians to choose either Islam or Christianity. Sago has been supplanted by rice, even though it requires more labour for a lower yield. Population has increased because cultural population controls have been abandoned. Hunting has become impossible in small and overexploited areas, and villagers are obliged to purchase their food, thus requiring either wage-work or handouts.

The Indonesian government, under Regulation 21 of 1970, defended the right of local people to continue traditional practices on concession areas. This was a considerable advance over earlier statements that turned native people into criminals when they cut a tree. However, the land-tenure system assigns real control to the Ministry of Forests, which has leased over 90 per cent of the total forestry area to logging companies; 1 per cent has been reserved as a protected area. Compensation to the villagers is not provided, even for the loss of their vital canoe-making tree, the sal. In Caufield's judgment: "The logging is not only depleting an important capital resource, it is also silting up the rivers, making navigation difficult and reducing the fish catch. Few Mentawaians have jobs with the logging companies. They are not considered good employees because they have no experience working to other people's timetables and are more concerned with tending their crops" (1984:95–6).

Recognizing some of the problems, the government developed a new policy, whereby Siberut was divided into three zones: a 200-square-mile nature reserve without logging, a 400-square-mile traditional-use zone, and a 1,000-square-mile development zone. This plan, worked out in conjunction with the World Wildlife Fund and Survival International, was intended to help the Mentawaians reclaim their traditional way of life and reduce their reliance on charity and cash employment. However, logging had not yet stopped in the traditional zone by 1984. Tribal people were still being forced to live in wooden huts and cultivate rice, and military force was used to ensure they did not return to their traditional longhouses on the coast.

During my own fieldwork in 1990, I encountered several Dayak resettlements along the river-banks in Sumatra and West Kalimantan. In Sumatra the native people are almost all resettled in confined quarters, far from traditional forest activities. In West Kalimantan the Dayak people still comprise some 80 per cent of the population, loosely grouped in over two hundred tribes. I witnessed about twenty men, women, and children arrive at a trading post, load up longboats with the detritus of western civilization – beer, soda pop, candies, canned foods – and then paddle up the river to a more distant and still permitted camp. West Kalimantan has poor soil and is not heavily infested with industrial logging operations. The more remote Dayak are still hunters. But their purchases sadly informed the onlooker that their main quarry was at the local store, no longer in the forest. Other Dayak were resettled near the trading post; their poverty was stark. Repeating the history of North America, Indonesia was destroying an indigenous culture and a people.

The governor of West Kalimantan informed me that the majority tribe of the island, the Iban, had abandoned nomadic patterns and shifting cultivation and had accepted government-provided housing and land grants; further, that intermarriage and integration were well under way, and assimilation would be the outcome. He observed that the primary impetus for granting land to Dayak, as to Javanese migrants, was to increase the labour pool. The resettlement plans under the Sorinoy agreement were unusual in acknowledging the existence of these people, but it assumed the legitimacy of obliging them to give up their itinerant existence to accommodate the new industry. To give the government credit, minimal housing and services have been provided for resettlement of the Dayak shifting cultivators, and attempts are made to locate income sources for people who, as forest dwellers, had not earlier been obliged to enter the cash economy.

As well – and this complicates the picture – the Dayak have already become enmeshed in the technology of the cash economy. By 1990 some were already travelling by motor boat, even when living on the rivers and practising shifting cultivation. They also benefit from modern medicine, which, combined with the loss of cultural controls on population, has caused their numbers to increase. With both additional population numbers and reduced territorial range, they practise shifting cultivation within territories too restricted and in a rotation too short for nature to regenerate the tropical forest. A rotation period of fifteen to twenty-five years is minimal for most tropical forest land; in a shorter time frame, regeneration does not occur spontaneously. A family would therefore need twenty-five times the land area available to it as it would actually use to live on for one year (personal interview with Janus Kartasubrata at Bogor Agricultural University; also Kartasubrata, 1987, 1990).

The objective of the social-forestry project mounted by the government in conjunction with Bogor Agricultural University is to enable the native people, and also the migrants under the transmigration projects, to develop more sedentary agricultural economies around cash crops such as rubber and fruit. Various projects are oriented toward educating villagers to manage forests in ecologically sound ways while obtaining fuel, food, and employment in them. Forestry schools themselves are becoming more "people-oriented" and include com-

pulsory courses in social forestry. A major organizer and scholar with
the social-forestry project, Junus Kartasubrata, noted that since 1984,
foresters, governments, and social scientists have attempted to diag-
nose the problems of forest management in Java. They have tried to
understand the tensions between the local population and the forest
industry by conducting sociological research throughout the forest
regions. Thirteen pilot projects have been established in areas where
a critical situation has developed. These were engaged in developing
agro-forestry systems (taungya) where up to a thousand trees might
be planted with sufficient space allowed for the growth of fruit-trees
or other crops. The projects have also been engaged in organizing
forest farmers' associations with their own elected officers who could
become a medium for the extension of agro-forestry. In the opinion
of Kartasubrata "it is difficult for people to express themselves because
they are used to being told what to do. People need to learn how to
organize themselves. Foresters do not know how." However, he mentioned the successful formation of a non-governmental organization,
the Binesweteia, in one region.

The intended solution is to establish integrated organizations for
land-use policies, with village heads delegated to monitor the shifting
cultivators and special village police to be used as backup. Although
the researchers believe that persuasion and a "psychological" approach
known as the "prosperity approach" will suffice to turn shifting cultivators into conservationists in agro-forestry arrangements, the policy
would use strong measures if required. The incentives would be cash
for commercial crops and employment in sawmills and pulpmills. The
number of shifting cultivators is not known. The distinction between
these tribal peoples (who are ethnically different from the Malays)
and the migrants from Java has become somewhat blurred.

Creative solutions for the indigenous peoples may not be the best
approach for frail tropical forests. If the indigeneous peoples
encroached on the conservation or protection areas, they could
destroy a fragile balance. When that balance is already threatened by
colonization (known as transmigration) from Java, and now as well by
clearcutting and afforestation of exotic species, there is little room for
native hunters to roam and no forest in which to practise shifting
cultivation.

The Landless Poor and Transmigrants

Encroachers on the forest land of the Indonesian islands include, as
they do in Thailand, poor peasants who seek agricultural land and
fuelwood; they also include transmigrants – people moved by the

government from overcrowded areas on Java and other densely populated regions to rural areas of Kalimantan, Sumatra, and the 13,000 outer islands. Transmigration was initiated in 1905 by the Dutch colonizers, but the movement of people vastly increased in the 1970s.

Encroachment on forest lands by settlers and farmers was a gradual process in most of the world, but in Indonesia and Brazil, where transmigration projects have been mounted by governments and promoted by international agencies since the 1940s, the encroachments were sudden. The stated objectives of transmigration in both cases were to provide land to the landless and to establish a bulwark against territorial claims by neighbouring states. The projects satisfied the international community that something was being done for the poor. The "something" avoided redistribution of good agricultural land, which would have increased agricultural productivity and, in the long run, the wealth of such nations, but landowners successfully resisted redistribution. In neither case was the resettlement successful, either for the majority of those who homesteaded or for the government. The clearing of land in the name of resettlement, however, had serious impacts on the forests; much of what was cleared failed to support agriculture and was abandoned; some of what was left was taken over for ranching, mining, or timber concessions.

John S. Spears examined the situation of encroachers in tropical forests in 1979, and he concluded that the expenditures then made by World Bank agricultural-settlement schemes – an important contributor to the transmigration projects – could have improved agricultural productivity more if they had been applied to non-forest areas (Spears, 1979). He argued, as did Val Plumwood and Richard Routley (1982), that the encroachments were unnecessary: they were not caused by a genuine land shortage. Years later, the same arguments are mounted with no more impact. Patricia Adams (1989) has contended that Indonesia's transmigration project, which by the mid-1980s had moved 3 to 4 million people, destroyed millions of hectares of tropical rainforest and ecologically sensitive wetlands, only to yield wastelands when the fragile soils could not support farming.

For many of the migrants, conditions in the new regions are a great improvement on the poverty of their lives on Java, and not all of their settlements have turned the land into waste. More often, the results are ambiguous. My field notes record a visit to a project on Sumatra. Here, on the road north from Lampung and through a series of towns beginning with Metro and including Toeagajah, Rumbia, and the native village of Surabaya Ilirr, the land was covered in thick forest until nearly the end of the 1970s. Native Dayak practised shifting agriculture and regarded the land as theirs. Some 50,000 people now live on transmigration land grants of 2 hectares per family, practising agro-forestry, backed by a military presence and the Ministry of Forests. Though in the more remote areas of the transmigration region, electricity and other services were not yet provided in 1990, families were housed, roads were in place, and there was reasonable evidence that the settlers would "take" to the land. Where electricity existed, television had become an important source of entertainment. Where it did not, the roadsides became evening meeting places; there people of all ages, dressed in their best clothes, enjoyed community events. There were no cars, but some bicycles. The region appeared to be a peaceful and self-contained community, even modestly prosperous. These villagers also provided a potential local labour pool for companies granted logging and clearcutting concessions along the Wai Seputih river. More important, they provided the labour for estates growing palm oil and rubber, both of which are labour-intensive plantation crops. Here, as in many similar developments, the forest had been destroyed by settlers and plantations; the developers had gained cheap labour, but the settlers had also gained a means of surviving, and the outsider should be wary of judging that gain lightly.

The cheap-labour aspect of transmigration was emphasized in West Kalimantan, where 5,000 families had been settled during 1990 and another 55,000 were to be settled within the following five years. These migrants received 2 hectares each (a quarter for housing, the remainder for planting). Each family was given a house, seeds, and sufficient food for a year and was obliged to plant according to a schedule established by the local government. The government has three companies, two in rubber and one in palm oil. These, plus one private company, purchase the migrants' produce (my field notes, 1990). They thus became labourers in a controlled market system, but at the same time they became owners of land and had a means of surviving. It would be going too far to say they had options.

One obvious aspect of the transmigration project is that it has substituted migrants for tribal peoples, pushing the tribals from their traditional lands in order to accommodate the poor from the over-crowded central island of Java or obliging the tribals to become smallholders. And it is true that many of the newcomers do not have the knowledge or survival skills to inhabit these forests without destroying them. Where they move to forest regions with poor soil, even when provided with up to 5 hectares of land and a year's supply of food, they frequently fail to sustain crops and finally abandon their homesteads or, worse, retreat into shifting agriculture practices that cannot support them in large numbers. It is also true that many people on the outer, more sparsely populated islands regard the policy as colonization or pacification: it brings an influx of Javanese to islands that are part of Indonesia only by historical accident (*Ecologist*, 1985).

Environmentalists and human rights activists mounted a sustained campaign against transmigration in Indonesia. The *Ecologist* published numerous critiques in 1985 and a special edition in 1986. It aimed its attack at the World Bank and other international agencies, which provide most of the funding for the project. Survival International, one of the contributors to the campaign, claimed success when the Indonesian government announced that only 1,000 instead of 100,000, families would be relocated in 1987 (Colchester, 1987). However, the reduction might well have been due more to recession and to a persistent problem with transmigration: about half the settlers return to Java within a few years. Even where the resettlement land provides a modest survival capacity, the rural lifestyle is disjunctive with the urban roots of Javanese settlers. The development of the pulp and paper industry and of pulpwood plantation economies adds another dimension to these problems.

FOREST MANAGEMENT AND PLANTATIONS

According to Indonesian government authorities, the rainforest is now well managed. That claim is contested by many ecologists and even by the World Bank, which reported in 1989 that actual exploitation levels were substantially above those recorded. Environmentalists contend that logging has not been stopped in the protected areas and that overcutting of the industrial sites has reduced regeneration capacities of the land. Wastage, poor logging practices, and general corruption in the granting and utilization of concessions are among the charges of the foreign critics. Nicholas Guppy notes that concession agreements are for twenty-year periods, providing incentive for neither conservation nor reforestation. In his view, logging, by 1984 already proceeding at a rate of 800,000 hectares annually, with follow-up by shifting cultivators who cut about 200,000 hectares of secondary forest, will destroy remaining stocks before the turn of the century (1984:943).

Replanting has taken place, even if the extent is debatable. By Indonesian government estimates, some 60 million hectares on concessions under the legislation of the 1960s has undergone reforestation. Most of the replanted areas are on Java. As well, the government established experimental pine plantations as early as 1970. By the 1980s, private companies could obtain concessions to thin in these areas. Some companies also established plantations. One of the pilot projects included the Japanese Overseas Consortium in plantations of

Table 10.1
Indonesia: average deforestation rate[1] by island, 1982–90

Region	Forest area (million hectares)		Annual deforestation (1982–1990)	
	1982	1990	Hectares	% Loss
Sumatra	21.32	20.38	367,700	1.8
Kalimantan[2]	39.62	34.73	610,900	1.6
Sulawesi	11.27	10.33	117,500	1.1
Maluku	6.35	6.03	24,300	0.4
Irian Jaya	34.96	33.65	163,700	0.5
Nusa Tenggara & East Timor	2.47	2.36	14,100	0.6
Bali	0.13	0.13	400	0.3
Java	1.00	0.96	16,100	1.7
Total	119.12	108.57	1,314,700	1.2

Source: Wahana Lingkungan Hidup Indonesia and Yayasan Lembaga Bantuan Hukum Indonesia, 1992, table 3, p. 15.

[1] Deforestation signifies conversion of any closed or open forest to any other land use, through shifting cultivation, crop tree estates, or forest clearing without further cultivation. It also includes succession of alang-alang grassland and fire-ravaged areas. It does not include forest clearing for commercial wood planations.

[2] Includes the 1982–83 forest fire, which burned approximately 3.6 million hectares.

eucalyptus, acacia, and albizzia on some 2,000 hectares on Sumatra in 1980. The Finnish company Jaakko Pöyry introduced a pilot project in 1984 in South Kalimantan known as Marta Puras. Foreign groups were providing mainly technical assistance; financing was largely from the Indonesian government.

The Industrial Forestry Act (1988) obliged concessionaires to plan, and it introduced stumpage fees. Financial assistance was provided for replanting above required levels. Concessions were extended for thirty-five-year renewable periods in place of the twenty years of the earlier concessions, and concession areas on the outer islands were large. The government established nurseries, with the aid of Jaakko Pöyry and Bogor Agricultural University. The university was to retain 40 per cent of the seedlings for further experimentation and development. A clearcutting allowance – indeed, requirement – depended on defining some land as irreparably damaged or as non-tropical forest land. Some land was indeed in one or other of these categories, and the development of plantations might well regenerate it or make it productive for the first time. Some was, however, merely slightly used through previous selective logging or shifting agriculture. Left to nature, it might regenerate a tropical forest. The problem was, Indonesia was in a hurry. Indeed, it was in such a hurry that it defined forest land as non-productive and therefore open for clearcutting where the productivity was lower than 20 CM per hectare, only 2 CM below the norm for tropical forests. The relationship between clearcutting and plantation afforestation is plainly drawn by the vice-president and director of Indah Kiat in a 1988 interview with *Pulp and Paper International*: "Our target is to get access to 150,000 ha of forest which will allow our mill to expand to one million tons/yr. We have 65,000 ha now; we are now in the process of getting concessions for another 65,000. Basically we are looking for forest which can be clearcut and replaced with eucalyptus and acacia" (Soetikno and Sutton: 1988:41).

Plantation *Pinus merkusii* will be the fibre source for two mills being constructed by Inti Indo-rayon and Kertas Kraft Aceh in Aceh province in northern Sumatra (Soetikno and Sutton, 1988:42–5). In Riau, Indah Kiat plants *Acacia mangium* because it proved to be better adapted to the high temperature and soils of Indonesian islands. It forms a canopy within a year and reaches commercial size within five to seven years. Prime harvesting age would be eight years, when trees attain an average height of 20–25 metres. It requires less fertilizer than eucalyptus, though unlike eucalyptus it cannot be coppiced to produce second and third crops.

New forests are thus in the process of creation, and clearly if they are successful plantations, they will affect the entire forest industry, as well as this nation of islands. The species now being planted are potentially productive, but not yet tested. They grow well in some soils, not at all well in others, and the pilot projects were planted only a short while before plantation forestry became a national fixation. Native hardwoods do not store well, are difficult to replant artificially, and have less commercial value than the exotics, but there is still much to learn about the exotics. Of the native hardwoods used for furniture, teak is the most valuable, but it needs water and care. Mahogany is not a native species – it came from South America about a century ago – but it has been in place long enough to have developed its own enemies and is susceptible to various beetles and disease. Eucalypts take a great deal of water from the soil and evaporate it. In growing this (and other) exotic species, much more research needs to be done to determine the effects on soil and water-table. So Indonesia has not planted as much as it intended; plantation forestry is under way, but not yet sustainable on any large scale as an annual harvest. Research is also under way, especially connected to Biotrop, the Southeast Asian

Regional Centre for Tropical Biology, established in 1968 with a major location at Bogor Agricultural University.

It is partly because of the high risks at this stage of plantation development that companies are reluctant to invest in them, especially when the investment first involves the clearing of alang-alang grasslands or the regenerating of degraded soils in forest areas. Instead, as clearly indicated in the quotation from the vice-president of Indah Kiat, they seek out productive natural forest areas where they can profit from the cut tropical wood and the not-yet-degraded land before planting the new monoculture species. Wahana Lingkungan Hidup Indonesia (the Indonesian Forum for the Environment) claims that in Indah Kiat's case, "timber estate creation is instead resulting in the destruction of remarkably productive lands. The natural forests which they are now clear cutting for *Eucalyptus* and *Acacia* plantations are so productive that by felling just a few of the largest trees, loggers can start a domino effect which clears a swath hundreds of meters long; an added advantage is that the loggers may claim the deer and pigs trapped as the forest collapses around them" (1992:26). Even with steadily diminishing natural resources and untested or not yet planted trees, pulpmills are going ahead with plans to build new facilities. The model is the Aracruz project in Brazil, to which representatives have travelled in search of information on species, soils, and nurseries.

Plantation in Lampung

The Ministry of Forests for the province of Lampung on Sumatra informed me in 1990 that its target was 81,000 hectares, to be planted at the rate of 10,000 hectares per annum. The first stage then was merely a year old, and 7,000 hectares had been planted. Prior to that, selective logging "in conformance with preservation and sustained yield conservation principles" had been undertaken. There were three concession holders in the province, all strictly regulated, according to the ministry. For plantations, the ministry locates non-productive areas and gives them to industrial forestry companies. These lands have small-diameter trees. The Putera group, which granted me access to its properties, applied for a concession. To obtain the licence, it had to offer an economic justification, be recommended by the local governor, create a forestry agreement with the ministry, and pay licence fees, royalties, and other percentage contributions from the harvest to the government. To the eyes of an outsider from North America, the Putera area was a meadow; a few big trees remain in patchy, but large clearings. The land is reached by motorized longboat, with native guides together with the local warden. Whether the warden warded people or animals is not clear, but when he accompanied me on the trip he carried a sizeable rifle.

THE PULP AND PAPER INDUSTRY

The pulp and paper industry is rapidly expanding on Sumatra, Kalimantan, and Java. Earlier wood-free mills are antiquated and inefficient, but new mills under construction at the end of the 1990s are state-of-the-art in every respect. They have the newest technologies for everything from debarking to chipping wood, and their emission control systems are, in general, superior to any mills in the northern hemisphere constructed prior to 1980. While labour costs are extremely low – estimated at an average of $2 per day in 1994 – new mills are providing regular wages, and some are providing housing and townsite amenities as well. Because of the labour costs and virtually free land, Indonesia's mills have about the lowest production costs of any country in the pulp and paper business. Even when the companies invest in plantations, they are operating at a cost advantage over other exporting countries.

There were twenty-nine paper manufacturers in Indonesia in 1980. Fourteen were integrated units with pulp-producing sections, but only 20 per cent of fibres were obtained from wood. Rice straw, bagasse, bamboo, and waste paper were the major raw materials. The main wood source was *Pinus merkusii*. One firm, P.T. Sabindo, located at Sesayap in East Kalimantan, used tropical hardwoods. Some other companies, including P.T. Banto Pulp and Paper, obtained pulpwood from South Kalimantan for a mill in East Kalimantan (Hunter, 1984:104). The chief obstacles to development were deficiency of capital and lack of technical expertise. Machinery for wood pulping had to be purchased from advanced industrial countries, and foreign technical aid was required to train workers in its use.

These deficiencies were overcome in the next ten years. By 1990, there were forty papermills in Indonesia, thirty-two on the island of Java and the remainder on Sumatra. Seven more were constructed or in the planning stage by 1994, and others were under consideration. Indonesia, formerly a net importer and still importing 124,000 tons of paper that year, was already exporting some 190,000 tons of paper. Imports of pulp at 217,000 tons still exceeded exports at 181,000 tons; as well, Indonesia imported 463,000 tons of waste paper in 1990 (Wahana Lingkungan Hidup Indonesia, 1992:24). New capacity was on line, and the government projected the establishment of fifty-six large-scale mills by the year 2010, drawing on thirty-three proposed plantations covering in total some 70,000 square kilometres. The

Table 10.2
Indonesia: approved domestic and joint ventures in the pulp and paper industry,
in the period 1967–91

	Domestic	Foreign	Total
Number of projects	72.00	10.00	82.00
Investment			
$ billions	7.80	2.40	10.20
Foreign exchange			
million tpa[1]	3.20	1.60	4.80
Pulp production			
million tpa	4.15	2.19	6.34
Paper production			
million tpa	6.07	1.45	7.52
Pulp export			
million tpa	0.20	1.45	3.44
Paper export			
million tpa	2.38	0.47	2.85
Number of employees	132,046	32,899	164,945

Source: Indonesian National Investment Coordination Board, 1991, as reprinted in Wahana Lingkungan Hidup Indonesia and Yayasan Lembaga Bantuan Hukum Indonesia, 1992.

[1] Tpa = tons per annum.

industry anticipated supplanting Japan as the world's top producer of paper and paperboard, drawing on the resources of neighbours and new producing regions. Despite considerable growth, these dreams are unlikely to materialize by the year 2000. Infrastructure remains undeveloped in much of the country. Most of the new paper and paperboard mills still rely on imported waste paper or non-wood fibres. Indeed, the industry's chief critic, Wahana Lingkungan Hidup Indonesia, is in accord with trade journals in concluding that there is insufficient domestic wood to sustain the industry through the 1990s, and planted areas, if successful, will not fulfil the requirements of existing mills (1992:20 and table 7; *World Paper*, June 1994). Even so, new mills are underway.

P.T. Kertas Bekasi Teguh–Djamzu Papan Holdings

The largest private pulp and paper producer in Indonesia is P.T. Kertas Bekasi Teguh, a division of the Putera group, owned by Djamzu Papan and Associates. Field notes written during a 1990 visit to the company's operations provide a fair indication of the stage of development.

Djamzu is a Chinese entrepreneur whose childhood was spent near Pontianak, West Kalimantan, where he proposes to expand his plantation and mill capacities. A self-made man, he now presides over a substantial group of businesses that include seven pulp and paper mills, a garment plant, and a vegetable and tea plantation on Java. His first business investment, in the 1960s, was in a packaging plant, for which he imported materials; he then established a corrugated box plant and in 1974 the first pulpmill in Bekasi, West Java, some 30 kilometres east of Jakarta. By 1990 it was producing about 120,000 tons per year of kraft paper (kertas) and strong brown corrugated paper (samsun). The linked box factory takes about 30 per cent of the output, and itself produces about 100 tons a day, or 36,000 tons a year, primarily for the domestic market. Mills were subsequently established in central Java (Sermara) and Jakarta, and more are planned.

Fibre sources are numerous for the Bekasi and other mills. Long fibre comes from old corrugated cartons imported from Singapore, the United States, and New Zealand and small quantities of unbleached kraft pulp from the U.S., South Africa, New Zealand, and Chile. Some 7,000 tons per month of waste paper is purchased from abroad, and between 10,000 and 15,000 CM of wood is bought locally from tree farmers to whom the company provides start-up costs and seeds. There are standing contracts with ten suppliers, with the company providing technical assistance to the farmer. The trees include *Albizzia falcataria*, rubber tree, *Pinus merkusii*, kapok, maeopsis, and other fast-growing tropical hardwoods, which are mixed with waste paper and recycled pulp to produce paperboard. Some 100,000 hectares of *Pinus merkusii* plantations in central Java were seeded by the government in the early 1970s. To obtain wood from this plantation, the company must agree to thin the stands every three years. The wood is classified: large-diameter woods go for furniture; small-diameter wood is for firewood, but can be purchased for pulp; and Kertas Bekasi buys about 3,000 tons of pine firewood each month.

In 1970 the company started planning a long-term plantation project with a Japanese partner, but in 1989 the project was cancelled because of land-protection policies in central Java. However, other plans have gone ahead, and the extraordinarily fast- growing species *Albizzia falcataria* was planted in several West Java areas. The company purchased 12,500 hectares in Lampung, Sumatra, with a concession of thirty-five years' (renewable) duration. An application by the company for another area of 50,000 hectares in South Sumatra (Selatan) near the border with Lampung province was still pending when I visited the company plantations in 1990.

Kertas Bekasi pays nothing except the cost of administration for the resource itself. The government says that it provides this kind of concession in order to increase employment, preserve the land, and gain export dollars. On the 12,500 hectares, the company was planning to begin planting albizzia, acacia, and eucalyptus in the summer of 1990, as soon as local forestry officers granted permission. Employment of 200 people was intended, and initially seedlings would be grown in a nursery in the open grassland area, while roads and bridges were constructed. The logs would be rafted down the river to Jakarta, a journey of about 200 kilometres taking about three days. The Selatan application included trees about 30 centimetres in diameter, which would be turned into lumber or chipped for the new pulpmill. The company argues that in the harvesting region (adjacent to the river), the rainfall is high and there is no fear of a reduction in the water level. The precise location of the new pulpmill is not yet determined. There are native groups and transmigrants in the region.

A measure of benefits for mills is employment; by this index Kertas Bekasi is a significant force. The kraft mill at Bekasi employs about 850 workers. Of these, 12 are engineers, another 40 or so in managerial and clerical positions, and the remainder in production jobs. The box mill employs just under 1,000 workers, of whom 750 work regular shifts on production. About 200 per day are on piece rates. The office staff of 150 is larger than is usual in mills because the company maintains its own sales staff.

Indah Kiat Pulp and Paper

The Sinar Mas group, held by the Widjaja family, owns six pulp and paper companies in Indonesia. It has 51 per cent interest in the Indah Kiat Pulp and Paper (IKPP) company at Perawang in Riau province, Sumatra, which was the first green-jungle (natural wood-based) mill in Indonesia. Chung Hwa and Yuen Foong Yu, both of Taiwan, between them have 25 per cent. Chung Hwa provided the mill equipment; Yuen Foong Yu is a major Taiwanese papermaker which purchases output. Sinar Mas also has interests in banking, palm oil, rubber, petrochemicals, and construction. Of its paper companies, Indah Kiat is the largest, with mills at Serang and Tangerang, West Java, as well as the mills in Sumatra.

The first Sumatra mill was built in 1984. Its fibre source consisted of over one hundred species of tropical hardwoods on a 65,000 hectare concession in Riau province. With two additions, IKPP had an annual capacity of 790,000 tons of bleached hardwood pulp and 254,000 tons of writing and printing paper at Perawang. Noted in news reports of the third mill was the source of wood: "virgin forest on the company's 300,000 hectare concession." Also noted was the anticipation of continued substitution of tropical hardwoods with plantation hardwoods (mainly *Acacia mangium*), "which are gradually replacing virgin forest in the concession" (*PPI*, Oct. 1994:24). The company was rushing to cut the forest in its concession mindful that the Rio Earth Summit Conference in 1992 recommended the year 2000 as the deadline for logging in tropical rainforests. As reported by M. Bayliss in *Papermaker*, the company was paying around us$5 per ton in royalty and taxes on roundwood from its concession at Perawang, and total wood cost per ton of pulp was under us$100 (Oct. 1994:32). Despite such cheap fibre resources and huge concessions, IKPP was charged with illegal logging in 1994, and there are questions about wood supplies for this mill, as well as the others in the Sinar Mas group.

The IKPP mill complex at Perawang provides employment for some 6,000 employees. The company estimates that another 10,000 are dependent on the mills. Labour was originally imported to the town, so Indah Kiat provided dormitory accommodation and built company houses, a library, recreation facilities, and a medical clinic. In short, this is the classic company town, a replica of towns created throughout British Columbia in the 1950s (except that BC companies depended on governments to provide townsite facilities).

Raj Garuda Mas

The world's largest pulpmill has been constructed in central Sumatra by an Indonesian industrial conglomerate, Raj Garuda Mas (RGM), privately owned by Sukanto Tanoto. The company, with interests in steel, textiles, palm oil, shipping, and forest products, established its first pulp and rayon mill on the island in 1988. The new mill, to be called Riau Andalan, will produce 750,000 tons of bleached kraft market pulp by the end of 1995. The fibre source will be mixed tropical hardwoods, which, like the forests used by Indah Kiat, are scheduled for demolition in favour of acacia plantations. The company says it might build two more mills, and it expects to export about 65 per cent of its product to China, India, and Taiwan (*World Paper*, June 1994).

The Riau Andalan mill is scheduled to consume some 512 roundwood deliveries per day (one lorry per three minutes according to *PPI*, 1 Sept. 1944:22). Concessions amounting to 257,000 hectares of forest land will supply about 70 per cent of the wood. The remainder will be purchased from external concessions. Some of this land has already been logged once, and the company contends that it is not going to cut down virgin forest but merely clean up an earlier mess and then replant. The mill plans to use only plantation wood by the year 2002, but for the half-decade between start-up and plantation self-sufficiency, it will use all the waste wood on its own and neighbouring lands. RGM is establishing research nurseries on the model of Brazilian plantation pulp operations (see chapter 11). Also like Brazilian prototypes, RGM is building a townsite for mill workers to include schools, church and mosque, housing, a clinic, a hotel, and recreation facilities. The company will employ only Indonesians in junior positions, though upper management posts will initially be filled by outsiders.

The RGM mill incorporates the latest technology, including a pulp-cooking system known as SuperBatch, manufactured by Sunds Defibrator, Finland. This system, already installed at Enocell, Finland, is state-of-the-art, and it is matched by other technological features of this and other new mills in Indonesia.

PT Kiani Kertas

Indonesian tycoon Mohamad (Bob) Hasan is the force behind a $1.2 billion pulp mill designed by H.A. Simons of Canada. Hasan, a close friend of the Suharto family, is the major shareholder in firms manufacturing cables and glass as well as plywood and kraft paper. The new mill, intended to be on stream by 1997, will produce 475,000 tons a year of hardwood pulp, initially from tropical forest lands in East Kalimantan. According to its management, it will switch to plantation eucalyptus early in the next century (*Globe and Mail*, 1 Sept. 1994). This particular project, the fourth new state-of-the-art, gigantic mill to be constructed in Indonesia within a few years, will replicate many of the technological features of the Alberta-Pacific mill and will compete with that mill for the same Japanese and other Asian markets.

Tjiwi Kimla

There are numerous other paper companies still relying on wood-free fibre sources or using small amounts of locally purchased wood mixed with other fibres and imported pulp. One of these, Tjiwi Kimia at Mojokerto, East Java, is already reported to be the world's largest producer of stationery products (*PPI*, Oct. 1994:27). These companies and the Sinar Mas group are seeking fibre sources elsewhere and markets in China and neighbouring countries in Southeast Asia.

The Critics

Wahana Lingkungan Hidup Indonesia notes that by 1990,

the policy induced forestry sector development has resulted in the operation of 578 forest concessionaires which occupy an area of 59.9 million hectares of production and conversion forests, 296 sawmills and 119 plywood factories. Both the sawmills and plymills require some 42 million cubic meter of logs annually, while the 39.3 million hectares of production forest can only supply some 22 million cubic meter of logs at the sustainable production level. It is clear that the situation has put too an heavy burden on the [*sic*] Indonesia's forest despite its longterm threat for the whole timber industry" (1992:3)

This group argues that only 62 per cent, or 90.2 million hectares, of the country's resources remains undamaged, and of this some 12.6 million hectares are scheduled for transmigration or other purposes. Of the total, in their estimation, less than 40 million hectares, or 51 per cent, have the potential for sustainable timber production. To determine whether such a precarious situation could be justified, if not sustained, the group examined government claims that the industry provided employment income, government revenue, foreign-exchange earnings, and other wealth essential to development. Its studies indicated that forestry-sector contributions to domestic income between 1983 and 1989 accounted for only 1 per cent of Indonesia's total GDP at both current and constant 1983 market prices. In terms of growth rates, at constant 1983 prices, forestry production grew by 1.4 per cent during the same period, a rate considerably below that for agriculture as a whole. In terms of employment, there has been growth, though most of it occurred early in the 1980s. For forestry in the 1980–87 period, overall sector employment growth amounted to 8.4 per cent, compared to employment in all industries of 4.1 per cent. These figures accord with earlier reports in this chapter.

The most significant role of the forestry sector has been in its contribution to foreign exchange. Exported timber products in 1989 accounted for US$3.5 billion in foreign exchange, 15 per cent of total export value from Indonesia. Wahana Lingkungan Hidup Indonesia asked, Then who controls the revenues? Are they used for further investment? It discovered that the central government's forest revenues were modest, never accounting for more than 0.2 per cent of total government domestic revenue during the 1984–89 period. Calculating the actual revenue and comparing this to the potential rent if the collection system were effective, the group concluded that the revenue actually generated amounted to less than 10 per cent of a reasonable rent appropriately collected (calculations based on FAO forestry sector reports, IMF, and IFS for data on Japan log prices). In their judgment, "The amount of rent uncaptured by government has become an implicit subsidy for wood-processing industry. It is this subsidy that has been boosting up the industry, while discouraging efficiency increase in production due to the very cheap input" (1992:21).

Table 10.3
Indonesia: economic rent in timber production; potential and collected revenue,
1989–90 (US$ million)

	1989	1990
Economic rent per meter log	99.24	94.66
Rent captured by government licence fee,	5.00	6.00
property tax, royalty reforestation fee	7.00	10.00
Total rent captured per cubic meter log	12.00	16.00
Log production (1000 m³)	31,215	26,000
Total potential rent	3,098[1]	2,461
Total collected revenue	253[2]	416
Percentage of rent captured	8%	17%

Source: Calculation made by Wahana Lingkungan Hidup Indonesia, 1992, table 2.5.
[1] Total amount if US$ 16/m³ rent could be realized.
[2] Converted from Rp 447 billion revenue.

In short, the industry provides employment and brings in foreign-exchange earnings. However, it is heavily subsidized, contributes a negligible amount to the domestic budget or government revenue, and has capacity in excess of existing timber reserves. It is not sustainable at current rates of timber extraction.

SUMMARY

Deforestation in Indonesia has been dramatic and rapid, due to a variety of causes including transmigration projects. In the late twentieth century the transmigration program is linked to the logging industry. Though logging on an industrial scale only began in the early 1960s, by the mid-1970s large tracts of land were deforested. Logging was extensive, scarcely regulated, and extremely lucrative, and the logs were destined primarily for export markets. After 1975, progressive bans on export of logs were introduced until, by 1985, no exports were legal. However, logging did not decline when plywood mills took higher quantities of the timber. Some estimates suggest that Indonesia will be substantially deforested by the early twenty-first century. In the 1980s the pulp industry expanded its use of wood fibres and began to develop pulpwood plantations on land formerly forested by native species. In fact, the industry was deforesting its concessions as fast as possible in view of the recommendation of the Rio Conference to ban all logging in tropical rainforests by the year 2000. The recommendation is thus used as a rationale for deforestation.

Japan's success has been critical to the development of neighbouring Asian countries. It has been consumer and joint financier, but above all, the model. Initially reliant on Japanese consumption of its wood, Indonesia has now developed independent momentum in world markets; in 1980 it became the world's top producer of plywood and by 1990 was outranking others in production of stationery. As well, where it was formerly a net importer of pulp, it is now more self-sufficient in its domestic market and supplies some pulp to other Asian countries, including Japan. Its paper industry is well ahead of that in Thailand, and more of its production derives from wood fibres than other plants, though imported waste paper and wood-free fibres are still the major components of pulp production. Plantation fibres are expected to become major sources of pulp and paper in the future.

Indonesia is still a very poor country; its unemployed comprise half the population. Forestry is a major source of foreign currency, and it provides employment; further, though foreign-owned companies, are everywhere engaged, Malay and Indonesian-born Chinese entrepreneurs are major participants. The victims of progress, if that is what this industry represents, are, as usual, tribal peoples and land-poor farmers or hapless transmigrants; they also include rice farmers and others whose claim on land has obstructed industrial growth in the past. The Bogor Agricultural University projects are designed to transform shifting cultivators and hunters into a labour force for industry, in order at the least to prevent the continuing decimation of forest by short-rotation cultivation. The beneficiaries of these developments are civilian and military élites and business investors in Indonesia, together with merchants and investors from neighbouring Southeast Asian countries and Japan, the consulting and engineering firms that supply the new mills, and Japanese and Taiwanese papermaking companies.

Map 10   Brazil, central to southern Atlantic coast: forest types

11 The Plantation Economy in Brazil: A Case Study

The public-relations officer at Aracruz mused over lunch about the character of Brazil. In university, he said, he had learned Marxist theories. These coincided with what he had earlier been taught by the Catholic Church. "Both taught us that poverty is virtue; wealth, sin." Now as an employee of a major corporation, he had come to the conclusion that the company was doing more to relieve poverty than either the Marxists or the church, and he no longer thought poverty was particularly virtuous.

A history of plantation-crop cycles characterizes Brazil and much of Latin America: brazilwood from 1500 to 1550, then sugar from 1550 to 1700, and following the gold rush of 1700–75, rubber and coffee from about 1850 to 1930. Cattle ranching has been dominant since the 1930s in much of the territory now being planted in eucalypts. In earlier phases, Brazil was under colonial rule and then it was a client state, captured by and serving the interests of large, external resource companies. The current plantation industry differs from its predecessors in an important respect: it is not dominated by external companies. Foreign ownership of shares in the pulp and paper companies is a significant source of funds and more important as a guarantee of external markets, but it does not reach levels that allow external investors to dictate the policies of the forestry sector.

Critics of pulpwood plantation crops are numerous. The companies are accused not of relieving, but of creating, poverty through displacement of landless peasants and Indians and through contractor employment practices. The debate is intense in Brazil, the more so because of the long history of forest destruction caused by various plantation crops. One might well question the claim that tree crops are any different. Critics see the eucalyptus phase as part of the continuum, with the corporations cutting down native forests to plant the cash crop and ignoring social impacts on the poor. Greenpeace spokespersons argue that pulpmill effluents have caused deaths and illnesses, and here, as elsewhere, there is concern about monocultures, soil impactment, microclimate changes, and rising temperatures.

Not fair, say the defenders. With the exception of pine forests in the mountainous region of Paraná, the new industrial forestry in southern coastal and near-coastal states – Rio Grande do Sul, Santa Caterina, São Paulo, Minas Gerais, Rio, and Espírito Santo – are located where the subtropical forest was substantially depleted long before companies began planting eucalyptus. Ranching, now being replaced by plantations, causes greater ecological damage. Decimation of the coastal araucaria forests was due to industrial forestry of an earlier era, not to modern plantations, and plantations are providing a new forest cover to the denuded lands. As to social concerns, companies point with pride to their employment statistics, to established townsites with medical facilities and schools for workers, to progressive social welfare policies. Land, they say, is purchased, not stolen, from private farmers; other farmers are paid for planting pulp trees under company supervision. Displacement of landless peasants is a larger process which began long before the companies became established, or land was taken over in the 1960s by the state from peasants without title – not their doing, though the state granted them much of that land. Pulpmill operations also distance themselves from the lumber trade, which damaged forests in the Amazon, just as it depleted the araucaria forests of the south (described in chapter 7).

The native species are not systematically utilized in the production of pulp because they are less suitable as raw material, but they may be used in the start-up phase or for lumber, if plantations are planned for land already forested. The issue concerning the pulpmill economy is not that it uses the native trees, but that it displaces them. Jari pulpmills are located on the Amazon; to date, the most extensive invasions have been by plantation forestry, using an imported tropical species. The planned Celmar pulp project for the southern Amazonian state of Maranhão, involving the companies Ripasa, Simão, and Campanhia Vale do Rio Doce, and the Japanese trading company Nissho-Iwai, if it goes ahead, would be even more contentious. The Celmar group plans the development of 80,000 hectares of eucalyptus plantations and a 420,000-tons-per-year bleached eucalyptus pulpmill

Table 11.1
Brazil: total area reforested by state and species, 1991 (1,000 hectares)

State	Eucalyptus	Pine	Araucaria	Other	Total
Amapa	4,621	84,922	0	0	89,543
Bahia	178,593	49,438	0	0	228,031
Espírito Santo	106,373	157	0	0	106,530
Maranhão	5,217	0	0	0	5,217
Mato Grosso do Sul	21,584	0	0	0	21,584
Minas Gerais	110,262	3,740	738	1,833	116,574
Pará	17,636	36,591	0	9,174	63,401
Paraná	45,153	203,223	14,581	3,097	266,005
Rio de Janeiro	4,472	0	0	0	4,472
Rio Grande do Sul	54,117	14,592	1,542	3,876	74,129
Santa Catarina	16,041	119,828	3,387	281	139,539
São Paulo	250,760	45,555	530	2,320	229,166
Total	814,831	558,049	20,778	20,583	1,141,244

Source: Associação Nacional dos Fabricantes de Papel e Celulose, 1991: table 6.04.

(PPI, Jan. 1990:23, May 1992:17). New projects in Bahia are also claimed as potential destroyers of the remains of a subtropical forest.

But these projects, planned for Amazonia, are not yet typical of the plantation economy. Most of the mills are located in the south and central coast states. The early boosters (e.g., Beattie and Ferreira, 1978; Berger, 1979; Neves, 1979; Nogueira, 1980) envisioned plantations displacing marginal and unproductive land. They saw large pulpmill complexes as the means of providing rural employment and industrial development through multiplier and trickle-down effects. In fact, some positive effects have been achieved in little more than a decade. The export trade and this use of land resources have considerable economic benefits for Brazil in employment, townsites, technology transfers, education and training, and improved trade balances. The country's plantation companies are well organized, mindful of their public image, excited about what they have done and can do, and eager to inform the outside world about their remarkable success. The infrastructure of transportation and communication services is well developed. The country continues to reel from inflation, international debt-service payments, painful poverty in urban and northeastern rural areas, millions of abandoned children, and political corruption, but the plantation forest companies are part of the other Brazil, an industrial powerhouse. They represent a new beginning, they say, an opening up of new economic opportunities.

Table 11.2
Brazil: pulp production in selected years, 1950–91 (tons)

Year	Chemical & semi-chemical			Mechanical[1]	Total
	Long fibres	Short fibres	Total		
1950	38,367	1,592	33,959	55,400	95,359
1955	50,182	22,986	73,168	72,900	146,068
1960	80,329	119,908	200,237	86,200	286,437
1965	116,211	203,862	370,073	201,500	571,573
1970	278,156	385,907	664,063	113,206	777,269
1975	358,768	830,840	1,189,608	162,578	1,352,186
1980	755,572	2,117,124	2,872,696	223,569	3,096,265
1985	1,058,310	2,345,154	3,403,464	312,513	3,715,977
1990	1,174,456	2,740,232	3,914,688	436,455	4,351,143
1991	1,212,464	3,134,056	4,346,520	431,596	4,778,116

Source: Associação Nacional dos Fabricantes de Papel e Celulose, 1991: table 2.01.
[1] Includes mechanical, thermomechanical, and chemi-thermomechanical processes.

GOVERNMENT TAX INCENTIVES FOR
PLANTATIONS AND GROWTH OF THE
PULPMILL INDUSTRY

Wood-based pulps have been produced in Brazil since about 1870. By the late nineteenth century and early twentieth century, the initial groupings that became Klabin, Simão, Santista at Cubatão, and Melhoramentos de São Paulo were established. A burst of new activity occurred in the 1940s and 1950s, including the expansion of Klabin, the construction of an Olinkraft mill, the growth of Suzano, and the establishment of Champion. Suzano and Champion (a subsidiary of the U.S. firm Champion International) were beginning to produce eucalyptus pulp in the late 1950s. A sufficient industry had emerged by the mid-1960s to justify the creation of a pulp and paper association. While these precursors to the modern industry were producing wood products much earlier, the rapid growth of the pulp and paper industry began in the late 1960s with tax incentives, technological changes, and the introduction of biotechnology.

The Codigo Florestal (law no. 5106), declared in the mid-1960s, provided for tax-free income from plantation-based forest products and total deduction of costs for forestation or reforestation. This law, combined with subsequent enabling legislation and financial support by state agencies, provided the base for growth of the industry. After the 1973 oil crisis, the incentives were expanded to induce investors to plant eucalypts suitable for charcoal as well as pulp. Particularly important to the development of industry, the National Bank for Economic Development (BNDE) was reorganized in 1971 as a public-sector enterprise designed to support public and private firms in such areas as forestry. Among its subsidiaries is the Special Agency for Industrial Financing (Finame), which provides the finances for Brazilian-manufactured machinery and equipment purchases.

Most earlier reforestation was undertaken for charcoal production, mainly in southern Brazil, especially in the state of Minas Gerais; as well, pulp plantations had been established in the state of São Paulo. The land was poor and unsuitable for agriculture, and it was cheap and plentiful. While there was considerable debate within Brazil about the utility of continued reforestation, and investors had to cope with inflationary trends and difficulties of access into world markets, new plantations and mills quickly came on stream from the mid-1970s through the 1980s. In the mid-1970s it was estimated that eucalypts would require twenty-one years to reach commercial size and pine twenty-five years. The fiscal incentives were augmented by increasing attention to research and training of foresters in the 1980s.

Under the generals' regime of the early 1970s, Brazilian pulp production was shaped largely by external paper companies combined with internal investors. Borregaard of Norway, for example, held 31 per cent of the shares in the Riocell unbleached hardwood kraft pulpmill established in 1972; Klabin and the BNDE owned most of the remaining shares. Foreign investors in other enterprises included Westvaco, Olinkraft, Wiggins Teape, and Scott Paper. The National Pulp and Paper Program was introduced, with a targeted production of 4.2 million tons of pulp by 1980. Foreign companies were invited to participate, to obtain kraft pulp for their manufacturing firms elsewhere, and to manufacture newsprint for an expanding domestic, as well as an export, market.

What had been a relatively small activity related to mining suddenly became a potentially huge activity when tied to forestry. Size requirements and other provisions for investment favoured corporate investors; indeed, they precluded small business participation. Sebastiao Kengen (1985, 1991a) argues that in Brazil, tax savings easily offset the costs of planting the trees and maintaining them for their first three years. Companies in Brazil would pay 30-per-cent tax on taxable profit, but could divert 25 per cent of this tax for reforestation if they chose to do so. The minimum area for a reforestation project was 1,000 hectares as decreed by a 1976 law (no. 1503, 12 December, article 13). Says Kengen: "The response was spectacular. In a decade, the rate of planting leapt to over 400,000 hectares a year as investors, many of them without previous experience or links with industry, hurried to obtain effectively free plantations on whatever land they owned or could buy cheaply" (1991a:10).

Investors talk about risks and complain that companies bear the costs of developing plantations, constructing towns, and building roads. Even if such rhetoric is well founded, companies in Brazil are posting substantial profit margins as they benefit from low land and labour costs. The profits continue to increase as research and biotechnology pay off. The early 1980s had slow growth because of Brazil's foreign debt, decapitalization, and political instability. But by mid-decade, despite an actual 600-per-cent yearly inflation rate, pulp capacity and exports were both growing. Brazil as a whole now has a capacity of some 2.2 million tons of new pulp annually. Its manufacturers anticipate continued growth because, they say, "eucalyptus is now a desired fiber worldwide." As the director of the Brazilian forestry organization FICEPA stated:

The forestry resources [have] a rapid growing rate when compared with the Northern Hemisphere, offer low wood cost, high productivity per unit area, great land spaces ready for reforestation, and biomass resources as an energy source. The pulp demand in the international markets along with increasing prices and low labor cost are promoting a leading technology for eucalyptus pulp and paper manufacture as well as more efficient mills ... There is the possibility to transform a part of the foreign debt loans to risk capital, and the participation of interested groups in existing and new projects. (Sagarra, 1988:25; see also Zobel, 1988)

Pulp production capacities have increased at a rapid pace. Most mills began adding capacity almost as soon as they were built; Aracruz, the outstanding, but not singular example, doubled its capacity in a decade (Hall, 1988:67). The rapid increase in chemical pulp and paper production represented 3.6 per cent of world production by 1985, ranking Brazil as the seventh largest pulp producer in the world at that time (Paoliello, 1988:70). Because of local demand, the Brazilian share of world pulp-export markets decreased between 1980 and 1985, but since then new mills have added vastly more capacity. While most producers concentrate on pulp production and ship to paper manufacturers in Japan, the United States, and Europe, several companies have also entered the paper manufacturing business. Most of the produce goes into the domestic market, but export sales have also increased and the range of paper products has grown substantially since the early 1960s. The Klabin group of companies produces over 16 per cent of the total, most of this at its Paraná division. The Suzano group accounts for just under 10 per cent of the total, with

Table 11.3
Brazil: geographical distribution of pulp production by states, 1991 (tons)

State	Total	Long fibres	Short fibres
São Paulo	1,322,529	50,892	1,271,637
Espírito Santo	816,458	–	816,458
Santa Catarina	571,204	571,204	–
Paraná	528,323	370,967	157,356
Minas Gerais	392,954	–	392,954
Rio Grande do Sul	306,291	20,874	285,417
Pará	270,004	100,216	169,788
Pernambuco	57,434	23,402	34,032
Bahia	41,002	41,002	–
Maranhão	23,194	17,448	5,746
Paraiba	11,933	11,933	–
Rio de Janerio	4,526	4,526	–
Ceara	668	–	668
Total	4,346,520	1,212,464	3,134,056

Source: Associação Nacional dos Fabricantes de Papel e Celulose, 1991: tables 2.03 and 2.04.

most of its production concentrated in Suzano de Papel e Celulose. Champion produces just under 7 per cent; the Ripasa group, about 6 per cent. Smaller amounts are produced by the Simão and Trombini groups and other independent mills. Wrapping and cardboard products are also exported, and by 1991 Brazil had become self-sufficient in production of these and sanitary papers.

The country still imports about half its newsprint and exports very little, but its domestic consumption has steadily grown, as has its consumption of other paper products. The near doubling of newsprint production in 1985 may be attributed to the construction and full development of capacities in new pulpmills using thermomechanical and chemi-thermomechanical pulping technologies. As the pine plantations come on stream and production capacities increase, Brazil may be expected to become more self-sufficient in newsprint production, even with climbing consumption levels.

SOCIAL AND ENVIRONMENTAL IMPACTS:
A STUDY IN MINAS GERAIS

In the southeastern state of Minas Gerais, charcoal forest reserves ranging from 145,000 to 150,000 hectares were established by the 1950s. These were owned by the pig-iron and steel firms, of which the state-owned CVRD (see Cenibra below) was the largest. Expansion of these plantations in the 1960s included some 60,000 hectares of eucalypts. The incentives scheme introduced in 1967 induced further expansion, and by 1981 the approved plantation projects totalled 1.4 million hectares, of which about 1.2 million were finally planted. Sebastiao Kengen reports that a 1980 survey showed many of the investors had no clear idea of the end use of the wood they planted; some 43 per cent had no identified market use (1991a:12).

One-third of the plantations in Minas Gerais were in the Jequitinhonha Valley, where Kengen examined the outcomes relative to the arguments advanced for the tax incentives. In the decade 1971–80, 354,000 hectares were planted, and the valley, which had been a stagnant rural region, was suddenly populated with newcomers. Both plantation forestry and cattle ranching expanded. The majority of the peasants who had resided in the valley for many generations did not have legal documents to prove ownership of their land. The state government agency Ruralminas chose to classify their land as owner-less, thus owned by the government, and subsequently sold it to plantation investors. Companies also purchased land from farmers. Much of this land was of low agricultural productivity, but it did provide some sustenance and medicinal plants. Kengen reports that "the general level of services in the Valley was extremely low." The plantations provided some infrastructure services for their workforce, but health problems "typical of more urban and industrial centres increased" (1991a:13). His study continues, since it is too early to know whether other industries will enter the Jequitinhonha Valley or whether it will simply produce ever more plantation crops at the expense of a displaced peasantry.

Companies do not discuss the displaced peasantry; they deem it to be a responsibility of government. As far as they are concerned, the land was obtained fairly through government offers and reasonable incentives. In what follows, I have provided the company perspectives and information on major enterprises; all of this should be read within the context of the history of military and oligarchic rule in Brazil (which benefited large companies), the tax incentive laws, and the eviction of peasants without land title. The plantation economy provides undeniable benefits, but they do not come free, and they still do not extend to the majority of the population.

THE JARI PROJECT

Before Jari was in production, B.J. Meggers (1971) argued that every large-scale effort to develop the Amazon had ended in "ignominious failure," and he called the "myth of boundless productivity ... one of the most remarkable paradoxes of our time." The history of failures did not dissuade Daniel K. Ludwig, founder of the Jari project near the mouth of the Amazon. He is said to have anticipated a world shortage of wood fibre for pulp and believed pulpwood could be produced on a continuing basis in the Amazon. He began the project in 1967 with the $3 million (U.S.) purchase from a Brazilian consortium of some 1.5 million hectares near the mouth of the Jari River in the state of Pará. Trials had indicated that *Gmelina arborea* would take to the soil and climatic conditions there. The selected site had a huge land area, a moist or wet tropical climate, and a deep-water port. His American company, National Bulk Carriers, established Jari Florestal e Agropecuaria in 1970. Ludwig held loan guarantees and promises of tax-free imports of machinery from the Brazilian government. He apparently believed that the government was stable and would eagerly facilitate foreign investment and that the Amazon region had unlimited potential for sustained yield of commercial crops. In 1978, Ludwig arranged for a complete pulpmill, with capacity for producing 500 tons per day, to be towed from Japan to the mill site at the mouth of the Amazon River. This 20,000-kilometre journey was undertaken in order to save time constructing a mill in Brazil (Jari's history is documented in numerous sources, including Guppy, 1984:953; Russell, 1987; Smith, 1990).

Gmelina arborea, indigenous to Thailand and Sri Lanka, had been successful in some African and Asian plantations. Trials in Costa Rica and Panama were encouraging. Ludwig bulldozed and burnt the tropical forest and then commenced large-scale planting of gmelina and Caribbean pine in 1969 at Monte Dourado. The trees did not grow as expected, the scheme went bankrupt, and the state-controlled Banco do Brasil was obliged to assume its $180 million of debts. The fragile soil had been severely damaged by heavy machinery. Massive bulldozing and clearing had proceeded despite evidence that the trees grew more rapidly on sites where the slash and topsoil was left in place. Experimentation with site-preparation methods did not improve the outcome: production of gmelina at Jari was 40 per cent below projected rates. Other species were tested, but Jari under Ludwig's management never quite succeeded as a plantation.

It may be argued, however, that Jari was successful in another sense. It provided a model for regional development, as distinct from single-industry development. Whether or not one approves of the ecological costs, this form of development had more benefits than others for imported labour. Ludwig built a town – complete with schools, hospitals, supermarkets, and dormitories – a railroad system, roads, outlying housing for workers in the remote areas, an airport, and the pulpmill, imported complete from Japan. Ancillary industries were also established, not only sawmills, but also rice-farming operations, cattle and water buffalo farms, and a processing facility for kaolin, which could be mined on the Jari property. Companies established later and farther south copied the Jari model, building into their projects some of the same social infrastructure.

The Ludwig empire sold its interests in 1982 to a consortium of twenty-seven Brazilian firms for $280 million (Fearnside and Rankin, 1985), far short of the $1 billion invested in it, and Ludwig's associates blamed government regulations and political problems in Brazil. The government, in turn, blamed poor management and profligate spending by the company. Not frequently mentioned at the time were low soil fertility and disease, though several studies produced strong evidence that these were the fundamental causes for the poor performance of the plantation (Russell, 1987). Smith (1990) claims that since Ludwig's withdrawal, some hillsides have spontaneously reforested. While that claim may be contested, new owners have had some success with reforestation.

Under new ownership and with the benefit of experience, Jari eventually became modestly successful. The June 1985 issue of the *International Society of Tropical Foresters News* claimed that Jari was among the top one hundred agribusiness firms and reported that it had enjoyed a $2.1 million (U.S.) operating profit in 1983. As Charles Russell points out: "This amounts to a return of 0.2% on the initial investment of $1 billion. On the $280 million Brazilian investment, it is a return of 0.8%" (1987:89). That dour note notwithstanding, Jari was reported to have recorded its best-ever sales at $175 million in 1990, with a record production of pulp amounting to 252,035 tons. Production in 1991 was higher still, and Cia Florestal Monte Dourado, as the new company is known, accounted for 6.21 per cent of the total production in Brazil. In 1992 it announced an intention to double capacity by the end of the decade (*Paper*, Feb. 1992).

ARACRUZ

The first bridge over the Rio Doce near the coast in Espírito Santo was constructed in the 1940s; not until then was exploitation undertaken north of the river. The northern highway between Vitória and Salvador was completed in 1973. Coffee plantations had some success before the collapse in prices; cattle ranching was less successful; small farms were mainly unsuccessful because of poor-quality soil. The population, descendents of slaves who had intermarried with predominantly Italian immigrant settlers, was poor. This is why Aracruz spokespersons, such as the one whose words opened this chapter,

Table 11.4
Brazil: pulp production for major companies, 1991

Company	Pulp (tonnage)	% Total
Aracruz Celulose SA	816,458	18.78
Klabin Group	783,473	18.03
KFPC-Paraná Division	399,343	9.19
Riocell SA	272,658	6.27
Papel e Celulose Catarinense SA	111,472	2.56
Cia Suzano de Papel e Celulose	427,865	9.84
Celulose Nipo-Brasileira SA (Cenibra)	375,924	8.65
Champion Papel e Celulose Ltda	291,211	6.70
Cia Florestal Monte Dourado	27,004	6.21
Ripasa SA Celulose e Papel	261,206	6.01
Igaras Papeis e Embalargens Ltda	230,382	5.30
Inds de Papel Simao SA	182,023	4.19
Rigesa Celulose, Papel e Embs Ltda	137,264	3.16
Total	3,775,810	86.87
Others	570,710	13.13
Total	4,346,520	100.00

Source: Associação Nacional dos Fabricantes de Papel e Celulose, 1991: table 2.02.

speak of the company's role as a wealth creator. To them, the company has provided employment and income where there was not even hope before.

Aracruz* (named after the district where the plant is located) originally purchased charcoal-producing plantation land from a Vitória steel company. By the 1990s, Aracruz had become the largest single forestry project in Brazil, and it retained its position as the world's leading producer and exporter of bleached eucalyptus pulp when it more than doubled its mill capacities in 1991. The $1.2 billion expansion project provides the capacity to produce over a million tons per year of eucalyptus pulp – about 20 per cent of the world's total supply – from extensive plantations in the states of Espírito Santo and Bahia. Eighty per cent of its production is exported, representing an annual income of approximately US$300 million in 1990 (Aracruz Celulose, 1990).

The original investment amounted to US$540 million, of which some $200 million came from the company's own capital, with much of the remainder supplied by the national development bank, BNDE.

* Unless otherwise noted, information on Aracruz is from my fieldwork in July 1992, which involved interviews and site visits, and from company reports and other in-house documentation.

The feasibility study was prepared by Sandwell of Vancouver, Canada, in 1969, and the principal consultant for the mill's construction was Jaakko Pöyry of Finland. Billerüds AB of Sweden provided technical experience in eucalyptus pulping (from its Celbi mill in Portugal) and marketing aid (*PPI*, Dec. 1977:63). The first mill was completed in 1978. By 1990, total mill production was 501,000 tons. The following year a 59 per cent increase in production was recorded, for a total volume of 816,000 tons, with sales of 799,000 tons. Prices dropped that year because the industry was afflicted with excessive capacity; as well, a sales tax on exported pulp and monetary variations in connection with financing of the expansion project resulted in the company posting a net loss of approximately US$95 million (*Wall Street Journal*, 21 May 1992). Even so, it was poised to capture new export markets, and its losses did not leave it devastated.

The consolidated company has five separate divisions: Aracruz Celulose, Aracruz Florestal, Aracruz Trading, Aracruz International, and Portocel. Ownership in 1990 was distributed as follows: Souza Cruz, Lorentzen, and Safra, each 27.97 per cent; BNDE, 12.47 per cent; other, 3.62 per cent. This represents a declining share by the BNDE since the company's establishment in 1970 and an increasing share by private investors, especially the chairman of the board, Erling Sven Lorentzen. Lorentzen, still a Norwegian citizen though his family has been in Brazil since the 1890s, has headed the otherwise Brazilian investment group since the beginning of the project.

Plantations and Environmental Concerns

Aracruz Florestal began planting eucalyptus in the states of Espírito Santo and Bahia in 1967. The current area for mills and plantations is 203,000 hectares, of which 134,000 hectares are reserved for industrial plantations (there is a slight discrepancy in the figures cited, as new land continues to be acquired from other private owners). By 1990, 122,000 hectares were planted or planned for eucalypts, with a planned rotation age of seven to eight years.

Company researchers E. Campinhos and Y.K. Ikemori note that "the plantation areas were originally covered by tropical rain forest; the valuable timber was harvested and the rest converted to charcoal" (1989:169). These are provocative words in a heated debate between the company and environmentalists about whether Aracruz itself was responsible for the harvesting of the rainforest. Greenpeace-Brasil-"Agapan" argued in a brochure distributed at the Rio Earth Summit Conference in 1992, "Although they claim otherwise, Aracruz have cleared tens of thousands of hectares of rainforest in the State of

Espírito Santo in order to establish plantations" (1992:9). In another publication for the conference, Greenpeace argues that "analysis of an environmental impact report presented by Aracruz to expand its pulp production shows that at least 30 per cent of the municipality of Aracruz (around 20,000 hectares) had naturally generated secondary forest which was replaced by eucalyptus plantations" (Greenpeace International, 1992:27).

Company brochures emphasize that the plantations were situated on previously degraded land. The company is a member of the (UN-sponsored) Business Council on Sustainable Development and was cited in its publication *Changing Course* as a model forestry company (1992: case 17.4). Aracruz took a leading role in coordinating the business participation at the Rio Conference. Its citation formed the basis of correspondence between the general manager of environmental and public affairs for Aracruz and Simon Counsell, a tropical rainforest campaigner with Friends of the Earth International, in which the Aracruz representative states that most of the 4.5 million hectares in Espírito Santo were forested at an earlier time; today, the forest covers only 8.3 per cent, or 378,000 hectares. It continues to be destroyed at a rate of 40,000 hectares per year. The land-use distribution includes 3.4 per cent for all eucalyptus plantations (not all owned by Aracruz) and 8.8 per cent in Atlantic forest or other reserves. The remainder is in crops and pastures, is unused, or is otherwise used. The deforestation prior to 1967 was due to many and varied causes: coffee plantations, cattle ranching, low-intensity agriculture, fuel, and logging for furniture and civil works. Further, the Aracruz spokesman notes that the IBDF (Brazilian Institute for Forestry Development, now called IBAMA) undertook land-use planning and allocated previously deforested land for plantations under a management plan. He claims that Aracruz plantations in Espírito Santo amount to 1.8 per cent of the state's area, are not continuous, and are intermixed with conservation areas. Indeed, conservation areas on Aracruz land constitute 23.7 per cent, compared to the 8.8 per cent in the rest of the state. As well, areas in the state of Bahia, purchased more recently, were photographed prior to planting. Aerial and terrestrial photographs showing that the vegetation was poor are available for viewing. For the two states, the ratio between the conservation area and the planted eucalyptus area is 2.4:1 (132,000:56,000 ha).

Company claims were, however, questioned in an environmental and social impact study conducted by the Instituto de Technologia of the Universidade Federal do Espírito Santo in 1987. The institute noted that the information available did not permit it to conduct as thorough an investigation as should be undertaken, but on the basis of visits to the plantations, tests on nearby waterways and soils, and study of the scientific literature, it found many cases where the claims of the company were not supported (Instituto, 1987:4–6). On the basis of aerial photos taken in 1970–71, the institute argued that at least 30 per cent of the municipality of Aracruz was covered by native flora at the beginning of the 1970s (1987:6).

There is concern here, as elsewhere, about biodiversity. The company claims that its maintenance and enrichment of natural preservation areas, interspersed and sometimes integrated into eucalyptus plantations, meets these concerns. It has planted about a hundred indigenous species to enrich the food supply for wildlife, as well as to reclaim degraded areas. The nursery has produced some 1.5 million seedlings of indigenous species each year. One hectare of indigenous vegetation is now planted for every 2.4 hectares of eucalyptus. In the 1987 impact study, the institute observed that the substitution of heterogeneous native plants by homogeneous eucalyptus produced a great environmental impact, reduced ecological diversity, and compromised the survival of many populations of different species of flora and fauna (1987:6).

Aracruz argues that much of the land has low-fertility soils. It was seeded initially by Brazilian species, but exotic species from South Africa, Zimbabwe, Australia, and the island of Timor were subsequently obtained to produce better results (Campinhos and Ikemori, 1989:169). Instead of burning leaves and branches after harvesting, the company uses these as nutrient and organic matter to reduce soil erosion. Soil fertility and biomass production are carefully monitored, and experimental fertilization has been introduced. Tractors are used on flat land, but they have "specially sized tires to minimize soil compaction." In addition, the company argues that soil erosion has been reduced through spacing and fast growth of eucalyptus, resulting in soil covering and protection before the first year of growth.

Aracruz is widely acknowledged as the innovator in the cloning process that provides higher-quality yields in shorter time spans than experimental seedling technologies. The objective of the cloning technique, introduced in 1979, is to increase the density of wood per tree and per hectare, while reducing susceptibility to pests, diseases, and fungi. The amount of wood required to produce a ton of pulp has been steadily reduced, providing for gains of up to 22 per cent in operating costs for the mill. In recognition of this pioneering work in forestry, Aracruz was awarded Sweden's Marcus Wallenberg Foundation award in 1984 and has since won other awards for research and plantation innovations. Awards notwithstanding, cloning does raise environmental questions. Monocultures could be extremely vulnerable to pests and diseases. The company counters that it uses a large number of different genetic bases for cloning and tests clones extensively for adaptation to diverse soil types to offset the danger. To date there have been no outbreaks of pathogens in company plantations. In addition, outbreaks of minor diseases have been naturally controlled by biological interference. The Instituto de Technologia study argues that the monocultures have provoked a disequilibrium in the basic structure of a vast region, and it recommended that expansion of forestry should be restricted (1987:7–8).

New varieties of wood are still being tested, and in 1991 a new methodology was developed to enhance the rooting of cuttings from clones. Less reliance is now placed on chemical fertilizers, and experiments are in progress with intercropping eucalyptus and legumes as an alternative means of weed control. Intercropping also improves ecological biodiversity. Forest planning was advanced by the creation in 1990 of a forestry-resource database that contains information on operational areas, research, forest inventory, topography, and economic variables. It allows the company to plan three decades into the future and streamlines decision-making about planting, harvesting, equipment purchases, and land acquisitions.

In 1991 the planting program covered 11,700 hectares. Some 38.8 million seedlings were produced. Of these, nearly 20 million were reserved for the seedling inventory and company forest plantations, the remainder being distributed to "partners." Partners are local farmers who are given the seedlings and a guaranteed sale of timber at market prices when they mature. Training is provided by company and government agencies. This form of "contracting out" is now widespread in Brazilian plantations.

Employment and Community Development

Aracruz has a company town, Coquieral, with a total population of 38,000; as well, it draws on a labour force from five other communities and employs in total about 7,500 workers. Education and health systems are subsidized by the company, and it has been a significant force in the opening of twelve schools, eleven clinics, and four sports complexes throughout the region. A child-care centre has been established at the mill site; as well, there are regular recreational, cooking, and crafts lessons in company facilities. Nutrition classes are also provided for householders. Some programs are designed specifically for retarded people living within the sphere of the company.

As with the ecological issues, the social issues are fiercely debated. Aracruz spokespersons say that their top priority is to reduce the poverty of the population in Espírito Santo and Bahia. According to their interpretation of history, the native people of the region, the Tupinguinus, were already dispersed and their lands long since taken over for coffee plantations and cattle ranching or agriculture. About forty Guarani Indians from Paraná or Paraquay came into the region of Guarapari south of Vitória in the spring of 1979, and a year later they moved into the Aracruz region, where they were absorbed by the Tupinguinus. They were helped by a group from the Catholic Mission for Indigenous Peoples. In the 1980s, Aracruz gave an area of 2,800 hectares to the Fundáçao National do Indio (FUNAI, the official government authority for the area). FUNAI settled the Indians on this land, with agreement by Aracruz that it would not discuss whether the settlers were Tupinguinus or not. The situation is complicated by differences in cultural heritage between the two groups: the Tupinguinus no longer have a distinctive language or customs, and there has been much intermarriage with non-Indians. The Guarani, however, have retained their distinctive culture and language. They have moved onto the reserve land.

Opponents claim that native people have been dislodged and make no distinction between Guarani and Tupinguinus. They also claim that the company's contracting operations are ways of avoiding long-term commitments to the labour force. Contractors, rather than Aracruz, pay benefits, and unions continue to be weak because of the diversity of employers and conditions created by the major corporation. The institute's impact study noted that the extreme dependence of the population and municipality on a single employer was a serious problem, and it recommended diversification. It contested the claim, made by Aracruz, that use of the land for plantations showed a more balanced cost-benefit relationship when compared with alternative uses, since no studies of alternative uses of the land had been undertaken. Further, the institute study contested claims by Aracruz about the financial returns to the municipality. Indeed, on this score, the study raised questions concerning several company claims about public benefits (Instituto, 1987:16–17).

CENIBRA

Cenibra* was established in 1973 and began operations in 1977 in the municipality of Belo Oriente, which is near the town of Ipatinga

---

* Unless otherwise stated, information on Cenibra is from my fieldwork in July 1992, which involved interviews with personnel, and from company reports and in-house promotional materials and newsletters.

on the Rio Doce in the state of Minas Gerais. It is a joint venture between a consortium of eighteen producers of paper and pulp in Japan, the Japanese government's Overseas Economic Cooperation Fund, the trading company C. Itoh, and the Brazilian government's iron-ore mining company, Campanhia Vale do Rio Doce (CVRD). Cenibra is an abbreviation for Celulose Nipo-Brasileira. The Japanese consortium includes Oji Paper, Honshu Paper, Daishowa Paper, and Mitsubishi Paper. Since its establishment, Cenibra has become a partner in several other mill developments, and CVRD is a major presence in both mining and forestry throughout Brazil.

Cenibra owns over 143,000 hectares of forest land, of which about 85,000 had been planted by 1992. The remainder is either in natural forest preserves or is undergoing soil preparation. Between 6,000 and 7,000 hectares, with 1,667 trees per hectare, are planted each year in the states of Minas Gerais and Espírito Santo. Much of the land now under cultivation was used earlier for cattle ranching. Land near the mill was originally planted in eucalyptus for charcoal, used in the iron-ore mines. CVRD was among the plantation owners. With the phasing out of wood charcoal use, the same lands were gradually turned over to other eucalyptus species useful in forestry. Some of the local lands still belong to Belgo Mineira, a Belgian steel producer, and Acesita, a Brazilian mining company. Anticipating the needs of a double-sized mill, Cenibra continues to purchase farmland within a radius of 150 kilometres of its mill in a region that encompasses thirty-seven cities. In addition, CVRD has other interests in Bahia and northern Brazil. A Cenibra spokesperson, quickly pointing out that the Amazon lands were not part of Cenibra, also noted that they had the advantage of flat terrain and possibly less stringent environmental restrictions.

In addition to planting its own lands, Cenibra, like other Brazilian companies, provides free seedlings to local farmers and guarantees purchase of the wood at market price. Farmers are given training sessions by the State Forestry Institute. A typical practice is to grow the trees interspersed with corn, so that the farmer has an annual harvest of a marketable and edible crop while building up the eucalyptus for cash sales. Local residents are permitted to enter some of the plantations to obtain fuelwood. They are not allowed to bring in tools and can only take away scraps they can carry, but the practice inhibits the raiding of plantations.

Funding for the Cenibra mill included US$66 million from company capital, $78 million in foreign loans, and $76 million in domestic loans. The foreign loan came from the Japan-Brazil Pulp and Paper Resources Development Company (JBP), which obtained money from Eximbank and Japanese financial institutions. The domestic loans came from the

BNDE and its associate company, Finame. CVRD owned 50.6 per cent of Cenibra's stock, the remainder being held by JBP. By 1992 it produced 370,000 tons and was beginning construction of new mill capacities to double output to 700,000 tons per year, or 1,200 tons per day. Expansion costs were estimated at $800 million (U.S.) and the company was seeking loans with IMF approval. Pulp is shipped by train to the Portocel terminal in the state of Espírito Santo, of which Cenibra is one of the shareholders. Eighty per cent of all production is exported. By the terms of the original agreement, half of the output must go to Japan and in some years nearly 60 per cent is shipped to that country. The United States and Europe provide other export markets.

Plantations and Environmental Concerns

Cenibra is virtually self-sufficient in fibre sources, using primarily *grandis* and *salignas* species of eucalyptus. Plantation trees near the Rio Doce are afflicted with a fungus (*seca de ponteiro*) that dries their top branches and stunts growth; attempts to control this disease have involved cross-breeding of species. The native forests are not afflicted with the fungus.

Cenibra boasts of its environmentally protective policies. It has planted 150,000 seedlings of native species along 20 kilometres of company-owned lands on the banks of the Rio Doce in hopes of sustaining a mixed forest and regenerating some of the original native forest ecology. It has created a 1,000 hectare Atlantic forest reserve on the banks of the Rio Doce in the district of Cartinga. In agreement with the city of Belo Horizonte, it cooperates in the reintroduction into the reserve of several wild species of birds threatened with extinction. It has also cooperated in the development of a major city park at Belo Horizonte. Altogether, the company protects some 40,000 hectares of permanent reserve and preservation areas spread across the districts of Minas Gerais against forest fires through a system of watch-towers, computerized weather stations, school campaigns, meetings with owners of neighbouring rural areas, and a modern radio-communications system. It is carrying out a survey of the local fauna and flora throughout the region of Belo Oriente and the Rio Doce basin. These protective policies should be situated in a context, one that includes both the loss of diversity as entire regions become plantation monocultures interspersed by small natural forests and the historical and continuing loss of diversity from other causes. These regions were not pristine before Cenibra established its plantations.

Among the company's industrial projects are several to classify and improve the quality of effluent from the mills. It complies with the state's environmental regulations in its new activated-sludge treatment system for water, a modern oxidation system to control the emission of odorous gases, and improvements in the unbleached pulp washing process to reduce organic matters. Its gas emission values are below those required by regulatory agencies. The company is increasing the proportion of pulp produced without elemental chlorines, thereby reducing the formation of organochlorinated compounds. It uses the most modern computer equipment to monitor water and air effluent. And it is constructing a new sanitary landfill in compliance with strict quality standards to provide for the disposition of solid wastes (Cenibra, *A Cenibra e o meio ambiente*, n.d.). In terms of contemporary standards for pulpmills, Cenibra's claims are not disputed. The mill utilizes technology and applies standards that are the equal of any modern mills in the world.

Employment and Community Development

While Cenibra in its current incarnation has much to boast about, the record of land take-overs is less auspicious. The Japan Tropical Forest Action Network (JATAN) (1993) contends that Cenibra obtained its land through violence and deceit. In a study of eucalyptus plantations involving Japanese investment, JATAN states that during the 1970s, "Floresta Nippo Brasil (Flonibra), established by Japanese general trading companies and Brazil's Companhia Vale Rio Doce, carried out eucalyptus planting in southern Bahia. From early on, Flonibra used whatever methods were available to acquire land for its plantations, sometimes they purchased land at above market prices, at other times residents were chased away with violence. It also often resorted to deception." Deception took the form of ostensible third parties pretending to mediate land conflicts between planted associates and local farmers. JATAN says that some "landowners subsequently took court action against Flonibra and won, forcing it to return their land" (p.4). After that, Cenibra took over from Flonibra and traded the southern Bahia territory for land near Ipatinga, a move made possible through political arrangements and brokers who pressured farmers into selling their land. Says JATAN, "The displaced farmers and their families moved to nearby cities and, since they generally lacked employable skills, most of them joined the growing ranks of slum dwellers" (p.8).

By the 1990s much of this is buried history. The imported managerial and professional research staff are unlikely to be, or have reason to become, aware of earlier land acquisitions, and local production and service workers (who might actually be migrants from some distance away) are unlikely to want to inquire further. They are the beneficiaries of plantation and pulpmill development.

Cenibra employs about 1,200 workers in the industrial sector and another 5,000 directly in the woods, plus about 4,000 on contract arrangements in forestry projects. The bulk of the labour force comes from the local region. According to company sources, 70 per cent comes from Ipatinga, 30 kilometres distant, and the remainder from Governador Valadores, about 70 kilometres from the mill. The company provides in-house training and employs university graduates for its more technical operations. It also provides hot lunches, uniforms, bus service to and from the towns, low-cost medical services (every worker is required to undergo an annual examination), dental services, and free AIDS examinations. Uniforms – described by one worker as "like having more in your salary because it saves the cost of clothing" – are issued on a regular basis: three complete sets of suits, shoes, and belts on the first day of employment and a new outfit every six months thereafter. Workers in direct employment wear khaki-coloured outfits; service workers and others on contracts wear blue ones.

Woodworkers are employed all year. From August to December, they prepare the land and plant trees. From January to July, they are engaged in silviculture and maintenance, as well as logging. They use chainsaws, and at some sites they operate Massey-Ferguson tractors and grapple yarders. Harvesters and heavy tractors, however, are hard on soils and of limited use on hillsides. Many workers are on service contracts, usually at a fixed salary plus incentive pay and benefits that include food (a hot lunch delivered on site every day), uniforms, health insurance, and school materials for their children. At the time of my visit in July 1992, salaries ranged from Cz 441,049 for the basic forestry work and Cz 764,483 for chain-saw and tractor operators to Cz 1,057,856 for Massey-Ferguson grappler operators and Cz 1,291,381 for those using harvesting machinery. Incentive pay could improve these figures by up to 45 per cent. The cruzeiro had an exchange value of about 4,000 to one U.S. dollar at that time, so the average pay of Cz 750,000 was roughly equivalent to $187 per month. (Because of rapid inflation, salaries must be renegotiated monthly.)

The mill and woodworkers are unionized. There are eleven separate unions for woodworkers, splintered by locality as much as by trade. About 70 per cent of them went on strike for a short period in 1989. The pulp workers at Cenibra and Impasa are in a union known as Sinticel, but it is not a member of any of the three central unions. They have never mounted a strike. The unions bargain over wages, food, social services, time, working conditions, and what is known as the "licence" of maternity and paternity – the law requires that fathers have five days off at the time of birth; women get four months' maternity leave. Logging is male dominated; the nurseries and planting work, is done by both men and women. Most labouring work in the mills is done by men, but there are, in general, more women on the floor of the mills in Brazil than on those of Canadian mills, and many women in professional, managerial, and technical positions.

KLABIN

Klabin,* a group of eighteen plants throughout Brazil, including its main paper and newsprint mills at Monte Alegre in the state of Paraná and Riocell at Porto Alegre in the state of Rio Grande do Sul, is Latin America's largest forest products company and is ranked fifty-seventh among major pulp and paper companies by *Pulp and Paper International* (1990). Until 1991, it was Brazil's largest pulp manufacturer (Aracruz moved into first place when it expanded its mills). Klabin remains the largest paper manufacturer. The group produces over 18 per cent of the country's total pulp supply, 16 per cent of its paper supply, and 16 per cent of the corrugated container supply. In 1990 it added a disposable tissue producer, Companhia de Papeis (COPA), to its stable in order to sustain leadership in this market (Klabin, 1990). COPA was purchased from Scott Paper and Caima.

The original company was established in 1899 and the present company in 1934. Unlike most other companies in southern Brazil, Klabin established its mills and plantations in a natural forest. It began logging the forest in the 1940s, originally for timber and more recently to clear land for plantations. Its mountainous location at Monte Alegre provides it with native pine species suitable for paper manufacturing.

Klabin was and is a Brazilian company which trades on the Brazilian stock exchange. The Paraná mill was constructed in 1941, and although parts of the mill are old, much of it has been modernized in recent years. The newsprint machine was installed in 1962 and the paper machine in 1979. The pulp line receives some 2 million tons of pulpwood per year, of which about 30 per cent is eucalyptus and 70 per cent pine. Most of this comes from company plantations. The Paraná mill sells output in the Brazilian market, but it supplied traditional European markets with export earnings in 1990 of US$51 million. Other Klabin mills are smaller, and several specialize in the

* Unless otherwise stated, information on Klabin is from my fieldwork in July 1992, which involved interviews with personnel, and from company reports and in-house promotional materials and newsletters.

Table 11.5
Brazil: paper production by company, 1991

Company	1991 (tons)	% Total
Klabin group (7 divisions)	788,840	16.05
Suzano group (3 divisions)	469,523	9.55
Champion Papel e Celulose	341,859	6.96
Ripasa group (3 divisions)	297,377	6.05
Igaras Papeis e Embalagens Ltda	246,039	5.01
Simao group (2 divisions)	230,048	4.68
Trombini group (4 divisions)	204,065	4.15
Rigesa Celulose, Papel e Embs Ltda	198,918	4.05
Pisa, Papel de Imprensa	151,898	3.09
Fca de Papel Sta Therezinha	80,000	1.63
Santa Maria Cia de Papel e Celulose	72,022	1.47
Iguaçu Celulose e Papel	54,004	1.10
Papirus Ind de Papel	51,966	1.06
Cia Indl de Papel Pirahy	46,013	0.94
Other	1,681,541	34.22
Total	4,914,113	100.00

Source: Associação Nacional dos Fabricantes de Papel e Celulose, 1991: table 3.02.

production of corrugated, tissue, or other papers. One at Recife uses only waste paper. A plant at Camacari in the state of Bahia formerly produced pulp from sisal and has recently modified the technology to produce bleached pulp from eucalyptus. The most important of the other mills is Riocell, which is dealt with separately below.

Plantations and Environmental Concerns

Klabin has 297,000 hectares altogether, of which 195,000 are in pine, eucalyptus, and Paraná pine (*Araucaria* species), through the states of Paraná, Santa Catarina, and Rio Grande do Sul. The remainder, about 102,000 hectares, are in preserved native forests. (The law requires that 15 per cent of the land be sustained as natural forest, so Klabin's current reserves are above the legal requirement.) In the Paraná division alone, 113,000 hectares are in plantation and 73,000 in natural forest. Universities, research institutes, the Curitiba zoo and natural history museum, and the national environment protection agency are all active in preserving and studying the fauna and flora of the forest regions.

The plantations include loblolly and radiata pine, both exotic to the region. These are thinned after seven years and clearcut at twenty years. Elliotii pine is also grown and is used for resin as well as pulp.

The resins are distilled for turpentine and glue. The Paraná pine, is thinned at ten years and at intervals thereafter to thirty-five years and is used for construction lumber and furniture. It produces very little resin. A small quantity of loblolly pine and slash is sold in the domestic market to nearby sawmills. The *E. dunnii* is grown in preference to other species of eucalyptus because it is resistant to light frosts, which occur in the mountainous region of the Klabin mill. *E. globulus* is also planted. The eucalypts yield about 36 CM, the pines about 28 CM, and the araucaria about 18 CM per hectare.

Research at the Paraná mill nurseries in Monte Alegre includes work on biotechnology, silviculture and nursery techniques, and industrial products. Klabin collaborates with researchers in Brazilian universities and other countries to improve the stock. As well, the project known as Cuncor at North Carolina University provides seeds of endangered species, which Klabin plants in the natural forests. That project is not yet far enough advanced to determine its success.

Since logging of native forests was undertaken by the company, critics are on stronger grounds in charging Klabin with deforestation than most other southern Brazilian companies. However, much of the logging preceded the development of ecology groups and international concern for native forests in Latin America, and the company's present practices are more attuned to current ecological concerns. The range of objections to monoculture plantations are as apt here as elsewhere, except that these plantations include both several pine species and eucalyptus, and native forest reserves intersect the plantations.

Employment and Community Development

Altogether, Klabin provides 17,800 direct jobs in its industrial and woodlands activities throughout Brazil. The Paraná company employs about 3,000 workers. Some 200 of these, the technical and management personnel, live in the company town called Harmonia. This town has a Klabin-funded hospital, and another company hospital is located at the nearby town of Telemaco Borba. Schools are public, but because there is insufficient funding for them, the company provides some portion of the costs. A recreation area with swimming pool and other facilities has been built by the company. Woodworkers live in seven villages nearby. They are provided with the noon-hour meal and have free medical and dental service, 50 per cent off prescription drugs, and free rent. The villages are equipped with schools, banks, supermarkets, and company housing. Half of the transport workers are on contracts, but Klabin does not use contractors for logging, and all others are on direct wages paid biweekly. There is a set salary per worker. The minimal woodworker's salary, for cutting trees, is $110 (US) per month.

Mill workers live in a nearby town and have free hot lunches in a cafeteria large enough to serve 4,000 employees, plus much the same range of social services and benefits as the woodworkers. According to the mill manager, "the whole community depends on us; we cannot disemploy workers. The workers are reasonable in their demands, and we try to offer a reasonable arrrangement, too, but it's not easy; it's really not easy." The main impediment is inflation; labour costs per unit are lower, but there are far more workers in this mill than in a modern mill in Canada of comparable capacity. The company employs contractors in service jobs and the woods, but its contract system is less extensive than at Riocell.

Phytotherapy

As a minor, though interesting, footnote to this description, Klabin sponsors the research of a pharmacologist who is studying the chemical properties of plants known to native peoples and used in native medicine. Many of these are now understood to be effective for precisely the ills identified by native peoples, and Klabin has provided a small pharmacy in the woodworkers' town to sell packages of over 130 of these at nominal cost. They include everything from camomile tea to aphrodisiacs. None are sold beyond the town. According to the forest manager, 70 per cent of local prescriptions are for natural forest extracts, and these include natural antibiotics.

RIOCELL

Rio Grande do Sul is known for its colourful gauchos, vast ranchlands, German settlements, shoe production factories, and the massacres of the Guarani indigenous peoples in the area of São Miguel four centuries ago. The immigrants of an earlier time cut the original subtropical forest; only 3 per cent of it remains today. Eucalyptus plantations are relatively recent.

Riocell,* situated some 20 kilometres from the capital city of Porto Alegre, was originally owned by a most unpopular firm, Borregaard of Norway. Its labour policies were the primary source of its unpopularity, but as well, the local population objected to its pollution of the

* Unless otherwise stated, information on Riocell is from my fieldwork in July 1992 and from company reports and in-house promotional materials and newsletters.

Guaiba River and its policy of producing only raw pulp to be processed in Norway. Forceably closed in 1975, the mill was taken over by the military pension fund, which marketed its output until Riocell took over the plant in 1977. Riocell built a new bleach plant and took the operation into the market pulp network of Abacel as a participant with Aracruz and Cenibra (*PPI*, Dec. 1977:66). Riocell is one of the Klabin group of companies, but operates as a separate and independent entity. Seventy per cent of its shares are owned by a combination of Klabin, the Banco Iochpe, and the Varig Investment Group, Votorantim, with Klabin holding majority control. The remaining 30 per cent are held by the national bank, BNDE. All directors are insiders, and all are professional directors (there are no family members). The mill produces some 300,000 tons a year of eucalyptus pulp, 10 per cent of which is converted into papers. About 85 per cent of its pulp and 15 per cent of its paper are exported, with Southeast Asia, Europe, and the United States as major customers. A second mill is under construction, and by 1995, production is expected to increase by 140 per cent to 700,200 tons annually.

Plantations and Environmental Concerns

The company has planted eucalyptus (*salignas, grandis,* and *tereticornis*) and *Acacia mearnsii* on 51,200 hectares. In addition to these, it has 13,000 hectares in natural preserved forests within its territory. It plants 5,000 new hectares annually. It also enters into "partner" agreements with local farmers who plant eucalypts on their land, with guarantees of market-price sales to the company as they mature. Some 17 million seedlings per year are given free of charge, and municipal officials provide training, distribution of seeds, and supervision; all the pros and cons for Riocell's partnerships are identical to those for Aracruz and the other Klabin plantations.

Riocell's commitment at the time of buying the mill was to clean up the river and establish a state-of-the-art effluent plant. It has built a water-clarifying plant and added an ogygen delignification line to precede its bleaching unit and improve the pulp-washing system and the chlorine-substitution rate. It has achieved an 80 per cent elemental chlorine reduction and now sells a line of chlorine-free market pulp. It boasts that a study recently undertaken by Tübingen University in Germany showed Riocell's sludge and treated effluent had no influence on the Guaiba River system. In other areas, Riocell is experimenting with all waste materials to discover ways of turning them into marketable products or inputs into its own product lines. As in other mills in Brazil, sawdust and barks are used for fuel, but Riocell is breaking new ground in turning bark and sludge into fertilizers after bio-composting. It leaves a large part of the branches and bark in the forest to provide fertilizer for new crops. It also turns dregs and grits into soil-acidity correctors for agricultural applications; fly-ashes from the coal boiler, into components for the cement industry. Recycling of office and laboratory garbage is well established, and the company has ceased to use the word "residue" for many by-products; it now uses the phrase "new products" for these.

Agro-forestry is under way, essential where local farmers need food crops while growing the tree cash crops for the company. Riocell undertakes extensive experimentation on crop combinations. It also has large nurseries and outstanding research laboratories that undertake the study of every phase of the operation from seeds (hybrids, clones, and new varieties), through planting and silviculture, pests and environmentally friendly means of dissuading them from ruining crops, and diseases and fungi, to new products from wood. Because the company's plantations are spread over 150 non-contiguous farms and there are other crops (rice, cattle farms) in the region and because the company uses several clones, it has so far avoided monoculture diseases and has no insect problem. The chief research forester claims that in fifteen years he has never seen a problem in a cloned forest that was not already present in the seedlings; the objective, then, is to do the environmental planning before planting.

The company's commitment to research is signficantly greater and more visible than that of competing companies even within the Klabin group. One of the differences is its library. This was established at a cost of $400,000 per annum. The chief librarian is a chemical engineer, and the operation is set up as a service contract firm. The library was a busy location when I visited, and its research materials were up to date and extensive.

Employment and Community Development

About 1,100 workers are employed directly on salary or wages; another 900 or so are on service contracts. Contract employees do janitorial, computing, and much of the clerical and library work and forestry-woodlands jobs. When the company started, it had some 2,500 direct employees; the change in labour organization has thus been substantial in a short time. There are now 165 companies providing services to Riocell. Each company pays its own taxes, has its own offices, and hires its own workers. Riocell uses the term "partnership" for this organization and notes that mutual audits are allowable. If it believes that a small company charges too much or does not pay enough to employees, an audit can be undertaken; theoretically (but no one offered an example), the small company can request an audit of Riocell for complementary reasons (it does not pay enough for the contracts or offer sufficient technical support services, etc.)

In the woods, there are only 100 direct employees. All silvicultural and harvesting tasks are performed by contract firms. Most jobs in the nurseries and research labs are likewise undertaken on contract, though in all cases, key positions are held by direct employees on salary. Of the total direct employees, some 35 per cent are university graduates. The management rejects the argument that this form of labour organization offloads responsibility for the labour force and either destroys existing unions or inhibits the formation of new ones. At Riocell now, fewer than 500 workers are in unions. The company argues that these contract arrangements provide some elasticity when pulp markets or government monetary policies alter its economic condition; thus they allow the whole organization to survive in bad times as well as good ones. Service contract companies are free to develop other business in the community, and several of the small firms have already done so, thus reducing their dependence on Riocell. They are, the management argues, much more innovative when they are running a small business, much more engaged, and individual workers are, as well as feel, more important to the success of the enterprise.

The average monthly salary according to management personnel was about US$275 in July 1992. Blue-collar workers – company managers use this terminology – earn more than they can in other occupations in the region; managers and professionals make somewhat less. Working conditions for professionals, however, are extremely pleasant, and the company is regarded as a good employer. Hot lunches, social services, medical care, and various other benefits are provided to both managerial and labour workers and, via the contract firms, to their workers.

RIPASA

Ripasa,* an integrated pulp and paper company, was founded by the Zarzur, Derani, and Zobgi families in 1958 at Limeira 130 kilometres northwest of São Paulo city in the Piracicaba river valley. The group's holdings include five large mills, most importantly those at Cubatão

* Unless otherwise stated, information on Ripasa is from my fieldwork in July, 1992, involved interviews with company personnel, company reports, and other company documents.

(Santista), São Paulo (Embu), and Limeira, which produce pulp, writing and printing papers, duplex and triplex boxboard, groundwood paste from pine pulpwood, and special-purpose industrial papers. The Limeira mill produces 750 tons per day of pulp (266,250 tons per annum) and 610 tons per day of paper (216,550 tons per annum). The pulpmill is fifteen years old, and production is labour intensive. The papermill has two machines, one old, the other a sparkling new, computerized Voith Duoformer especially designed for the company. A modern purification plant removes sediments from the mill effluent before water is returned to the Piracicaba River. The domestic market absorbs whatever pulp is not consumed by company mills – about 25 per cent of the total. Sixty per cent of the paper is exported to Europe, the United States, and, in lesser amounts, Africa and the Middle East. About 1,000 tons per month of writing-grade paper is sold to stationery outlets in the United States. The company maintains a sales office in Belgium and a trading company (Rilisa) in São Paulo.

Plantations and Environmental Concerns

Ripasa acquired land via the National Pulp and Paper Program, and it now owns some 64,000 hectares and rents another 20,000 from local farmers in eight locations in the state of São Paulo, with the centre at Limeira. The company maintains a stock of some 82 million trees in these locations and is acquiring new plantation land in northern Brazil (discussed later on). Reforestation was greatly expanded in the late 1970s. The plantations are mainly *Eucalyptus salignas* and *E. grandis*, together with *Pinus caribaea*. Research laboratories are located at Araraquara City, and nurseries there and at three other locations. Research is done jointly with the universities, some of which have formal joint contracts with Ripasa and other companies.

The company operates an environmental education program for employees and local communities, and it boasts of land-management techniques for soil conservation and water-resource protection. It has a biological pest-control program for insects making use of their natural enemies and thus obviating the need for chemical pesticides. In 1991 the Brazil Society to Encourage Environmental Management awarded its Green Seal certificate to Ripasa for its environmental projects. (Among the other twenty winners were Riocell and Aracruz.)

Employment and Community Development

In the total operations, there are over 5,000 mill employees. The Limeira mill alone employs 1,500, 60 per cent of these in pulp production, 30 per cent in paper production, and 10 per cent in professional and clerical positions. Most of these workers live near the mill in the towns of Americano, Limeira, and Nova Desa. According to the mill manager, workers in the papermill who handle the new computerized machinery are required to have completed middle school; the mill teaches them computer skills. Other mill workers have less education. While the pulpmill and the first-machine section of the paper mill are labour-intensive operations, the new papermill has relatively low labour input, and many of those who are on the floor of the mill are trained in tradeswork. Women are employed in the packaging section; otherwise they do not work in the mill.

There are about 1,800 workers in the forest section. When the company purchased land near the Limeira mill, it did not evict local inhabitants. "We invited them to work for us," is how the mill manager describes the relationship. But finding and keeping woodworkers is a major problem, and turnover rates are around 60 per cent per annum. The nearby sugar cane plantations compete for labour. Ripasa provides a transportation system for workers, cafeteria food, and aid to the local schools. Its manager claims that working conditions are now much better than on the sugar plantations, and more of the workers can now be employed on a year-long basis on the eucalyptus plantations. Social services are more extensive for mill workers than forest workers. Mill employees receive annual wage increases and bargain via a fairly strong union. Forest workers are also unionized, but have been less successful in sustaining their membership base and improving their incomes. The company offered forestry workers the same increase in 1990 as was given to mill workers and claims to have thereby reduced the turnover rate from 100 per cent to 60 per cent.

OTHER GROUPS

In addition to the very large companies described in detail above, there are other substantial operations. One of these is Industria de Papel e Celulose Arapoti (Inpacel), owned by the Simão Racy family of São Paulo since the 1920s. In 1992, 48 per cent of its stock was purchased by the industrial conglomerate Votorantim. Brazil's fourth largest papermaker, Inpacel produces 250,000 tons per year of paper at four mills in São Paulo state and began producing another 200,000 tons per year in an integrated CTMP papermill at Arapoti in Paraná state in 1992. Not including production from this mill, the company already made 16 per cent of the country's printing and writing paper and about a third of all its speciality papers. It produces banknote and fax papers, claiming to be the only producer of both products in the southern hemisphere (Knight, 1992:74; *Paper*, Oct. 1992)).

Another sizeable operation is Papel de Imprensa (Pisa), half owned since 1988 by New Zealand's Fletcher Challenge company. Other shareholders are Brazilian newspapers, the World Bank, and the Brazilian development bank. It produces 150,000 tons per year of newsprint and groundwood papers, for which its own 50,000 hectares of pine plantations provide the fibre source. Its integrated mill complex is located at Jaguariaiva in Paraná state, about 400 kilometres south of São Paulo.

PROJECTS IN BAHIA STATE

The soils are poor, the land is degraded, the remaining trees are still the major source of domestic fuel, there are few business enterprises, and poverty is widespread, but the state of Bahia is the new location of choice for several companies. One of the major industrial projects in Brazil during 1992 was the construction of a mill to produce printing and writing papers at Mucuri in Bahia. The project, called Bahia do Sul Celulose, involves Suzano and CVRD. The state conglomerate already owned 100,000 hectares of timber within 60 kilometres of the mill site, and this provides the initial fibre source. Suzano has a 38 per cent share; CVRD, 30.8 per cent; BNDE, 27.45 per cent; and the World Bank, 3.67 per cent. The Banco do Noreste do Brasil was also involved in the financing (*Paper*, Feb. 1992).

Copene and Riocell are two partners in plans for a $340 million, 420,000-tons-per-year pulpmill to be financed through a debt equity swap at Entre Rios in Bahia (*Paper*, 13 Aug. 1991). The project is known as Norcell. A group dubbed "Odebrecht" by Brazilian newspapers, which is currently involved in construction, petrochemicals, mining, and equipment assembly, has signalled its intention to invest $1.4 billion over the next decade in a pulp project in southern Bahia under the subsidiary name Veracruz Florestal (*Pulp and Paper Week*, 30 Jan. 1992). The fibre sources will include 47,000 hectares of land near Eunapolis leased from the CVRD subsidiary Floresas Rio Doce. The company anticipates the purchase of another 75,000 hectares (*Paper*, 5 May 1992:20). As mentioned, Klabin has a mill at Camacari in the state of Bahia, which was transformed from a sisal to a eucalyptus mill. So far it has been besct with technological problems and has produced very little, but when it is fully operational, the company anticipates an annual capacity of 90,000 tons of pulp per year.

PLANTATIONS IN NORTHERN BRAZIL

Jari is the major company in the Amazon; none of the other industrial forestry companies has a similar northern presence. But while companies in the southern states of Brazil have, in the main, taken over or inherited lands already deforested and have generally sustained whatever remains of native stands and even replanted subtropical species in protected areas, several companies have also taken out options on acreage in the tropical regions of northern Brazil. They do not boast about these, rarely mention them during tours of their operations, and may even disclaim ownership if pressed (usually formal ownership belongs to a separate company established as a subsidiary or joint venture with third parties).

In addition to Jari, among the operations situated in various parts of the north is the Klabin mill Papelao Ondulado do Nordeste, located at Goiana in the state of Pernambuco. It produced some 20,000 tons of corrugated containers in 1990, plus 13,000 tons of packaging paper and moulded pulp. It is being expanded to capture markets in the north and northeast. The Celmar pulp project in Maranhão state of southern Amazonia has been mentioned earlier. Tree plantations of up to 250,000 hectares will be established to feed the new mill anticipated for start-up by the turn of the century (*PPI*, Dec. 1991:43–5; *Pulp and Paper Week*, 20 Jan. 1992).

While projects in the south may claim innocence in the despoilation of native forests – most have inherited degraded land rather than created it – the same cannot be said of new projects starting up in the Amazon region. Tropical wood itself is not of primary value to the pulp industry, though it is exploited for roundwood and lumber by the sawmill industry. The reason for moving north is land, and the moves indicate that cheap land is becoming scarcer along the southern coast. These projects are cause for concern on environmental grounds. They are also of concern because the few remaining native groups in Brazil – with an estimated population now at not more than 220,000 individuals – are resident mainly in the Amazon region.

In addition to native peoples, there are thousands upon thousands of homeless and impoverished people in Brazil. They come from the northeastern sector; they also come from Paraná and other southern states. Of the migrants from the south, Guppy notes that 40 per cent of the smallholders of just one municipality of Paraná departed in 1982, victims of government policies that favour large operations over small landholdings (1984:939). He also argues that there is sufficient cultivatable land in Brazil, outside of Amazonia, to give 2.3 acres to each individual Brazilian, or ten acres per family; yet concentrated landholding has resulted in 43 per cent of all farm land being owned by 1 percent of farmers, and large tracts of land are owned by industrial and plantation organizations. Settlement schemes of the 1960s to 1980s lured individuals to low-fertility land in the north. Unsuccessful at farming such land, they roam in search of employment. The pulp-mill industry may well be viewed by them as a means of survival and by the Brazilian government as a means of sustaining the population. It is now a cliché, but true none the less, that the ecological problems in the Amazon will be solved only when the poverty problem is solved; the question is, Can that problem be solved only by land-ownership patterns that deprive the poor of alternative means of subsistence?

BACKWARD LINKAGES AND INTERNATIONAL TECHNICAL ADVICE

Brazil, unlike many other countries including Canada, has managed to create a substantial industrial sector to service the pulp and paper industries. Mill equipment is manufactured in the country under licence from parent companies, and new equipment is being developed for logging in fragile terrains. Domestic companies have generally had preferential treatment, but with persistent inflation and changes in tariff laws, competition from elsewhere has occasionally interfered in this arrangement. The Bahia do Sul operation, for example, had intended to purchase all of the $355 million worth of equipment domestically, but with lower tariffs introduced in 1988, offshore bids became more competitive. Mill owners demanded that domestic suppliers submit lower bids.

Jaakko Pöyry of Finland and Sandwell of Canada have had a prolonged run in preparing feasibility and technical reports for the Brazilian pulpmill expansion of the last decade. They have been engaged by virtually every company at various stages of planning and construction. As well, international banks and the World Bank have been involved in financing most of the projects.

SUMMARY

Brazil has become an industrial giant in pulp and paper production. Investors have benefited from advantageous tax laws, subsidies, and other incentives, cheap land, fast-growing and highly productive eucalyptus and pine trees, and cheap labour sources. Backward linkages have developed, and the manufacturing of machinery for the industry is now well established. International investors have benefited as well, though domestic sources of capital are dominant in the industry. The World Bank and private banks have been active, and the national bank of Brazil is a central financier for most projects. The industry provides employment and makes useful products for the domestic market; it generates foreign-exchange earnings for an indebted country. These are accepted as "facts" by most critics; all else is contested. At issue most particularly are claims to ecological sustainability and innocence of causing the displacement of peoples.

As land was claimed for plantations, both natural trees and people were moved out. This process is now under way in the Amazon region, but most of the original plantations farther south were not connected to forestry, and contemporary forest companies have become the indirect beneficiaries of earlier land clearances. Labour policies may be viewed as exploitative from the North American perspective, but in the context of rural Brazil, employment at pulpmills or in the woods confers benefits not easily obtained elsewhere. One continues to muse, with my informant from Aracruz: Will it be Marx and the Church or the giant corporations that come closest to reducing poverty?

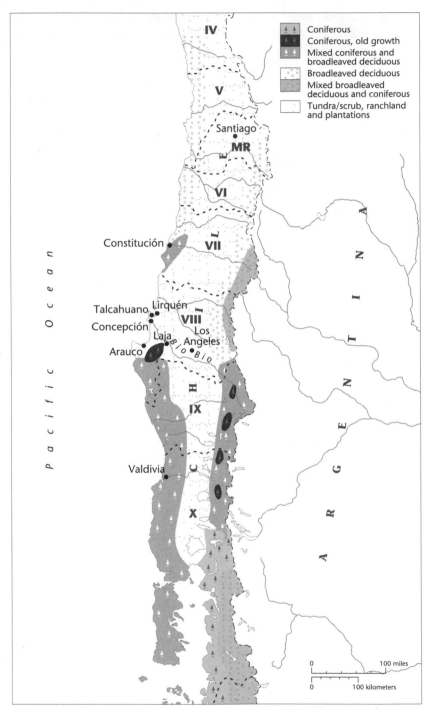

Map 11   Chile, central to southern Pacific coast: forest types

12 Chilean Temperate and Plantation Forests: A Case Study

In August 1991, forty-one Chilean university faculty members issued a statement calling for immediate political action to conserve native forests ("Urgencia," 1991). The group, consisting of individuals at different universities throughout the country, asked for the termination of massive subsidies for the planting of monoculture exotic species that displace native stands. They argued that government policies had facilitated the destruction of native forests by encouraging foreign and domestic investment in plantations. In April 1992, the president of the republic introduced a bill in Parliament for the "Recovery and Promotion of the Native Forest" (Chile, Ministerio Secretaria General, 1992). Between these two dates, and for some time prior to the circulation of the petition, debate on native forests and plantations had been intense throughout Chile. The proposed bill will not satisfy critics. This chapter outlines the structural reasons for the debate.

CHILE AND BRAZIL COMPARED

Apart from the fact that they are on the same continent, Brazil, a giant, sprawling country bordering the Atlantic coast, and Chile, a thin strip of mountainous terrain along the Pacific coast, have little in common except that both have become players in the global forest industry. Their separate developments are often lumped together by outsiders, but they are not alike. Much of the Chilean plantation area was forested in either primary- or secondary-growth native species before the development of a large-scale forest industry. The native forest is in the temperate zone and contains both softwoods and hardwoods. Brazil, by contrast, was dealing with already cleared lands in much of the central and southern coastal region, and these are located for the most part in a subtropical zone. In Chile most of the plantation fibre consists of softwood (radiata pine), and thus the country is poised to produce newsprint and a range of other papers not, or less well, produced from hardwood fibres. However, companies are now increasing their investment in eucalyptus in an attempt to offset the impacts of market fluctuations typical of the newsprint industry.

Since 1987, Japanese pulp and paper companies have provided a large market for wood chips. These chips are produced in native, as well as plantation forests, and the wood-chipping business is profitable. Chilean law since 1975 has permitted the export of both wood chips and roundwood. Brazilian conditions have not raised this issue in connection with plantation regions, and the international outcry over destruction in the Amazon, as well as the diversity of species, has inhibited the growth of a wood-chip business in the tropical zone.

The two countries are positive in their attitudes toward economic growth. Both are eager to establish a "free market" economy, and their leaders tend to equate economic growth with development. However, where Brazil has managed to maintain a fair degree of domestic control over the industry and its companies retain some independence, post-Pinochet Chile has leaned heavily on foreign investment and the industry has become highly concentrated. One of its major pulp and paper mills is a joint venture with external funding and management participation; the largest group is currently under domestic ownership but is linking up with foreign investors in new projects. Its two principal holding companies control a very high proportion of the total output, and most other companies are in one way or another affiliated with these two.

Chilean government and industry brochures for potential foreign investors make much of "the economic climate," by which they mean a positive version of the legacy of the Pinochet regime. But the legacy is less attractive than the brochures portray. Chile has numerous problems to overcome as it tries to create a free-market economy with an elected government. Among these are the deficiencies of infrastructure: roads and communication links are poor compared to those of Brazil, though ports have been much improved and road construction is under way in the 1990s. Less obvious and more difficult to document, but palpably present even so, is an ambience of distrust. The outsider senses a reluctance to undertake serious discussion or debate on politically sensitive or economic-policy subjects. Whereas Brazilians engaged in lively debate while the military controlled the country, Chileans manifest signs of a people too long repressed by a brutal military regime. Although democracy became official in March 1990, the impact of nearly two decades of repression and the loss of potential leaders for opposition movements, is painfully evident. As well, the repression has not ended; its more violent aspects have diminished or are veiled, but the population knows it is not yet over. This notwithstanding, of course, investors are willing to take advantage of government incentives, and foreign investment is not subdued by history.

The general reluctance to express criticism openly is not shared by an organization called CODEFF – the Comité Nacional pro Defensa de la Fauna y Flora – which provides a nexus for critics of forest policy and publishes both academic and popular analyses. More criticism is mounted on ecological, than on social grounds, but social criticism is often implied. (Most university centres of social analysis, such as departments of sociology and anthropology, are either missing or relatively small in post-Pinochet Chilean universities.) It should also be noted that the Chilean forest service, CONAF, has several times attempted to change regulations harmful to native forests; its lack of success has been associated with underfunding of the service and superior power in the ministries of finance and economic development.

NATIVE AND PLANTATION FOREST ACREAGE AND LOCATION

The forest regions of Chile are ranged in a vertical strip along the Pacific coast south of Santiago (the northern regions have desert climates). Divided into administrative districts, these are labelled regions V through XII. Native forests are concentrated in the most southern regions, IX through XII. Plantations are now most plentiful in regions VII–IX. By 1990 the plantations covered some 1.46 million hectares.

Official statistics published by the Instituto Forestal–Corporación de Fomento de la Producción (INFOR-CORFO) classify nine million hectares in 1990, or 12 per cent of Chile, as forested, with an estimated volume of 1.1 million CM of wood. Of this, 7.61 million hectares in 1990 were classified as native forest, with a volume of 915,000 CM of wood. The native forests are temperate and include southern beeches (*Nothofagus* spp.) and other broadleaf trees, as well as *Araucaria araucana* and other species of conifers. Of the 1.46 million hectares in plantation forest, 1.2 million (86 per cent) were planted with *Pinus radiata*, and most of the remainder with eucalyptus (Chile, INFOR-CORFO, 1991, caudro 2.2).

Table 12.1
Chile: area planted, radiata pine and other species, 1978–90

Year	Radiata pine		Other species	
	Ha	%	Ha	%
1978	65,413	84.5	11,958	15.5
1979	48,869	93.6	3,357	6.4
1980	60,086	83.3	12,078	16.7
1981	88,529	95.4	4,252	4.6
1982	61,637	89.9	6,949	10.1
1983	63,884	83.7	12,396	16.3
1984	76,982	80.5	18,620	19.5
1985	80,630	83.7	15,648	16.3
1986	55,058	83.2	11,137	16.8
1987	55,386	84.6	10,055	15.4
1988	61,841	84.8	11,103	15.2
1989	65,587	75.6	21,118	24.4
1990	61,310	65.1	32,820	34.9
Total	845,212	83.1	171,491	16.9

Source: Chile, INFOR-CORFO, 1991, as published in Lara and Veblen, 1993, table 9.4.
Note: Figures include afforestation and reforestation.

FOREST POLICY

Early predations against the forests were widespread and included massive clearing and burning for colonial settlement, agriculture, and pasture land. The first law restricting forest exploitation was passed in 1872, but by then deforestation in the Lake District (regions IX and X) was extensive. (Veblen, 1983, and Lara and Veblen, 1993, claim that it was "one of the most massive and rapid deforestations recorded in Latin America" prior to 1980.) Radiata pine, introduced originally as an ornamental species, was discovered to be useful as a source of timber. Following legislation in 1931 that more effectively restricted the exploitation of native forests and provided tax exemptions for reforestation, pine plantations were encouraged. The government agency for promotion of primary and industrial production (CORFO), established toward the end of that decade, stimulated private planting through low-interest loans and partnerships (Lara and Veblen, 1993).

During the 1965–73 period, when the Chilean government was following policies of more direct participation of the public sector in the economy, publicly financed nurseries and two major pulpmill complexes (Arauco I and Constitución) were created. CONAF was established to oversee the development of the forestry sector, which was now perceived as strategic to Chile's development, both because it supplied housing and employment for labour and because it could become a leading export sector. CONAF undertook a system of afforestation partnerships, the "convenios de forestación," in the early 1970s, by which small landowners provided land and CONAF provided the management and logging for plantation species. The value of the products was to be divided on a 75:25 per cent basis in favour of the public sector. CONAF was responsible for over 90 per cent of new plantations in 1973, and forest production overall increased at an average of 8.5 per cent per year between 1960 and 1974 (Lara and Veblen, 1993).

The military government's forest policy, beginning in 1974, vastly expanded the plantation area and privatized much of the forest economy. Decree Law 701, promulgated in 1974 and modified in 1979, provided up to 75 per cent in subsidies for afforestation costs, together with tax exemptions, for plantations established between 1974 and 1994 (Chile, Ministerio de Agricultura, 1974, 1979). In some years the subsidies rose to 90 per cent. Antonio Lara, in a 1992 study of the subsidies and policies, observes that the private sector received US\$71.1 million in afforestation subsidies and that further subsidies and grants were introduced in 1978 and 1983.

In addition to the forest laws, other government legislation, subsidies, and supports of various kinds were critical to the growth of the plantation economy. The overall thrust of government initiatives was to encourage the growth of a few very large private-sector corporations in forestry as in other sectors. State-owned enterprises were privatized. They, together with land and forests in the public sector, were transferred at below-market prices and on highly favourable terms to private companies. CONAF provided planting subsidies and services to private owners. Other state ministries and agencies assisted in developing export markets, encouraging foreign investment, funding research, and providing technical training (Lara, 1985, Contreras, 1987). Although private plantations were the focus of the new policy and the government was ideologically wedded to a rigorous market-economy model, in fact CONAF continued to be active in establishing new plantations until the end of the 1970s and was again active in 1983–85 as part of a program to reduce unemployment (Lara and Veblen, 1993). It planted on private lands, thereby providing yet another subsidy to the companies.

In 1975 the previous export ban on unprocessed forest products was lifted, allowing for the export of roundwood and wood chips. Labour legislation enabled the companies to contract out most or all of the logging and woodlands activity and an increasing proportion of service-related work in the mills as well. These arrangements allowed employers to avoid payroll and labour costs and also inhibited the growth of powerful unions or other concerted opposition.

The forest law did not refer to native forests, and subsidies were not intended for their conversion to plantations. In the beginning, there were enough degraded and marginal lands to serve as plantation acreage, and where such was the case the law was potentially beneficial. However, authorities turned a blind eye to the creation of new marginal lands through clearcutting, exporting of wood chips, and substitution of native forests. No government supports were provided for management or recuperation of those forests. In 1980, however, another law, Supreme Decree 259, imposed restrictions on the silvicultural systems that could be applied to particular native forest types and slope angle classes of land. Logging of native forests in private estates required a management plan approved by CONAF. Illegal logging – that is, logging without an approved management plan or contrary to the restrictions of the 1980 legislation – incurred penalties in the form of fines and impoundments of product.

Restrictions and penalties notwithstanding, the logging of native forests and export of wood chips was a lucrative business. Illegal substitution was rarely prosecuted, and fines were weak penalties compared to the profits of the undertaking. A total area of 48,600 hectares, or about 6,000 hectares per year of native forests, was destroyed and converted to plantations in regions VII and VIII between 1978 and 1987. According to analyses by forestry economist Antonio Lara and his associates, 31 per cent of native forests in the coastal range of region VIII were converted to plantations in that period (Lara et al., 1987; Lara, 1992). In the early 1990s, a debate developed about forest lands with steep inclines: the forest companies wanted to restrict exemption to lands so steep that they could not be logged and planted economically, but to regain legal rights to log land short of such an incline. Of this debate, more is said further on.

The native and plantation forests produce wood for various purposes, but the native forests are especially exploited for wood chips, exported mainly to Japan. In 1990, twelve companies exported 1.1 million dry tons of wood chips from native forest species. This represented 52 per cent of the total wood-chip volume exported that year (Lara, Donoso, and Cortes, 1991). Clearcutting and high-grading are commonplace methods. These areas are then converted to radiata pine and eucalyptus plantations or simply abandoned. Thinning and silvicultural methods are rarely undertaken. Although private companies argue that public land and native forests are inadequately managed (with occasional references to the "tragedy of the commons" metaphor), critics point out that the forest service is not provided with sufficient resources to protect the lands from the predations of private companies. As well, under both the military regime and the democratic government elected in 1990, the overriding concern in government has been to attract foreign investment and encourage economic growth.

Logging for fuelwood is another cause of native forest depletion. Before the boom of the wood-chipping industry, Lara (1985) estimated that over 80 per cent of the annual volume cut from these forests was for fuelwood or wood-charcoal production. This logging occurs in part because there are many landless and impoverished villagers without alternative fuel sources, but also in good part because the production of fuelwood is a profitable business. Wood provides a major source of energy for industrial plants as well as private households.

CODEFF argues that official data on reforestation are misleading. Only a small proportion of the areas classified as national parks and other protected areas is actually forested (much is water, rock, and ice in the far south), and of lands ostensibly protected by the national forestry commission, many are open to hydroelectric projects and mining, if not logging (CODEFF and affiliated groups, 1992). In addition, most of the area under national parks is in regions XI and XII; elsewhere there are important species and ecosystems not adequately represented in national parks.

In summary, the forest policies of the Pinochet and post-Pinochet governments succeeded in increasing investment in forest industries and afforestation or reforestation of forest lands. They privatized the land and production facilities, and paid a high price in public subsidies. They created improved management, though logging practices in native forests continue to be detrimental. A fire management program has been introduced, and soil erosion has been reduced (Lara and Veblen, 1993).

In 1991, Chile was granted aid to promote sustainable forestry under a program sponsored by the United National Development Plan, the World Bank, the World Resources Institute, and other organizations coordinated by FAO. This plan provides for research and development undertaken by universities and consultants working together with the major companies under the project Investigación y Desarrollo Forestal. CONAF is again the major funding agency within Chile (*Chile Forestal*, 1991a).

THE PLANTATION ECONOMY

Some 1.06 million hectares of *Pinus radiata* plantations were created in the 1970s and 1980s, the largest single radiata pine holding of any country (Jelvez et al., 1990). These have a twenty-five-to-thirty-year growing cycle for sawlogs and fourteen to sixteen years for the production of pulp logs. The pulp plantations of the 1970s are now of harvestable age, and production will increase considerably through the 1990s and into the next century. Plantation pine is expected to cover 1.6 million hectares by the mid-1990s (Correa, 1988:152, 159; CMPC's *Annual Report* for 1988 estimates a total of 1.7 million hectares and 25 million CM of wood per year by 2000). Annual forestry exports are anticipated to grow by another 50 per cent before the turn of the century. The industry speaks of the development of backward and forward linkages to accompany this export industry: forestry equipment, pulpmill machinery, and a range of related materials.

With this resource base and modern technology, the forest industry is an important contributor to the Chilean economy. It accounted for nearly 10 per cent of total Chilean exports by 1990. Pulp and paper, lumber, wood panels, chips, logs, and firewood are all marketed, but from the mid-1980s, when it accounted for 48 per cent of the total, to 1990 when it accounted for 37 per cent, chemical pulp constituted the major export. Sawnwood products accounted for 21 per cent in 1990; wood chips, 13 per cent; and roundwood, 9 per cent of the total forest exports. These were exported to Japan (25.6 per cent), western Europe (29.3 per cent), Latin America (15.2 per cent), South Korea (6.4 per cent), and the United States (5.9 per cent) (Chile, INFOR-CORFO 1991; Lara, 1992).

For bleached softwood kraft pulp, wood costs in Chile are substantially lower than in the United States (in both the South and the Pacific Northwest), eastern Canada, Sweden, and New Zealand. Only Brazil has lower costs, and it is not a major producer of softwoods. Almost all plantations and native forests used for timber production are privately owned.

THE COMPANIES

Four companies control all of Chile's pulp capacity and 90 per cent of paper capacity, as well as about 40 per cent of the plantations. The two largest companies are Compania Manufacturera de Papeles y Cartones (CMPC) and Celulosa Arauco y Constitución (Arauco). The first of these is controlled by Chilean capital, the second by a combination of New Zealand and Chilean capital. Both have subsidiaries that link them with capital from other regions, including the United States, Sweden, and Japan. An earlier large firm, INFORSA, has since become part of CMPC.

In addition to these major enterprises, other multinational interests own about 10 per cent of the plantations and control up to 20 per

Table 12.2
Chile: megaprojects in the forestry sector planned for the period 1992–96

Project type	Number of projects	Total cost ($US million)	% of total projects
Pulp and paper	3	1250.0	79.0
Sawmills	6	39.3	2.4
Panel Board	1	50.0	3.0
Roads	1	150.0	9.4
Ports	4	98.8	6.2
Total	15	1588.1	100.0

Source: Chile, INFOR-CORFO, 1992.

cent of forest production and exports. The remainder of the plantations are owned by either Chilean or combinations of Chilean and
external capital, some connected already to pulpmills, some to planned
construction of pulpmills, and some entirely to other forest-product
industries, including wood-chip and roundwood exports.

Virtually every company and subsidiary has expansion plans. A
special printout given me by INFOR-CORFO showing construction plans
for 1992–96 lists fifteen megaprojects worth, in total, $1,588 million
(1992). While three of these in pulp and paper account for 79 per
cent of the total ($1,250 million), projects in other wood-products
sectors are not small. Of the megaprojects, foreign ownership in two
will account for 41 per cent of the total cost, and joint or mixed
ownership in six will account for 44 per cent, leaving 14 per cent for
purely national capital investment. The three major pulp and paper
megaprojects include one in region VII owned by the Swiss group
Attisholz in a project to be called Licancel (Celulosa Licanten); the
Forestal Pedro de Valdivia mill in region X, to be built by Copec and
Stora (see below), and an almost equally large mill, called Forestal
Anchile, to be built by Daio Paper and Itochō (formerly C. Itoh) of
Japan, also in region X.

Compania Manufacturera de Papeles y Cartones

CMPC is a giant company by the standards of any country in the world.
It owns major mills at several sites, specializing in paper manufacturing.
These include Laja, with capacity to produce 310,000 bleached softwood kraft pulp and 73,000 printing and wrapping paper; Puente Alto,
with capacity for 20,000 tons of mechanical pulp and 76,000 tons of
corrugated board; Valdivia, with capacity for 11,000 tons of mechanical
pulp and 34,000 tons of multi-ply board; INFORSA (Industrias

Forestales SA), with capacity for 86,000 tons of mechanical and 34,000 tons of sulphite pulp, plus 120,000 tons of newsprint and some speciality paper; and Celulosa del Pacifico (Celpac), with capacity for 315,000 tons of bleached softwood kraft pulp (*PPI*, Nov. 1993:37). Celpac is a joint venture with Simpson Paper of the United States, situated at Mininco near other mills in the Bío-Bío valley. CMPC has also started to invest in neighbouring countries. It entered into its first foreign investment in 1991, a diaper manufacturing plant in Argentina (*Chilean Forestry News*, interview with Ernesto Ayala, president of CMPC, April 1992:6–7).

CMPC refers to itself as a group. Its shareholders are other large companies. Forestal Cominco SA and Forestal, Constructora y Comercial, Del Pacifico Sur SA each held nearly 20 per cent of the stock as of 1988, when eleven shareholding companies were named as holding, directly or indirectly, more than 10 per cent of the stock; their relationships to one another were not divulged in the annual report. Neither was any history provided in that or other annual reports, apart from occasional statistics dating back to 1979. Yet the reports do refer to "the founders," and with reference to timberland, the 1988 report mentions "several decades of continuous effort." Informal history has it that the group is the current organizational form resulting from battles in Chilean society of the 1970s and 1980s (the warriors are sometimes referred to as the barracudas and the piranhas). A 1991 tour of the Laja mill, situated southeast of Concepción on the Bío-Bío River in region VIII, provided enormous technical detail of the variety one finds easily enough in pulp and paper magazines, but the engineer assigned the task was not conversant with history or social context.

By the end of the 1980s, the company's tree nurseries were growing some 25 million seedlings and held a 336,869-hectare timberland base. Of this base, 213,439 hectares were planted in radiata pine, 8,416 hectares in eucalyptus, and 7,348 in other species. Another 42,401 hectares were available for planting. The remainder was used for fire-breaks and agriculture or was unused (Annual Reports). Of interest are the changes in these statistics from year to year. The timberland base grew by nearly 27,000 hectares between 1988 and 1989, the respective areas planted in both pine and eucalyptus by about 3,000 hectares. The rest of the expanded base became available for planting, augmented by a small reallocation from agriculture. This same pattern has characterized the company's timber holdings from year to year.

Direct employees of the company have gradually increased in numbers from just over 4,000 in 1979 to nearly 9,000 in 1992. According to company annual reports, relationships with unionized production workers are "smooth and harmonious," and compensation and benefits for workers have increased over the years. The company boasts of training programs for production and management workers, provision for attendance at overseas conferences for professional staff, industrial safety programs, and assistance with home ownership for workers. At the Laja mill training programs are regularly undertaken, a townsite is fairly substantial and endowed with schools and medical facilities, and hot lunches are provided for workers at the mill cafeteria. Some 13 per cent of employees are university graduates. This mill – indeed, this company – like those in Brazil, is concerned with creating high-quality products for markets that are increasingly outside the country. To do this, it must have well educated and well trained personnel in very modern pulp and paper mills.

The problems for workers are not within the large-company, direct-employee sector. They are in the feeder sawmills and woodlands operations under contract to small operators. Since these operations are generally seasonal, large companies enjoy the benefits of arm's-length employee arrangements. They do not have to provide expensive housing and other benefits to seasonal workers employed by small contractors. That is precisely the problem, according to the critics: the big companies ultimately benefit from the steady erosion of alternative forms of subsistence for these workers, but they take very little responsibility for the welfare of the displaced and the poor, and the small contractors operate on such limited margins that they are even less likely to put workers' welfare high on the agenda.

Arauco

Arauco was established in 1972 by the Chilean state and was subsequently transferred to the private sector. In 1979 the two major mills, Constitución and Arauco I, merged. The current ownership arrangement of Arauco is complex. Carter Holt Harvey (CHH) of New Zealand owns a 50-per-cent stake in Inversiones y Desarrollo Los Andes. This company in turn holds 60 per cent in Chile's largest industrial company, Compania de Petroleos de Chile (Copec). Carter Holt Harvey has, in all, a 59.1 per cent share in Copec. Copec's affiliate is Celulosa Arauco y Constitución.

In February 1991, CHH announced its decision to sell its holdings in Chile (*Paper*, 19 Feb. 1991), but a few months later it reconsidered. The original decision had been made when a major shareholder in CHH, Brierley Investments, sold half of its 32-per-cent stake to International Paper. However, after CHH had reviewed Copec's financial performance and considered the potential markets when the new

Arauco II pulp mill comes fully on stream, it decided its pulpmill holdings in Chile were worth retaining, and it made other sacrifices instead. Copec was reported in 1992 to have increased its profits to NZ$91 million (US$53 million).

The International Finance Corporation arranged a US $95.5 million loan to Arauco for a new bleached kraft pulpmill in 1989. It provided $55 million of this itself, and a group of European banks lent the remainder. Other financing included a local bond issue and credits and loans from suppliers, plus a $56 million loan from the West German development agency DEG (*Paper*, 22 Aug. 1989).

Arauco II, thus financed, was started up in 1991 on the same site as Arauco I in region VIII. It has a capacity of 350,000 tons per year. Three hundred kilometers farther north, at Constitución, a third mill has a capacity of 265,000 tons per year of unbleached kraft pulp. In total, then, the three mills have a combined capacity of 805,000 tons per year. In 1991, pulp and paper sales accounted for 66 per cent of all sales; wood products, 13 per cent; and logs, 21 per cent; for a total sale in U.S. dollars of $346.2 million.

The Swedish papermaker Stora Koppärberg Berlag AB purchased 50 per cent of Forestal Pedro de Valdivia, a subsidiary of Arauco, in 1989. The agreement was to construct a kraft mill similar in size to Arauco I and II, about 500,000 tons per year, at Valdivia, 400 kilometres south of Arauco. The mill was planned for construction over the decade of the 1990s, and 58,000 hectares had been planted with radiata pine and another 3,000 hectares in other species by the end of 1991 (Arauca, *Annual Report*, 1991). However, in 1993 Stora and Arauco announced a delay in plans. Arauco finally bought out Stora's holdings and planned to build the mill itself.

By 1991 Arauco had altogether some 437,200 hectares of forest land, of which some 310,500 hectares were planted primarily in radiata pine (*Annual Report*, 1991). As do all the large companies throughout Latin America, Arauco boasts of its concern for the native forests. Of its total 437,000 hectares, 83,000 are said to be protected native forest land. According to its *Annual Report* for 1991, the company has spent some US$40 million in preserving the forest. (This figure is stated as Chilean $14.980 millions, equivalent to US$40 millions in the report.) This claim is difficult for many forestry specialists to believe. For comparison's sake, the annual budget for the forest service to fight fires is about US$3 million.

Arauco accounts for 70 per cent of Chile's pulp exports. The larger part of these (47 per cent) go to Asia; Europe takes 31 per cent, and the United States 16 per cent. Of total forest products, Asia is the destination for nearly 50 per cent. The domestic market takes 33 per cent, Europe 12 per cent, and the United States 5 per cent (in each case, the remaining 1 per cent is unspecified).

Finally, Arauco, like other large forest companies on the continent, provides education and training for its workers, medical facilities, and financial aid for accommodation in towns adjacent to the mills. It also contracts out its logging, with the same issues arising from that practice as for CMPC.

Celulosa Santa Fé (Papeles Sudamerica)

An integrated, 140,000-ton-per-year pulp and linerboard mill based on radiata pine was planned in the early 1980s by Papeles Sudamerica at Nacimiento near Los Angeles in Chile's region VIII. The company, half owned by Inforsa at that time, went bankrupt in 1985, and eventually North American co-owners took it over. The co-owners are Shell (60 per cent), Scott Paper (20 per cent), and Citicorp (20 per cent). The name was changed to Celulosa Santa Fé. Scott Worldwide has the rights to 80 per cent of Santa Fé's output. In March 1991 a new mill, now based on eucalyptus, started up, with production capacity of 230,000 tons per year of market kraft pulp (*PPI*, March 1992:38). The mill will be first in Chile to use eucalyptus as its major or exclusive fibre source, although CMPC and Arauco are both planting more eucalyptus wherever the soil is suitable. If the end product is market pulp for most paper products, and as long as the market prices hold, a faster return on investment can be obtained from the hardwoods than from the pines. (For further studies of companies within the Bío-Bío region, see Moraga, Eugenia, and Rodriguez, 1989; Hochfarber, 1990.)

Papeles Bío-Bío

Papeles Bío-Bío has a capacity to produce 110,000 tons per year of newsprint. This was part of the CMPC empire until 1987, when it was sold to Fletcher Challenge of New Zealand. FC sold its 49 per cent share in the company in 1993 to a group of U.S. institutional investors (*PPI*, Nov. 1993:32). The reasons appear to lie in Fletcher Challenge's financial condition at the time rather than in deficiencies at the Chilean plant, since newsprint is a successful paper grade produced in Chile. The domestic market, split between INFORSA and Bío-Bío, is about 70,000 tons per year; as well, these two companies exported about 155,000 tons in 1993, with destinations in Europe, Asia, and the U.S.

Daio

Daio Paper announced plans in 1989 to construct a $520 million, 350,000-tons-per-year pulp and newsprint plant at Osorno in southern Chile, to be in operation in 1998. Fibre sources would be eucalyptus and radiata pine. The Japanese market would absorb the output. Daio, having 84 per cent of the shares in partnership with Itochō, confirmed the purchase of 22,000 hectares of forest land (*Paper*, 17 Aug. 1989), and details are provided by INFOR-CORFO in its listing of proposed megaprojects. Daio, through Forestal Anchile, had already started to clear land for plantations, indeed had announced in Japan that it had "purchased 60,000 hectares of old-growth forests, which will be cleared and the wood sold off to cover the plantation costs" (JATAN, 1993:10) before the Terranova incident (reported further on). Daio has since changed direction and is purchasing abandoned agricultural and grazing land. It says that it is now limiting its logging of native forests.

THE WOOD-CHIP INDUSTRY

Chipmills, exporting woodchips from both native and plantation forests to Japan primarily through Mitsubishi, Itochu, and Marubeni corporations, are very profitable enterprises in Chile. Operating mills have required investments under US$15 million. On that investment and operating at full capacity, they may produce up to US$20 million per annum, reaping as much as $8 million a year in profits (Lara, 1992). These companies are not generally owned by the pulpmill complexes, though they may purchase plantation and native round-wood and on occasion sell chips to the domestic mills. There were four major exporting wood-chip firms in 1991 and numerous smaller producers who sold chips to these exporting firms.

Mitsubishi owned one of the four exporting companies, and the other three were owned by Chilean interests; another three were in the planning stages. One of the latter is Terranova, now co-owned by the Chilean group CAP and Marubeni. CETEC Engineering Company of Canada is reported to be a third partner in the proposed expansion (Lara, 1992). Terranova owns land throughout region X and near the towns of Talcdahuano, Chillan, and Cabrero. Of its activities, CODEFF has critical insights, which are discussed later. Here it should be noted that Terranova has announced plans to construct a $50 million, 100,000-ton-per-year sawmill in region VIII (INFOR-CORFO, 1992).

Japanese paper companies Sumitomo and Sanyo-Kokusaku became become active partners in a purchasing arrangement with a new wood-chip mill in 1992. This is located at Punta Arenas at Chile's southern tip and utilizes natural forests consisting mainly of lenga trees. The fragile ecosystem of the region is a cause for concern amongst environmentalists.

Other Japanese interests

The connection with Japan is made partly through the exports of logs, chips, and pulp in ever-increasing quantity. In addition, the consortia of companies associated with Itochu includes some of Japan's major enterprises – Mitsubishi, Marubeni, Sumitomo, Oji, and Honshu. In Chile, as increasingly elsewhere, Japanese companies secure raw materials through five-to-ten-year marketing contracts and minority shareholding where necessary. This strategy allows them to avoid long-term landholding obligations, local opposition to either foreign ownership or ecological damage, and commitments to labour. As well, provided there are various sources for raw materials, Japanese corporations can stay aloof from political changes in supplier regions and can retain considerable control over international markets (this argument is also made in Lara et al., 1991).

Chilean Forestry News reported a decrease in exports of radiata pine roundwood to Japan over the late 1980s (April 1992:14). In its interpretation, a combination of economic changes in Japan and increased competition for the Japanese market by New Zealand logging companies accounted for the larger part of the decline. Chilean roundwood exports have now entered more diversified markets in Korea, China, and Turkey. As well, more domestic demand absorbs the production today than in the 1970s. What is of interest in this report is a note to the effect that Japanese imports have involved increasing proportions of sawnwood. In 1960, thirty times more timber was imported in log form than as sawnwood; the amount was down to four times more by 1990.

Though Japan may be importing more sawnwood and fewer logs, the demand for raw material for pulp and paper mills continues to be strong. The expanding relationships with Chilean investors bring yet another dimension to the debate on Chilean native forests. According to critics (discussed below), the government, and especially those sectors of government concerned primarily with economic growth and foreign investment, is unwilling to withstand or incapable of resisting the pressures from Japanese companies to increase the plantation acreage. Since the only possible way by which this increase could be made at this stage is through substitution of native forests, Chilean ecologists identify Japanese companies as major actors in the Chilean forest drama.

SOCIAL IMPACTS

Officials of the Chilean forest service (CONAF) whom I interviewed in connection with fieldwork in Chile expressed concerns about labour shortages experienced by forest companies and the need to move workers from elsewhere to provide the labour. They noted that small contracting firms who do the logging are highly competitive with one another and pay low wages to their workers. The displacement of local farmers constitutes a social problem, but it has the positive effect that it moves them off marginal and degraded land used for agriculture. In the officials' view, industries have provided some local and regional development aids, but local governments have not focused on rural development, and local populations are not well informed of available aid and tools.

The development of plantation forestry, again in the view of officials, has provided jobs and foreign-exchange earnings. Where previously, native forests were selectively logged by companies who put nothing back into the rural economy, the pine plantations have provided reforestation, silviculture, and logging work, and have regenerated degraded land. Work in the native forests diminished when the plantation forests were instituted. An important source of deforestation in native forests has been cattle ranching, mostly under small owners without land title.

Forest service officials also discussed the issues connected to Indian land claims. Reservations were assigned to Indians at the end of the nineteenth century, but various subsequent events, such as the agrarian-reform laws of the 1960s and early 1970s, reduced native land rights. In the view of the officials, the national government had been active in trying to stop the take-over of native lands. It had purchased land from private companies and other owners in order to give it to native groups. INFOR-CORFO was, they said, engaged in developing small projects for the benefit of Indians and small farmers. These relatively mild assessments of social conditions connected to plantation forestry are in sharp contrast with the views of critics. It should be noted, however, that internal conditions shift in such changing political conditions as Chile has experienced, and socio-economic research is now being undertaken within CONAF.

A very different perspective is advanced by organizations representing native peoples, especially in region IX, where various Mapuche groups have suffered expulsion at the hands of private claimants to traditional lands. The Mapuche traditionally inhabited a large territory extending southwards from the Bío-Bío River. Never subjugated by the conquistadores, they resisted integration and reservation status. The Pehuenche, a hunting-gathering tribe of Mapuche, continued to inhabit the araucaria forests, eating herbs and seeds, until the plantation economy, combined with planning for a hydroelectric dam disrupted their lives (Sears and Bragg, 1987).

The ownership issue has moved through various courts. In one particular case, that of the Pehuenche people at Quinquen who contested a private claim in 1989, the various courts and finally the Supreme Court supported the private owners against the native peoples. CODEFF launched a public campaign, and President Aylwin, while still a candidate for office, gave a public commitment to fight for the rights of the Quinquen people. When his government was installed, it paid the private claimant and created a national reserve for the Pehuenche. However, this is a rare event and may be the only one where the native peoples, supported by environmentalist groups, won a land-claims case (private correspondence with A. Lara, 1993). Plans for a hydroelectric dam on the Bío-Bío River are still in the wings; if it goes ahead, ancestral lands will be flooded.

The critics point out that rural areas have been taken over by plantation timber companies with little or no compensation or help to the displaced populations. Many larger (in excess of 1,000 hectares) and middle-sized (between 100 and 1,000 hectares) farms have been sold to the companies, and land previously held by peasants on customary rights without land deeds has been expropriated (Rivera and Cruz 1983; Cavieres and Lara, 1983; Cavieres et al., 1986; Otero, 1989; Lara et al., 1991; Lara, 1992). Impoverishment of small landowners, expulsion of forest dwellers, and buy-outs have caused widespread migration. Clearly these groups have not benefited from the plantation economy. Cavieres and colleagues (1986) and Otero (1989) have documented the relationship between acreage planted and reduction in the rural population. Many migrants end up in rapidly growing rural towns where poverty, unemployment, and deficient service systems depress all inhabitants.

There has not been an increase in employment in the forestry sector proportionate to the increase in exports. Between 1973 and 1990, exports increased almost twenty-two times. In the even longer period from 1968 through 1988, full-time jobs in forestry increased by 30 per cent (Contreras, 1987; Chile, INFOR-CORFO, 1989, 1991). As well, the sector is characterized by low salaries, job instability, and lack of hygienic conditions and comfort in labour camps (Otero, 1984). Although contractors may have specified requirements in their contracts with the big enterprises regarding the provision of food for loggers and there may be penalties for cheating, contractors frequently take advantage of poor and often illiterate workers. The officially reported accident rate is 20 per cent for forestry workers, twice that for industrial workers in other sectors. These accident rates are related to growing mechanization in the woods. According to a university researcher who studies the impact of mechanization on rural workers, most are relatively small in stature, and they are working with large power-saws. The same situation pertains in sawmills, where accident rates may be even higher (Elias Apud, Concepción University, interview notes). Unions were disbanded under Pinochet and have not yet been re-established, though there appears to be some nascent organization developing (GEA, 1990).

Elias Apud contends that some portion of rural workers and many more of the unemployed migrants are Mapuche Indians, displaced from their forest lands because they could not prove ownership. Other academics in forestry are less sure that Mapuche Indians have gained any employment in the forest industry. The cause of the Mapuche, and aborginal land rights more generally in Chile, has become front-page news (e.g., *El Mercurio/El Páis*, 10 julio 1992). Demonstrations are now attracting non-native supporters and international attention, as is true for native peoples in northern Brazil.

ECOLOGICAL IMPACTS

As noted above, there are numerous critics of the destruction of native forests. CODEFF is the organization through which many academic and other critics produce analyses of forest policy. CODEFF reports, for example, that in 1990 the Chilean company CAP and the Japanese company Marubeni, under the title of Empresa Forestal Terranova, applied to CONAF for authorization to plant eucalyptus on an area of between 14,000 and 23,000 hectares then forested in region x (called the Venecia Estate and owned by the company). They had to apply for permission because current regulations added to Decree Law 701 prohibited the replacement of native evergreen forests. The company wanted to replace all forests on slopes less than 45 degrees. Because of strong opposition, including petitions such as that noted in the opening paragraph of this chapter, the government finally established a commission of international experts through FAO and a Chilean interministerial commission. Both reports recommended against the replacement of native forests with plantations. According to CODEFF, "These reports were delivered to the Government and accepted by the Ministerio de Agricultura (Ministry of Agriculture) but not by the Ministerio de Economia y Hacienda (Ministry of Economy and Finance), in whose opinion the reports limited the performance of the private sector and endangered foreign investment" (CODEFF,

1992:2). Numerous reports later, a National Forestry Commission (CONAFOR) recommended substitution in all forests situated on slopes less than 45 degrees. Bills were proposed, and what might be perceived as compromise was achieved by mid-1992 in an initiative approved by CONAF to permit forest owners to replace 25 per cent of native forests situated on slopes less than 30 degrees. This is a compromise only from the viewpoint of private forest owners, since the ecologists, the Chilean opponents, the FAO, and the original national committee all objected to any substitution.

Various writers on Chile note specific environmental impacts, such as the loss of species of wildlife, trees, and shrubs. The list of endangered species keeps growing. Habitat destruction and herbicide impacts are frequently reported. More problematic in the long run is the destruction of whole ecosystems, not just of specific species. After 1990, a couple of endangered tree species were put on the "protected" list and could not be cut, but saving particular trees while destroying the forest around them does not save biodiversity or ecological systems. Soil erosion has also been a serious concern. Lara reports on a study in press that documents massive soil losses following the logging and burning of radiata pine plantations in region VIII. If these reports are accurate, the future of the plantations would be jeopardized, and the companies, as well as the ecologists, should be worried. In addition, watersheds have suffered negative impacts, and streams have become clogged (Lara, 1992; Otero, 1990).

ALTERNATIVES AND THE 1992 PRESIDENTIAL INITIATIVE

The critics of forest policy are not arguing against the existence of a forest industry or against plantations situated on already degraded or marginal land. Their dispute is with the logging of native forests by the wood-chip industry and the displacement of these forests by exotic plantations. They have made some positive arguments for reform. They contend that native forests are potentially productive and profitable units if put under sustainable management practices. In their opinion, there is adequate research data supporting such a conclusion; the research is ignored because short-term gains are greater from the current destructive practices. Schmidt and Lara (1985) estimated yields of 9–14 CM per hectare per year for thinned second-growth forests, compared to 20–30 CM per hectare per year for radiata pine and eucalyptus plantations. Although the yields are lower, native timber prices are two to four times higher than prices for radiata pine (Lara, 1992).

The plea for the preservation of native forests, or for sustained management of them, is rooted in the belief that native forests in and of themselves are vital to the health of the planet and are essential to the sustenance of endemic species of flora and fauna. As well, where areas have been deforested and are now eroded, reforestation is necessary. Thus the argument goes beyond preservation of what is left to a plea for national attention to regeneration of degraded land, preferably by native species. Where degradation is far advanced, the land could be turned over to exotic plantations. Another line of argument has to do with biodiversity. Monoculture plantations are subject to pests and disease. Interspersing them with native forest areas and planting various other species in addition to radiata pine and eucalyptus would reduce the potential for disaster.

These recommendations, advanced through CODEFF and individual forestry experts and academics, are opposed most vigorously by the association of wood-product companies (particularly in the wood-chip business) and would require legislative changes. Subsidies for current practices would be reduced, and incentives introduced for the reforestation of native forests. The Association of Foresters proposed something along this line in 1979 (Cavieres and Lara, 1983), and CONAF prepared a draft policy in 1984. Further drafts have been advanced more recently. A proposed law for the "recovery and promotion of the native forest" followed the debates so far reported. Advanced by the president in the spring of 1992, it offered small correctives on current practices and provided little comfort for the reformers.

The new law has two stated objectives. One is to establish incentives to regenerate and manage native forests. The other is to change regulations governing their exploitation. A subsidy is to be introduced to allow companies to manage native forests so that the number of intermediate-stage trees can be increased. That is, mature trees would be phased out, and selected species would be planted and grown as plantation trees. As noted in an earlier chapter, this practice saves the forest after a fashion, but does not save the overall ecological system.

Other incentives somewhat modify this scenario. They provide for complementary planting of "valuable commercial native species" of the same forest type beneath the tree canopy and improvements in the density and composition of forest masses. Afforestation with native species on terrains classified as appropriate for forestry is to be encouraged (*Chilean Forestry News*, May 1992:14–15). Subsidies are to be provided for silvicultural practices – pruning, thinning, and general management of the forest. These subsidies are directed specifically to small forest owners in order to reduce their marginalization. Small owners are to receive 85 per cent of the costs incurred in these activities, while larger landowners will receive 75 per cent (i.e., the same subsidy as pertains under Decree Law 701).

Substitution regulations are modified in the new legislation, though they would not meet the demands of many critics. An environmental-impact study for all substitution projects covering more than 500 hectares will be required. Regulations in the law aimed at protecting watersheds and streams will be enforced. Substitution will not be permitted on slopes steeper than 30 degrees (but by implication will be permitted on lesser slopes). Certain native forest species are to be protected, including araucaria and alerce, and logging will be prohibited in habitat areas of endangered or rare flora and fauna. Substitution will be permitted of up to 25 per cent of the native forest on each private property, but the part to be substituted must be that with the poorest quality forests. If an owner can demonstrate that a larger proportion of the native forest on his/her property is degraded, other options are permitted. In short, this law, though heralded as a major breakthrough on behalf of the native forests is modest indeed.

SUMMARY

The plantation-pulpmill economy has been established in Chile, but it is more controversial than in Brazil; critics include foresters and environmentalists, who argue that forest policies have favoured companies at the expense of ecosystems and indigenous or poor people. Native forests are in decline. Investment, including foreign investment, has increased, and afforestation by plantation species has taken place, but public subsidies have been expensive.

13 Conclusion: Sustainable Forests and Communities

There are currently many plans for sustainable use or sustainable development that are founded upon scientific information and consensus. Such ideas reflect ignorance of the history of resource exploitation and misunderstanding of the possibility of achieving scientific consensus concerning resources and the environment. Although there is considerable variation in detail, there is remarkable consistency in the history of resource exploitation: resources are inevitably overexploited, often to the point of collapse or extinction.

Ludwig, Hilborn, and Walters, 1993:17

Deforestation is a central problem of our time: it impoverishes the earth, threatens the survival of all species, reduces biodiversity, destroys subsistence cultures and human, as well as other creature, habitat, and robs humanity of a spiritual home. Even so, consumption of wood products continues to grow. It grows because people need fuel and lack other sources even more than because they turn wood into houses and advertising copy. The demand for fuel wood grows out of poverty; the massive waste in excessive buildings and unnecessary newsprint results from affluence. Can we halt deforestation while still sustaining human communities?

As the three scientists quoted above argue, the question cannot be answered in technical terms. We have knowledge and science enough to save forests: we need social, economic, and political capacities to accomplish that goal. In light of the evidence presented throughout this book, let us now review the issues introduced in chapter 1 and consider the possible solutions.

POPULATION GROWTH

Although halting population growth is an essential undertaking for the earth's survival, we have not found evidence that population pressures are directly responsible for most deforestation. When people lived in forests, their appetites were limited to its fruits, and conservation was essential to survival; they could not afford large numbers of dependent children. When they were dispersed and the cultural controls on population growth were removed, they lost the sense of conservation. Populations have steadily grown since the 1950s, but the regions first showing severe signs of deforestation were not then over-populated; further, some of the regions now most endangered, such as the Amazon, have very low population densities. Finally, some of the population pressures are artificially created by government trans-migration or settlement policies, whereby poor people are (sometimes forcibly) shifted away from fertile agricultural regions, where good land is already privately owned, to poor forest land, where they are out of sight and less dangerous. Population controls are essential for long-term sustainability of all resources, but in the immediate future, what is more desperately needed is land reform.

ECOLOGICAL CONCERNS

Trees are essential components of the earth's cycle of carbon dioxide and oxygen. They conserve energy and mineral nutrients of the soil, moderate micro-temperatures, and provide necessary controls on the water-table. Although there are debates about precise effects on macroclimates, most world experts agree that deforestation has very serious impacts. In addition to climatic impacts, deforestation causes the leaching of nutrients from the soil, drying and compaction, general erosion, desertification, and the flooding of downstream areas.

While deforestation agents are numerous and long precede the development of industrial forestry, the industry is responsible for much of the deforestation occurring in the second half of the twenti-eth century. This occurs in temperate rainforests, in boreal forests, and in subtropical and tropical forests. Papua New Guinea, Myanmar, and northern Alberta are following the fates of the Philippines, Thai-land, and British Columbia. Brazil is repeating its own fate, transform-ing the Amazon through ranching and plantations today, just as its colonial élites transformed the northeastern coastal forests to plant sugar cane. The northeast no longer supports agriculture, and its people have fled south.

Forest Reserves

Forest reserves are among the proposed solutions to deforestation of natural stands (e.g., UNESCO, 1972; IUCN, 1980; US, Dept of State, 1981; McNamara, 1982; World Bank, *Report*, 1980). UNESCO's Man in the Biosphere program proposed biosphere reserves and a World Heritage Convention (Vernon and Christy, 1981, reviews the proposals). The International Union for the Conservation of Nature and Natural Resources, together with the World Wildlife Fund and the United Nations Environment Program, articulated several principles for a "world conservation strategy" in 1980. These included preservation of genetic diversity and maintenance of essential ecological processes. They also included sustainable development of rural areas in developing countries. At the present time, logging within reserves takes place even where the principles have been adopted by governments, and it is justified by reference to the surrounding rural poverty (see IUCN, 1980; Guppy, 1984:956–7). To make reserves and the principles of preservation work, we need to establish alternative means of subsistence. As observed by the Brundtland Commission, sustainability of the environment is inextricably linked to the reduction of poverty in the Third World.

Developed countries could provide much more of the financial support for tropical reserves. Ira Rubinoff (1982) proposed a strategy for the preservation of "a sample of the world's remaining tropical forests" in the form of an internationally financed system of tropical moist forest reserves. Under this plan each country with tropical forests would be paid an annual amount based on the area protected for custodial duties. Maintenance of the reserves would be monitored, with payments adjusted accordingly, so that incentives to protect the land would be sustained. The funding would cover the additional costs of intensification and diversification of agriculture in non-forest areas, so that pressures to provide new agricultural land could be met by developing countries, and payments would be made for public forestry education programs. Reserves would include about 100 million hectares of the world's tropical moist forests, or roughly one-tenth of the remaining land. While this plan would not stop deforestation, it would at least create a safety valve, preserve some diversity, and provide some options for the future. Rubinoff gives a detailed plan for financing by developed nations with high per-capita gross national products. There is already an example of such forest reserves in Costa Rica, with experimental and research stations financed by international funds and participating universities in the developed countries. Eco-tourism provides a secondary source of income for maintenance of the reserves.

Plantations

Plantations are preferable to the logging of remaining natural forests, but they do incur ecological costs. The ecological concerns include the inability of plantations to provide either habitat for wild animals or grazing for domestic cattle. They are monocultures, and the restricted gene pools are of considerable concern. Pines and other coniferous species are less controversial than hardwoods. Eucalypts and similar species are problematic because the trees have little top vegetation and do not modify the impact of rain on the soil or check soil erosion. The evidence is not consistent regarding their impact on the water-table, though many knowledgeable observers contend that their net effect over time is to deplete it. They definitely affect micro-climates, but their impact on macroclimates is yet to be determined.

Ecological impacts of plantations are still under debate, but there is little doubt that plantation forestry can be a great boon to developing countries and may save remaining tropical forests. Much depends on the location of the plantations. Because it is much more expensive (and sometimes impossible) to rejuvenate marginal lands or lands previously destroyed, companies will not undertake that task voluntarily. Instead, many will choose to cut growing natural forests if they can get away with it or plant on land that might yet be regenerated in natural stands. This propensity has actually been intensified by the Rio Conference resolution to ban all logging in tropical forests by the year 2000. In Indonesia and Malaysia, the stated intention of companies and government is to log out the natural forest so that land is freed for plantations. The Rio declaration gave an incentive to do so before the turn of the century.

Logging Bans

Bans on logging would work only if they began immediately, all countries agreed to them and had enforcement capacities, and all importing countries refused to buy contraband wood. These conditions are not likely to be met in a world that wants wood and, even with all the warnings about catastrophes linked to deforestation, is not dedicated to saving natural forests. Incentives for planting on poor land or prohibition of planting on good forest land would be more effective, since these approaches would reduce the advantages of cutting tropical forests. If the issue is viewed in market terms, we might suppose that when land itself becomes more expensive in plantation regions, companies will more readily begin land reclamation projects. However, by then they may have destroyed the remaining forests.

Two decades of creating eucalyptus and acacia plantations have not resulted in much objective literature on their ecological impacts. Many experts on plantations are embedded in the industry and have a self-interest in its expansion. The same international consultants conduct feasibility studies everywhere, each time reaching conclusions in favour of logging natural forests and establishing plantations. They consult with companies and governments. Feasibility is determined by resource supplies, costs of production, and markets, not by ecology, let alone spiritual, cultural, and social values. If the World Bank and other international agencies can finance plantation development, then the same and allied international bodies should establish guidelines for feasibility, environmental, and social-impact studies. The guidelines, which are the rules of the game, should be discussed at the village level, as well as with companies and governments, and should be announced publicly.

Pollution Control

Ecological concerns are not restricted to deforestation; they include the pollution generated by industrial expansion. Pulpmills have downstream impacts far distant from the mills, and acid rain is caused in part by pulpmill emissions into the air. Although the debate on dioxins is not yet concluded, what is known is that something in pulpmill effluent causes mutations in fish. In the past, environmental-impact assessments were not conducted, and even more recently where they have been done, recommendations have often been ignored or requirements not enforced. In British Columbia, evidence was presented that many mills failed to invest in effluent controls during economic boom periods. Developments in Alberta went ahead despite assessments that recommended against them on the grounds of potential damage to northern rivers and terrain.

In general, pulpmills constructed in the late 1980s and 1990s included much more effective effluent-control systems than earlier mills, and it is reported to be now possible to reduce effluent to zero through recycling of wastes in mills. State-of-the-art mills in Brazil are ahead of older northern mills in their recycling and effluent-control methods. Thermomechanical mills are less environmentally dangerous than chemical pulp processing, but the choice of pulping process depends on such other conditions as available energy sources and water and the relative costs, and opportunities, as well as on fibres. While regulations and taxation rules could favour the technologies best able to protect the environment, and world standards might improve operations, it would be unrealistic to demand that all kraft papermills or all bleaching of pulps be abandoned. Advanced effluent-cleansing operations could be required, but again, the implementation of these would depend on available capital for installation or renovation and maintenance. Consumer habits may change as information on pollution becomes more widely understood. Unbleached paper, for example, would be acceptable for many purposes, and recycled papers are already gaining markets (although the net effect of the technology of recycling is not shown to be less damaging to water than the effluent of ordinary mills).

PROPERTY RIGHTS

Is there clear evidence that one system of rights is less disastrous than any other? Unfortunately, generalizations about huge social systems are as unwieldy as those about large ecosystems. There are simply too many variables to pronounce on the relative merits of diverse private and public property systems. The evidence appears to favour a combination of small woodlot owners, local log markets, and separation of manufacturing from land ownership. State governments in such systems can effectively oblige landowners to engage in sustainable forestry practices and force manufacturing companies to restructure in the interests of the larger society. This approach has succeeded in reforesting Sweden and Finland, though second and third growths have less diversity than the original. The Japanese model is close to the European one, but it is complicated because so much of Japan's raw material comes from external sources. Internally, the cooperative system, combined with national forestry regulations and the retention of long-standing cultural traditions in rural regions, has protected land and regenerated the forest.

The American model includes enormous corporate estate owners, as well as state lands; sometimes land is regenerated on either private or public areas, sometimes both are mismanaged. The disadvantage of ownership by large companies is that there is no capacity to oblige private companies to operate in the interests of the public during economic downturns, such as when log sales to Japan will generate more profits than employment of local workers in milling the same wood. The Canadian model involves public ownership, but private control of forest land, and it does not seem to enable governments to reforest or influence private companies to undertake the task. It has the disadvantage that public subsidies are high, yet de facto control does not rest with the public. The huge corporations that dominate North America appear to be inferior instruments for sustaining forests.

In addition to their deficiencies in other respects, huge integrated corporations avoid market prices for their resource when they create internal pricing systems between their logging and manufacturing units. Where such corporations are dominant, there are no effective log markets, and valuation of logs is done in the interests of the manufacturing firm.

Developing Countries

Developing countries face a different set of issues. The land may be public in the sense that no individual has formal title, but may be held in common by villagers or indigenous peoples who assume the rights of ownership. As forestry becomes profitable or plantations look feasible, governments take over the land and then sell or lease it to companies. Private property rights may also exist in such systems in the form of estates representing earlier land-grabs by colonial powers, local gentry, or military juntas. Control of good agricultural land is typically held by the same élites, so that those dispossessed by plantations are unable to move elsewhere to develop subsistence or cash crops.

Even were they so inclined, governments in developing countries face constraints in caring for the poor. At the internal level, they are obliged to strive for political and economic stability, to be achieved, if at all, by the creation of industries and employment. Externally, they typically face market forces which they cannot control, accumulated debts which they must repay, and controls by international governments (the IMF, the World Bank) which they cannot evade.

International funding agencies have devised "community (or social) forestry" projects ostensibly designed to help the dispossessed. The objective is to engage village populations in planting or other tasks close to or in plantations. They sell their produce to nearby mills or on a local market. This approach may provide some benefits to villagers where villages are actually communities with recognized leadership and labour arrangements. But today many villages are not communities of long standing; they are groups of poor people forced to live together because their alternatives have been extinguished. They may or may not share a culture, language, and history, and members may work together or go off in all directions in search of income and employment. Community projects sometimes founder because they rest on benign assumptions about the villages their proponents want to help. As well, foresters may establish plantation projects for reasons that are not shared by villagers and may then be disappointed by village responses; or they may evaluate project success or failure along different dimensions than those used by villagers. The record of failures in these projects is substantial, and the root cause is that much of what external advisers value or their notions of how things should be done is not consonant with the realities of village life or the aspirations of villagers.

Since the need for social forestry is created because most of the land has been pre-empted by wealthy landowners, companies, or governments, social forestry tends to be about salvaging leftover plots for the dispossessed. If the objective were to enable the poor to survive, land redistribution would be more appropriate. Social forestry and agro-forestry may be the most that the unpropertied can hope for, but if international agencies are going to ease the world's conscience about macro-forestry, they might at least create opportunities for the development of effective community decision-making processes for social forestry projects.

International Struggles, Local Communities

Despite its transnational organization, the plethora of international agencies backing it, the global consulting and engineering firms urging it forward; despite military regimes and client-state governments; despite jail terms for opponents and exile for some; the global forest industry has failed to silence environmental activists. Often working with organizations of indigenous peoples, activists continue to mount resistance to the control of forests by global corporations and corrupt governments.

Whether it is the Mapuche of Chile, the Dayak of Indonesia, or the Penan of Malaysia; whether it is Greenpeace or the World Wildlife Federation; in British Columbia or in Thailand; the fight over forests is no simple struggle over trees. It is a fight about who owns the earth, who has the rights to its fruit; it is a fight about survival. The allocation rules are at stake, and those who have, in every country of the world, arrogated the earth's resources are under attack. What the critics ask for is the overturning of private rights in forest lands; that is, the communalization of property, a reversal of historical trends. They are claiming that the rights of people who live in a region should take precedence over the rights of capital. In their arguments, the capacity to decide on the fate of a forest should not rest with a company or an individual whose sole claim is the payment of cash (let alone the traffic in influence). In this view, forests, as common property, are inalienable.

Countering such a view is the argument that common property is not well managed because no one takes it as his/her personal task to protect it, and there are no private benefits involved in conservation. As well, argue the proponents of private property rights, the market society based on private ownership has survived precisely because it works: the "invisible hand" that depends on each individual maximizing his/her own benefits does end up creating optimal social outcomes. Central government controls do not work; they inevitably end up being control by an authoritarian, bureaucratic élite or dictatorship. Community ownership would likewise end up reflecting the interests of the few, while the many are effectively excluded.

Arguments such as these are often imperceptive or ill informed about the very wide range of conserver activities that have occurred over many generations in every society where small groups share land or fishing areas and manage these as common property. Even in an industrial society, the Japanese cooperative system includes well-tended common properties formally owned by communities and shrines.

However, cooperation is not built into large-scale market-driven societies, where populations share few interests and survival rarely depends on collective action. There are groups with strong ties to a geographical location who may be able to accommodate both external markets and internal systems of common property management, as some native bands in British Columbia say they will do if land claims are negotiated in their favour. But such intentions are at odds with market influences, and community cohesion is difficult to sustain when individuals are tempted by market opportunities. Even where individuals are prepared to put community survival above their own interests, external markets may diminish community capacities to survive: such is the case in rural Japan.

Without romanticizing small societies or ignoring market realities, we can observe that some communities, even in industrialized societies, are sufficiently cohesive that they could sustain resource-management responsibilities. Moreover, more communities would form and stay together if that option were available. If those who defend tropical forests really want action, they might turn their attention and funds toward establishing communities as owners and managers of tropical forests and plantations, where the rules are negotiated, explicit, and fully agreed to before the transfer of land and funds takes place. The rules would have to provide a delicate balance between conserving natural resources and ensuring that the community has sufficient room to make decisions.

Privileges once gained are rarely given up voluntarily, and those who own and control land in forested or agricultural regions show no inclination to redistribute it so that the dispossessed can survive. Militant struggles are inevitable between the privileged and the poor unless the international community creates capacities for change through incentives and rewards. As long as international agencies simply feed funds to investment groups, even where the money is ostensibly to save tropical forests, they contribute to growing inequality and deforestation.

MARKET CONTROLS

Taking as the objective the creation of sustainable forest practices, would the world achieve that goal more readily with totally free markets or with market restrictions?

Import Restrictions

One form of market restriction is for importing countries to impose conditions on the entry of products. Austria introduced a compulsory labelling scheme in 1992, and Holland announced that after 1995 it would allow imports of tropical timber only from sustainably managed forests. These moves had the effect of publicly warning exporting countries that improvements in forest management are expected by importers, but they are controversial as mechanisms for change. Southeast Asian producers strenuously objected to such limits on free trade. The London-based Environmental Economics Centre argues that rewards to good forest-managing countries would be preferable to trade bans, and such rewards might take the form of preferential access to import markets for those who engage in better management (*Economist*, 14 Nov. 1992:40).

During the public outcry about the conditions of native peoples in Sarawak, European and American governments imposed bans on imports and attempted to pressure the Malaysian government to alter its behaviour. Since most of the logs from Sarawak went to Japan, only the closure of the Japanese market would have made a difference. Some groups advocated a boycott of Mitsubishi products because that company was the most visible of Japanese firms trading in wood from Borneo (Sesser, 1991:65). Log exports were finally stopped in 1993.

Log Markets

Log markets have the great merit of providing a mechanism for pricing logs. Where standing trees are not valued, logs are not valued either unless an economic value is attached by a buyer. Where there are numerous woodlot owners and manufacturing companies are obliged to buy logs at competitive market prices, there is a built-in incentive to utilize the resource carefully. Where there are huge companies with free or nearly free trees, the incentive is missing and much wood is wasted. So, at least, runs the theory of markets, and, indeed, we have seen plenty of evidence in its support. The theory is less persuasive in relation to two conditions: log exports from tropical or other non-replaceable stands and international markets dominated by very large corporations that still have access to cheap wood in at least one location.

Log-Export Bans

A second form of market restriction would be log-export bans. Thailand and Indonesia instituted such bans, though by the time they did so, much of their natural wood was already gone. British Columbia has maintained a ban on log exports and combines it with public ownership of forest lands. The United States instituted a ban on the export of logs from national forests but could not enforce restrictions on owners of private lands. These policies are all controversial for those who prefer unfettered market forces to prevail. However, if the objective is sustainable forestry, do these export bans work?

Log-export bans do not receive unmitigated approval. The journal *Science* published an economic analysis of the tropical timber trade which argued that the boom-and-bust pattern in the tropics is in part caused by bans (Vincent, 1992:1653; see also debate in FEPA newsletter: Uhler 1991, for similar argument about British Columbia). Advocates of log markets argue that exports would signal to domestic producers and forest owners (or governments) the real prices of logs and thus the appropriate levels of stumpage to be charged. International prices would reflect conditions of supply and demand; as supplies diminished because of excessive logging, prices would rise. Stumpage increases would then oblige users to develop more efficient utilization standards and would impose upper limits on the expansion of processing capacity; thus excessive logging would decrease. High prices would reduce demand and increase the likelihood of developing alternatives and improving technology for better utilization. In theory, an equilibrium would be reached, and sustainable levels of logging would be attained.

Let us carry this argument to its speculative conclusion. If there were a scarcity of logs on world markets, the price would rise, but a scarcity would be reached only if all forests were reduced to a level beneath the total demand for logs. Thus the pricing mechanism as a means of creating incentives for sustainable management would become effective only if total world supplies were diminishing rapidly while consumption levels were increasing. This may be a reasonable pricing mechanism if the article for sale is renewable. If it is not renewable at all, or not immediately, and even more, if it is essential to the earth's welfare or even rather important in that respect, then leaving its survival chances to the marketplace seems risky.

As well, international trade involves very large, integrated corporations that typically create internal pricing mechanisms between their own units. Their power to influence market prices is substantial, and in addition to skewing prices in their favour, they can deeply harm small woodlot owners trying to sell on domestic markets that have been infiltrated by corporations. The Japanese rural villages (see chapter 5) provide an example of this process. As long as participants in log markets are unequal in size and power, and as long as manufacturing establishments own much of their resource, markets are not satisfactory mechanisms for determining the economic cost of logs.

Apart from the problems of imperfect markets, there is the double-edged meaning of "value." Ecological, social, cultural, and aesthetic values are not easily measured in market terms: how much is a habitat or an ancient village worth? If we value forests for non-economic reasons, and if we accept the argument that forests cannot be replaced in much of the world, then we need to create mechanisms that allow for market or market-like economic evaluations of logs while inhibiting market forces that could destroy the resource altogether.

Controls through Stumpage

Almost everywhere that natural forests grow, there is a history of systematic undervaluation of its wood at the source. Governments have rarely extracted resource rents even remotely reflective of the costs of infrastructure and reforestation, let alone the social costs where indigenous people and peasant farmers are displaced. Market values of logs likewise fail to reflect such costs when none of the exporting countries insists on their payment by the logging companies. Systematic data and calculations of real costs are not available on a comparative basis. Occasional single-country studies are published, including that by Wahana Lingkungan Hidup Indonesia (1992; discussed in chapter 10), which found that only the foreign-exchange contribution was a positive benefit to Indonesia, and even that did not accrue to the society as a whole, but was received largely by the owners of plantations and government officials or the military. The net indirect benefits for society would theoretically derive from increased investment. Where the major investors are not resident, such reinvestment is not necessarily forthcoming. But even where it is, if the investment goes into more capacity in a sector such as forestry, where there is already overcapacity and the resource is being rapidly depleted, it does not have long-term benefits.

The Indonesia study demonstrated that during the 1984–89 period, total forest fees and charges had never accounted for more than 0.2 per cent of the government's domestic revenue and never more than 0.1 per cent of the total government annual budget. The explanation for this outcome is that economic rent is low, collection is ineffective, and the government itself is inefficient in pooling collected revenues. Presumably these conditions could be improved and stumpage could be increased, but the examples of conditions in more advanced regions does not allow us to imagine that the whole problem rests with an inefficient or corrupt government. Economic rents in British Columbia have been as controversial and problematic; there are solid grounds for arguing that these have been as inadequate and the collection as ineffective as in Indonesia.

The apparent solution would be increased stumpage rates where the land is public, vastly increased costs if land is sold outright, higher land taxes, and no subsidies. Governments might arrive at reasonable resource rents without market signals if they simply instituted adequate accounting systems that recognized the costs of regnerating forests (including the cost of a good forest service). An increase in stumpage costs across the board and simultaneously by all countries would go a long distance toward decreasing imports and wasteful utilization; obviously the downside would be reduced exports for countries dependent on them. Japanese construction companies might discover that perfect dipterocarp plywood is not really necessary for throw-away molds after all. As well, higher resource rents would go much further than theories about free markets and comparative advantage toward encouraging the development of substitutes and non-wood fibres.

Closures of Forest Land and Logging Bans

In addition to changing stumpage values, governments could alter habits simply by reducing harvesting rights in public land and closing off some land to any activity that might destroy a forest ecosystem. Sweden and Finland have already done so, and British Columbia is beginning to move in this direction. Among such developments are preserves, parks, and wildlife areas.

The Rio Conference recommended complete closure of tropical forests to logging by the year 2000, but as we have noted in the case study of Indonesia, this proposal has had the effect of accelerating the tempo of logging before that date. That example represents a lack of foresight on the part of the well intentioned, but not all closures have the same defects. Indonesia could, if its government chose to do so, establish genuine tropical forest reserves and could introduce incentives for replanting on non-forested land or greatly increase the cost of logging in natural forests. The chief problem is that there is no will to save the forests: companies and government alike treat them as an obstacle rather than a blessing, and the entire history of deforestation throughout Southeast Asia demonstrates the same attitude.

Producers' Associations and Cartels

Rational economic behaviour for tropical forest countries, if free-market arguments were correct, might lead to the creation of an artificial shortage through restrictions on exports from all source countries simultaneously, perhaps on the OPEC model. That, of course, is not what advocates of free markets have in mind.

Negotiations under the jurisdiction of the United Nations Conference on Trade and Development (UNCTAD) began in 1976 toward an international commodity agreement (ICA) for tropical timbers. A very limited conclusion was reached in 1983 regarding research and development, monitoring, increased value-added processing by producer countries, and marketing arrangements. Trade regulation was not seriously tackled (UNCTAD, 1983; UNCTAD, 1976–83). In view of the apparent inability of producers and consumers to reach agreement, Guppy argued in favour of a cartel of timber-exporting countries (1983b; 1984:958–65). UNCTAD, in particular, considered the proposal at length, but rejected it because members feared that artificially raised prices would induce the development of substitutes, and the market would collapse. Guppy observes, in response, that all the known substitutes for tropical timbers are either species from other countries or highly energy-intensive in production, such as aluminum, plastics, steel, and concrete. With regard to the first, if all forested countries combined, other hardwoods would not be available; in the second, petroleum and coal supplies would diminish, with a consequent increase in the demand for wood. He further observes that there is no real supply-demand market relationship, since mature-phase rainforests take up to two hundred or even four hundred years to become established. He argues, as do many ecologists, for the removal of such forests from the market system, but doing so, of course, would not solve the problem of developing countries for whom these trees presently provide a source of income. And on that score, the argument against an organization of timber-exporting countries, that all these countries could not be counted on to cooperate in a cartel, may have greater cogency.

Aid to Developing Countries

The Tropical Forestry Action Plan was essentially an aid package. Selected countries received an injection of funds to regenerate their forests, establish protected areas, and create plantations. Donors were not altogether disinterested, of course: they had interests in buying chips and pulp or selling expertise and technology to developing countries. Although the plan began with the declaration that poverty was the chief cause of deforestation, it did not come to grips with the existence of the poor or, more specifically, the need the poor have for forests. While money was to be spent in creating forest industries and plantations, there was no mechanism in place to really change the generally chaotic, arbitrary, and highly profitable log trades conducted in countries with remaining forest reserves by a privileged class in control of the state, or with the land monopolies owned by both external and internal investors in profitable plywood, lumber, and pulpmill industries. Indeed, as noted in chapters 7 through 12, efforts by the FAO and similar charitable organizations created an incentive to these groups to cut remaining natural forests and plant fast-growing commercial species.

Donor countries need to examine their priorities and their actions. If the priority is to enable a country to increase its self-sufficiency, appropriate actions by donor *vis-à-vis* proposed recipient might include bans on imports of raw material, but also preferred trade status for specific manufactured products, including those that compete with the donor country's own exports. Contributions might be in the form of technology transfers and full funding for the creation of sustainable industries that do not impinge on forests at all. Another objective might be to reduce land-use conflicts created by aid programs. This would mean not funding agricultural programs, hydroelectric dams, or mines that will, incidentally, reduce forest land or harm watersheds.

ECONOMIC DEVELOPMENT AND ITS SOCIAL COSTS

Economic development is the shibboleth of the twentieth century. It rarely means development carefully planned to provide long-term benefits to the whole population. It typically means growth, and growth designed by and in the interests of powerful groups. Sometimes the powerful groups are external to the region; usually there are combinations of external and internal movers and shakers. Land ownership is the essential ingredient of power in developing forested countries, and it is effectively achieved through logging concessions that permit (or even require) the concessionaire to replace a tropical forest with plantation species. Of the plantations which former FAO director Jack Westoby had earlier seen as essential for development he later commented that "nearly all ... were enclave developments; multiplier effects were absent; welfare was not being spread; the rural poor were getting poorer, and their numbers were increasing" (1983:2).

Our survey suggested that some plantation operations provided more benefits to surrounding populations. This was especially true in Brazil, where companies extend housing, medical services, schooling, and other benefits to the workforce and in some cases to whole villages near their plants as well. Because historically their plantation crops came on stream long after most of the non-tropical forests had been cut to make way for other cash crops or ranches, contemporary forestry companies south of the Amazon are not as burdened with the responsibility for displacing indigenous peoples as most companies elsewhere. On the other hand, most of them received land at rock-bottom prices from a military regime, and such land grants came at the expense of the poor, who might otherwise have had land and who were often displaced when they could prove no title. The companies uniformly claim their commitment to the new Brazil, to democracy, to helping the poor – such may be the spiritual legacy of Catholicism and Marxism! But if their words have substance, then perhaps it is time for the successful companies to invest in the reclamation of land in the Atlantic northeast, once forested but now so degraded that the migrants are even willing to try homesteading in Amazonia. Reclamation there would be costlier than opening up new lands in the tropical north, but the social benefits would be much greater, and the ecology of both tropical lands and the northeast would be improved.

The Role of Governments

Forestry development necessarily involves governments because they are the allocators of public lands and the controllers of taxation systems, import and export regulations, and much else that affects the capacities of regions to attract external investment or induce local companies to move into plantations. No state operates in a genuinely *laissez-faire* system, though much is made of the virtues of such a system. Nowhere is the state a passive participant in the economic process, merely establishing general rules for operation of private enterprise. Without government incentives and land grants or international banking support, the huge pulp and paper developments in place in North

America and now under way throughout the world would languish because free-enterprisers are generally more adept at expressing the credo than undertaking the risks associated with it.

Those who control states generally have a pronounced interest in controlling territory. Prior to undertaking economic development, they usually seek ways of ensuring that their borders are secure. Borders rather often have geographical contours that include rivers, mountains, and forests. Logging of the forests is thus not only an economic activity; it may well be a political assertion of sovereignty. Establishing settlements in remote forested areas has the same effect. These patterns were created in Europe several centuries ago; in the Amazon and Indonesia, they are current manifestations of nation-building.

Whether still establishing territorial control or now geographically stable, governments determine the rules for resource allocation and utilization. The rules are of paramount importance whether the basis is solely economic or whether aesthetic, community, and spiritual values are also important. As we have seen in the case studies, the basis in most countries is economic. The allocation is typically given to already wealthy investors, landowners, friends of politicians or military juntas, and large corporations.

Resource sectors are unlike manufacturing sectors in that raw materials are allocated by governments, but in other respects the forest industry is not unusual. Public funding for economic development is everywhere the norm, and it is not only the gestation period of trees that makes it essential, but rather a combination of the absolute cost of most large-scale developments and the unwillingness of private investors to take substantial risks or deal with uncertainties. Likewise, public protection for private enterprise is normal practice; what differs from time to time and from region to region is the sectors with most influence over the public decision-making process and the decision-makers. Where forestry is the dominant sector or has the potential to provide the largest overall benefits to the region (as judged by a solely economic standard and usually by those who stand to benefit most), it will gain public financial support.

The history of many a contemporary state can thus be read in the history of forestry. The past is evident in the statistics, but the future may look quite different in northern countries where regional governments are being released from the captivity of large companies. Obliged now to create new economic capacities in other sectors and encountering popular opposition to the continued control of land by companies, governments may finally treat the remaining forests as something more than economic resources. Or they may discover diverse economic resources in standing forests. Their history may, however, be repeated by governments in southern regions so eager to reap fast profits that they cannot see the ecological and social costs.

Employment and Restructuring

Though the chief rationale given for continued logging of northern old-growth forests is that it employs local workers, employment in northern-based forest industries has declined over the last quarter-century relative to production (and even in absolute terms) because of technological developments that have displaced labour. In the north – and we might anticipate a parallel development elsewhere in due course – the multiplier effects were temporary: an upward curve associated with initial development and expansion, followed by a sharp downward curve once the industry was firmly established. The upward curve for the northern countries lasted a good half-century, reaching its apex in the 1950s to early 1970s as demand for wood products increased.

Each company acts in its own interests when it acquires expensive machinery in order to reduce its reliance on human labour. Yet the combined self-interest of all companies acting in this way leads finally to societies where so many people are unemployed and where the capital costs of maintaining the industries is so high that the whole structure begins to fall apart. Maybe the situation could be salvaged if nothing else were to happen, but it is at precisely this stage that other things did happen in forestry: receding forest resources and insufficient supplies, new competitors, investment moving south. Now the companies tried to restructure the remaining labour force, asking it to save the companies from oblivion. Workers, their unions depleted by declining membership, could fight and risk losing remaining markets and finally their own employers, or they could bend to the demands and lose all bargaining power and control over their jobs.

There are few real choices once an economy moves into this phase. Investors – whether in the form of the same or new companies – are free to move capital to other locations with no entrenched unions or with greater resource supplies. But if companies stay, they must restructure, develop new product lines, and seek out new markets, if only so that they can pay for their expensive capital investments. In the free-enterprise system, these companies are not also required to provide lifelong support for the workers they have disemployed. Indeed, if other sectors of the economy were functioning well, those workers would be absorbed elsewhere, as many were during the 1970s.

The cost of these dislocations is borne by the larger society, as well as the unemployed worker, through unemployment insurance and welfare, and all the social costs associated with low-income families. In Canada during the five-year period 1986–90, forestry workers claimed Cdn\$6.17 (1991 constant dollars) for every dollar they put into the Unemployment Insurance scheme (Statistics Canada data reported in *Globe and Mail*, 2 Jan. 1995). This is a substantially higher ratio than for all other industries, including fisheries. These pay-outs are taken by both disemployed workers and workers temporarily laid off. Employers lay off workers knowing that they can draw Unemployment Insurance and remain ready for call-up when markets improve. Workers' insurance is thus a subsidy to companies, and to forestry companies most of all.

When all industries are simultaneously undergoing similar experiences, societies cannot absorb all the costs, and sooner or later governments begin to dismantle welfare systems. Disparities between the haves and have-nots become more transparent as the middle class experiences declining fortunes. As the industrial countries are discovering, democracy is fragile when large portions of the population are economically disenfranchised.

The hopeful aspect about restructuring is that it may release energies and allow for the consideration of alternatives that were repressed during economic booms. This possibility was encountered in the study of British Columbia, where new protected forests have been created, and the government has initiated changes in the process for forest-land allocation.

Alternatives for Economic Survival

There are many alternative sources of fibre for pulping, including hemp, kenaf, sugar bagasse, straw, and rags. In most regions, these are not economical on a large scale and are not profitable for global industry. They are economical on a smaller scale for domestic consumption and are already used in Thailand, India, and many other countries. China, for example, has created a substantial pulping industry using non-wood fibres. An objective for other countries might be providing for more domestic needs through these fibres, rather than relying on imported wood-based pulps.

Trees provide numerous gifts beyond fibres and construction wood, including oils and resins, nuts, implements, and ornaments. In Malaysia and many other tropical areas, wood exported as logs and transformed elsewhere into plywood, construction lumber, or pulp could be manufactured into furniture or wooden implements. In some areas, greater development of the pharmaceutical properties of trees could lead to small-scale, but profitable industries. These industries require infrastructure, marketing organizations, labour training, and research – in short, long-term planning and development. But with genuine aid from richer countries, developing countries could create the necessary conditions for them. Such applications would avoid the pollution and much heavier demands on wood supplies of the plywood and pulp and paper industries (see Plotkin and Famolare, 1992, for an examination of small-scale industries in tropical rainforests).

These alternative industries are not likely to provide incomes for workers or profits for companies on a scale approaching those for lumber and pulp. That is the chief reason regions such as British Columbia failed to develop their own furniture industries and did so little research on alternative products from trees. Indonesians and Brazilians have the same mind-set: create a mass-production industry and make big bucks in a hurry. To develop the alternatives, one needs a different mind-set: create a small-scale industry and make enough to live on without destroying the environment.

A Change in Consumer Habits

While there is no indication that consumers will stop demanding paper or wood-based products, such a change is not impossible. Substitutes are available now, electronic data processing is having an impact already, and markets are always subject to change as demographic conditions, taste, substitutes, and economic underpinnings are altered. These might reduce the demand, or at least the growth in demand, in developed countries. However, such changes would not solve the economic needs of the developing countries; indeed, having made substantial investments in plantations and mills, they would be seriously harmed by any swift change in market trends.

COMMUNITIES: ARE THEY WORTH SAVING?

Repeatedly we discover the dichotomy between local and global in the history of forests, as in the history of industrialization and urbanization more generally. Communities form in localized economies, where people band together to sustain themselves by hunting and gathering or small-scale agriculture. They continue when small-scale markets are introduced, but as the markets become larger, the communities are fractured. Capital does not circulate locally; independent persons become dependent workers; dependent workers leave communities to seek employment; goods cease to be priced according to local abilities to pay and become commodities for much larger and more distant consumer populations. Gradually the local culture is eroded, the local becomes a truncated replica of the urban, and finally, as national markets similarly turn into global markets, all communities become fragments of something else. No one quite belongs where he or she is; nor is that place complete unto itself.

Profits, the motive force of capitalist societies, depend on constant growth and change. Saturated with consumer goods, the northern industrial societies are now experiencing decline. The organizations that once employed them rush to other countries where new consumers and cheaper labour can be exploited and where, in the case of forestry, cheaper raw materials and willing governments provide new profits. Community stability, long praised as the natural effect of sustained-yield forestry, is now identified as either unrelated or in conflict with it. In any event, in much of the forested area of North America, neither sustained-yield forestry nor community stability has been achieved. The declining viability of forest-based communities in the north needs to be addressed, but at present it is squeezed into an artificial debate on logging versus "ecofreaks," while both jobs and communities disappear irrespective of the outcomes in the battle over values.

Human life may depend on the survival of communities as much as trees depend on larger ecosystems. There is a limit to human tolerance for being cut off from sustained interaction with other people. Being "an individual" loses meaning when one is lost in a crowd of individuals, each isolated, none having deep reasons to care about the others' welfare. Real economic development, if isolated fates are to be avoided, will depend on creating economic capacities that sustain, not just individuals, but more – viable human communities.

If some of the proposals outlined above were adopted – if, for example, trees were assigned an economic value more consistent with the real cost of logging them, if the developed countries undertook the financing of tropical reserves, if alternative small-scale industries utilizing wood more efficiently were established, if community forestry were strengthened as a genuine grass-roots development strategy, if even limited land reform were introduced in developing countries, if non-wood pulp industries were encouraged, if ecologically sound substitutes for paper were invented – there would be a marked decline in the rate of deforestation caused by the forest industry and an enhancement of community capacities for change. Deforestation from other causes would still have to be tackled, but removal of this cause would start the process.

Bibliography

Aborigines Protection Society of London. 1973. *Tribes of the Amazon Basin in Brazil, 1972*. London.

Adams, Darius M. 1993. "Long-Term Timber Supply Prospects in the United States: An Analysis of Resources and Policies in Transition." In Peter N. Nemetz (ed.), *Emerging Issues in Forest Policy*, 131–56. Vancouver: University of British Columbia Press.

Adams, Patricia. 1985. "The World Bank: A Law unto Itself." *Ecologist* 15(3):220–1.

– 1989. "Moving toward Disasters." In A. Schneider (ed.), *Deforestation and "Development" in Canada and the Tropics*, 35–6. Sydney, NS: Centre for International Studies.

Adas, Michael. 1983. "Colonization, Commercial Agriculture, and the Destruction of the Deltaic Rainforests of British Burma in the Late Nineteenth Century." In Richard P. Tucker and J.F. Richards (eds.), *Global Deforestation and the Nineteenth-Century World Economy*, 95–110. Durham, NC: Duke University Press.

Aditijondro, George J. 1983. "Problems of Forestry and Land Use in the Asia Pacific Region: The Irian Jaya Experience." In Sahabat Alam Malaysia Conference Proceedings, *Problems of Development, Environment and Natural Resource Crisis in Asia and the Pacific*, 53–69. Georgetown and Pinang: SAM.

Ahmad, Mohiuddin. 1987. "Bangladesh: How Forest Exploitation is Leading to Disaster." *Ecologist* 17 (4/5) (July/Nov.):168–9.

Aitken, H.G.H. 1961. *American Capital and Canadian Resources*. Cambridge, Mass.: Harvard University Press.

Allen, Richard C., Thomas J. Straka, and William F. Watson. 1986. "Indonesia's Developing Forest Industry." *Environmental Management* 10(6):753–9.

Amnesty International. 1988. *Brazil: Cases of Killings and Ill-Treatment of Indigenous People.* London: International Secretariat, November.

Amyot, J. 1988. "Forest Land for People: A Forestry Village Project in Northeast Thailand." *Unasylva* 159 (40):30–41.

Anderson, A. (ed.). 1989. *Alternatives to Deforestation.* New York: Columbia University Press.

Anderson, Robert S., and Walter Huber. 1988. *The Hour of the Fox: Tropical Forests, the World Bank, and Indigenous People in Central India.* Seattle: University of Washington Press.

Andreason, Ove. 1978. "Future Forest Management Trends in Sweden." *Forestry Chronicle*, Feb., 29–33.

Anpo, Y. 1981. "Japan and the Timber Trade: An Appraisal from That Side." *Forest Industries*, July, 36–8.

Apin, Teresa. 1987. "The Sarawak Timber Blockade." *Ecologist* 27 (4/5):186–8.

Appita. 1988. 41 (5):349–50.

Aracruz. Celulose. 1990. *Annual Report.*

– 1991. *Annual Report.*

– 1992. Correspondence between Carlos Roxo, general manager of environmental and public affairs, Aracruz Celulose, and Simon Counsell, tropical rainforest campaigner, Friends of the Earth, London, England, 16 April 1992, July 22, 1992. Copy provided by Aracruz.

Araman, Philip A. 1988. "U.S. Hardwood Trade in the Pacific Rim." In Jay A. Johnson and W. Ramsay Smith (eds.), *Forest Products Trade: Market Trends and Technical Developments*, 99–104. Seattle: University of Washington Press.

Arbhabhirama, Anat, Dhira Phamtumvanit, John Elkington, and Phaitoon Ingkasuwan (eds.). 1988. *Thailand Natural Resources Profile.* Natural Resources Series. New York: Oxford University Press.

Arbokem Inc. and Al Wong. 1987. "Review of Pulping and Papermaking Properties of Aspen." Edmonton: Canadian Forestry Service and Alberta Forest Service.

Argawal, Anil, Darryl D'Monte, and Ujwala Samarth (eds.) 1987. *The Fight for Survival: People's Action for Environment.* New Delhi: Centre for Science and Environment.

Ashton, Peter S. 1981. "Forest Conditions in the Tropics of Asia and the Far East." In *Where Have All the Flowers Gone?* 169–79. Williamsburg, Va: Department of Anthropology, College of William and Mary.

Asian Productivity Organization. 1990. *Forestry Resources Management.* Report of the APO Symposium on Forestry Resources Management, held in Tokyo, Japan. 7–14 Nov. 1989. Tokyo: APO.

Asian Wall Street Journal. 1981. "Weyerhaeuser Exits Indonesia." 26 Oct.

Associação Nacional dos Fabricantes de Papel e Celulose. n.d. "Fatos rele-
vantes." Unpublished history of the industry.
– 1991. "Relatório Estatístico." Sao Paulo.
Association of B.C. Professional Foresters. 1984. "Not Satisfactorily Restocked
(NSR) Lands in British Columbia: An Analysis of the Current Situation."
Vancouver, BC, February. Photocopy.
Atienza, Nelflor S. 1990. "Philippines." In Asian Productivity Organization,
Forestry Resources Management, Report of APO Symposium, Tokyo, Japan, Nov.
1989, 295–321. Tokyo: APO.
Atlas of the 3rd World. 1992. Ed. George Thomas Kurian. 2nd ed. New York,
NY: Facts on File.
Aurell, Ron, and Jaakko Pöyry. 1988. "Pulp and Paper: Worldwide Trends in
Production, Consumption, and Manufacturing." In G.F. Schreuder (ed.),
Global Issues and Outlook in Pulp and Paper, 3–19. Seattle: University of
Washington Press.
Australian Conservation Foundation. 1987. "Australia's Timber Industry:
Promises and Performance." November. Photocopy.
Backman, Charles A., and Thomas S.R. Waggener. 1991. "Soviet Timber
Resources and Utilization: An Interpretation of the 1988 National Inven-
tory." Working Paper 35. Seattle: CINTROFOR, University of Washington in
cooperation with FEPA Research Unit, University of British Columbia,
Vancouver.
Balée, William. 1991. "Indigenous History and Amazonian Biodiversity." Paper
presented at Tropical History Society meetings, Costa Rica, Feb. 1991.
Bandyopadhyay, Jayanta, and Vandana Shiva, 1987. "Chipko: Rekindling
India's Forest Culture." Ecologist 17(1):26–34.
Banijbatana, Dusit. 1978. "Forest Policy in Northern Thailand." In Peter
Kunstadtler, E.C. Chapman, and Sanga Sabhasri (eds.), Farmers in the Forest:
Economic Development and Marginal Agriculture in Northern Thailand, 54–60.
Honolulu: University of Hawaii Press.
Barber, Charles Victor. 1989. "The State, the Environment, and Development:
The Genesis and Transformation of Social Forestry Policy in New Order
Indonesia." PHD diss., University of California at Berkeley.
Barlow, Colin, and Thee Kian Wie. 1988. The North Sumatran Regional Economy:
Growth with Unbalanced Development. Occasional Paper No. 82. Singapore:
ASEAN Economic Research Unit, Institute of Southeast Asian Studies.
Barnes, T., R. Hayter, and E.N. Grass. 1990. "MacMillan Bloedel: Corporate
Restructuring and Employment Change." In M. de Smidt and E. Wever
(eds.), The Corporate Firm in a Changing World Economy, 145–65. Andover,
Hants: Chapman and Hall.
Barnett, Thomas E. 1990. "Commission of Inquiry Report into Aspects of the
Forest Industry in Papua New Guinea." Uncompleted draft paper for sym-
posium at Australian National University, July 1990.

Barr, Brenton M. 1988. "Perspectives on Deforestation in the USSR." In J.F. Richards and R.P. Tucker (eds.), *World Forests and the Global Economy in the Twentieth Century*, 230–61. Durham, NC: Duke University Press.

– and Kathleen E. Braden. 1988. *The Disappearing Russian Forest: A Dilemma in Soviet Resource Management*. London: Rowman & Littlefield.

Bayliss, Martin. 1977. "Thermo-Mechanical Pulp around the World." *Pulp and Paper International* 19(6):69–74.

– 1988. "Enter Elders with Big Plans for NZFP." *Pulp and Paper International* 30(11):83–5.

– 1994. "Sinar Mas. Exquisite timing." *Papermaker*, October, 31–5.

Beattie, W.D., and Ferreira, J.M. 1987. *Analise financeira e socio-economica do reflorestemanato no Brasil*. Serie: Estudos perspectivos para o period 1975–1985 (Brasilia: IBDF/COPLAN). Colecao: Desenvolvimento e Planejamento Florestal.

Beckley, Thomas M. 1994. "Pulp, Paper and Power: Social and Political Consequences of Forest Dependence in a New England Mill Town." PHD diss., University of Wisconsin-Madison.

Berger, R. 1979. "The Brazilian Fiscal Incentive Act's Influence on Reforestation Activity in Sao Paulo State." PHD diss., Michigan State University.

Bertram, I.G., and M. O'Brien. 1979. "Optimal Pricing Policy and Indigenous Forest Management in New Zealand." *New Zealand Journal of Forestry* 24(2):261–77.

Binkley, Clark S., Michael Percy, W.A. Thompson, and Ilan Vertinsky. 1993. "The Economic Impact of a Reduction in Harvest Levels in British Columbia: A Policy Perspective." Working Paper 176. Vancouver: Forest Economics and Policy Analysis Research Unit, University of British Columbia.

Blowing in the Wind: Deforestation and Long-range Implications. 1981. Studies in Third World Societies. Williamsburg, Va: Department of Anthropology, College of William and Mary.

Bodowski, G. 1988. "Is Sustainable Harvest Possible in the Tropics?" *American Forests*, Nov./Dec., 34–7.

Boomgaard, Peter. 1992. "Forest Management and Exploitation in Colonial Java, 1677–1897." *Forest and Conservation History* 36(1):4–14.

Boonkird, S., E.C.M. Fernandes, and P.K.R. Nair. 1984. "Forest Villages: An Agroforestry Approach to Rehabitiliating Forest Land Degraded by Shifting Cultivation in Thailand." *Agroforestry Systems* 2:87–102.

Bourgault, P.L. 1972. "Innovation and the Structure of Canadian Industry." Background Study 23. Ottawa: Science Council of Canada.

Boyhan, George E., Yong T., and Kao and Luigi Terziotti. 1975. "Newsprint from Bagasse – Can It Be Done?" *Pulp and Paper International* 17(11):44–7.

Brady, Michael. 1985. *Indonesia Forestry Project Working Paper*. Report No. 122/85 CP–INS–59 WP.1, October. Rome: FAO/World Bank Cooperative Programme Investment Centre.

British Columbia. 1993a. "Clayoquot Sound Land Use Decision. Key Elements" and "Background Report." Both dated April 1993.

– 1993b. "The Government of British Columbia Response to the Commission on Resources and Environment's Public Report and Recommendations Regarding Issues Arising from the Clayoquot Land Use Decision." 1 June.

– 1993c. "Scientific Panel for Sustainable Forest Practices in Clayoquot Sound: Backgrounder." News release, 22 Oct.

– Communications Office. "Victoria This Week." News releases.

– Ministry of Forests. 1976. Royal Commission on Forest Resources (Peter H. Pearse, Commissioner). *Timber Rights and Forest Policy.* 2 vols. Victoria: Queen's Printer.

– Ministry of Forests. 1978. *Forest Act.* Victoria: Queen's Printer.

– Ministry of Forests. 1983. Special Log Export Policy Committee. *Report on Legislation, Policies and Procedures on Log Exports from British Columbia.* Victoria: Queen's Printer, July.

– Ministry of Forests. 1984. *Report Fiscal Year Ended March 31, 1984.* Victoria: Queen's Printer.

– Ministry of Forests. 1991–92. *Inventory Branch Reports.*

– Ministry of Forests and Lands. 1985a. *Five-year Forest and Range Resource Program 1985–1990.* Victoria: Queen's Printer.

– Ministry of Forests and Lands. 1985b. *Bill 3 Forest Amendment Act, 1985.* Third session, thirty-third Parliament, 34 Elizabeth II, 1985. Legislative Assembly of British Columbia. Victoria: Queen's Printer.

– Ministry of Forests and Lands. 1986–90. *Annual* Report. Victoria: Queen's Printer.

– Ministry of Forests and Lands. 1987–89. News releases re: forest act amendments and log export policies.

– Ministry of Forests and Lands. 1988. "The Small Business Forest Enterprise Program." Information paper, July.

– Ministry of Forests. 1990. *Towards an Old Growth Strategy: Summary of the Old Growth Workshop, Nov. 3–5, 1989.* Prepared by Dr. Bruce Fraser, Salasan Associates, Victoria.

– Ministry of Forests. 1994. "British Columbia's Forest Renewal Plan." Victoria.

– Office of the Ombudsman. 1985. "The Nishga Tribal Council and Tree Farm Licence No. 1." Public Report No. 4, June. Victoria: Ombudsman.

– Vancouver Island Regional CORE Process. Forest Managers and Manufacturers sector (FMM). "Update." Numbered news releases to members, Nanaimo, various dates, 1993–94.

British Columbia Forest Products and Finlay Forest Industries. 1988. "Mackenzie Tree Farm License Application." 5 Aug.

British Columbia Forest Resources Commission. 1992. *Timber and Log Markets for the Forest Resources of British Columbia: Theory, Practice and Proposals in a Competitive Regime.* Vancouver: CCG Consulting Group Ltd., June.

Britton, J.N.H., and J. Gilmour. 1978. *The Weakest Link: A Technological Perspective on Canadian Industrial Development.* Background Study 43. Ottawa: Science Council of Canada.

Brosius, J. Peter. 1990. "Penan Hunter-Gatherers of Sarawak, East Malaysia." *AnthroQuest* 42:2–7.

Browder, J.O. 1985. *Subsidies, Deforestation, and the Forest Sector in the Brazilian Amazon.* Report to the World Resources Institute, Washington, DC.

– 1986. "Logging the Rain Forest: A Political Economy of Timber Extraction and Unequal Exchange in the Brazilian Amazon." PHD diss., University of Pennsylvania.

– 1987. "Brazil's Export Promotion Policy (1980–1984): Impacts on the Amazon's Industrial Wood Sector." *Journal of Developing Areas* 21: 285–304.

– 1988. "Public Policy and Deforestation in the Brazilian Amazon." In R. Repetto and M. Gillis (eds.), *Public Policies and the Misuse of Forest Resources,* 247–97. Cambridge: Cambridge University Press.

– 1989. "Lumber Production and Economic Development in the Brazilian Amazon: Regional Trends and a Case Study." *Journal of World Forest Resource Management* 4:1–19.

Brown, Richard, and Murray Rankin. 1990. "Persuasion, Penalties and Prosecution: Administrative v. Criminal Sanctions." In M.L. Friedland (ed.), *Securing Compliance: Seven Case Studies,* 325–53. Toronto: University of Toronto Press.

Brown, Sandra, and Lugo, Ariel E. 1981. "The Storage and Production of Organic Matter in Tropical Forests and their Role in the Global Carbon Cycle." *Biotropica* 14(3):161–87.

Brown, William H., and Donald M. Mathews. 1914. *Philippine Dipterocarp Forests.* Manila: Bureau of Printing.

Brunig, E.F. 1977. "The Tropical Rainforest – a Wasted Asset or an Essential Biospheric Resource?" *Ambio* 6(4):187–91.

Brunow, Berndt. 1989. "Western Europe: Supply Will Exceed Demand by 1.5 Million Tons by 1990." *Pulp and Paper International* 31(2):51–5.

Bryant, R.L. 1993. "Contesting the Resource: The Politics of Forest Management in Colonial Burma." PHD thesis, University of London.

Budiardjo, Carmel. 1986. "The Politics of Transmigration." *Ecologist* 16(2–3):111–16.

Bull, Gary, and Jeremy Williams. 1994. *Mattawa and Surroundings Communities Socio-Economic Impact Study: Final Report.* Commissioned by Mattawa Communities Socio-Economic Task Force. Toronto: Faculty of Forestry, University of Toronto, 30 March.

Bunker, Stephen G. 1980. "Development and the Destruction of Human and Natural Environments in the Brazilian Amazon." *Environment* 22(7):14–20, 34–43.

– 1981. "Impact of Deforestation on Peasant Communities in the Medio Amazonias of Brazil." In *Where Have All the Flowers Gone?* 45–61. Williamsburg, Va: Department of Anthropology, College of William and Mary.

Bunyard, Peter. 1985. "World Climate and Tropical Forest Destruction." *Ecologist* 15(3):125–236.

– 1986a. "Waldsterben and the Death of Europe's Trees." *Ecologist* 16(1):2–3.

– 1986b. "The Death of the Trees." *Ecologist* 16(1):4–14.

– 1987. "The Significance of the Amazon Basin for Global Climatic Equilibrium." *Ecologist* 17(4/5):139–41.

Burdin, N.A. 1991. "Trends and Prospects for the Forest Sector of the USSR: A View from Inside." *Unasylva* 42:43–50.

Burger, Julian. 1990. *The Gaia Atlas of First Peoples: A Future for the Indigenous World.* New York, Doubleday Anchor.

Burgess, P.F. 1989. "Asia." In Duncan Poore, Peter Burgess, John Palmer, Simon Rietberger, and Timothy Synnots (eds.), *No Timber without Trees: Sustainability in the Tropical Forest,* 116–53. A Study for ITTO. London: Earthscan.

Buschbacher, Robert J. 1987. "Deforestation for Sovereignty Over Remote Frontiers, Case Study No. 4: Government-Sponsored Pastures in Venezuela Near the Brazilian Border." In C.F. Jordan (ed.), *Amazonian Rain Forests,* 46–57. New York: Springer-Verlag.

– Christopher Uhl, and E.A.S. Serrao. 1987. "Large-Scale Development in Eastern Amazonia, Case Study No. 10: Pasture Management and Environmental Effects Near Pargominas, Pará." In C.F. Jordan (ed.), *Amazonian Rain Forests,* 90–9. New York: Springer-Verlag.

Business Council on Sustainable Development. 1992. *Changing Course.* Boston: Massachusetts Institute of Technology.

Buttar, James. 1980. Speech to conference of Council of Forest Industries, Vancouver, 18 April. Photocopy.

Byron, R.N. 1976. "Community Stability and Regional Economic Development: The Role of Forest Policy in the North Central Interior of British Columbia." PHD diss., University of British Columbia.

Cameron, John I., and Penna, Ian W. 1988. *The Wood and the Trees: A Preliminary Economic Analysis of a Conservation Oriented Forest Industry Strategy.* Hawthorn, Victoria: Australian Conservation Foundation.

Campinhos, E., Jr, and Y.K. Ikemori. 1989. "Selection and Management of the Basic Population *Eucalyptus grandis* and *E. urophylla* Established at Aracruz for the Long Term Breeding Programme." In G.I Gibson, A.R. Griffin, and A.C. Matheson (eds.), *Breeding Tropical Trees: Population Structure and Genetic Improvement Strategies in Clonal and Seedling Forestry,* Proceedings of the IUFRO Conference, Pattaya, Thailand, Nov. 1988, 169–75. Oxford: Oxford Forestry Institute.

Canada. Canadian Forestry Service. 1987a. *Canada-British Columbia Forest Resource Development Agreement, 1985–1990: A Progress Report.* Pacific Forestry Centre. Victoria: Queen's Printer.

– Canadian Forestry Service. 1987b. "The Continuing Challenge: Competition and New Products in the World Forest Products Markets." Resource

Information Systems, Inc., *Proceedings*, Annual Conference, Sept. 1986, compiled by Tom Steele and Tim Williamson. Edmonton, Alta.

– Canadian Forestry Service. 1991. *Canada-British Columbia Partnership Agreement on Forest Resource Development: FRDA II, 1991–1995*. Pacific Forestry Centre. Victoria: Queen's Printer.

– Department of Industry, Trade and Commerce. 1978. *Review of the Canadian Forest Products Industry*. Forest Products Group, Resource Industries Branch. Ottawa: Supplies and Services, November.

– Forestry Canada, Economics and Statistics Directorate. *Selected Forestry Statistics Canada*. Ottawa: Forestry Canada, Annual.

– National Forest Strategy. Canadian Council of Forest Ministers. 1992. *Sustainable Forests: A Canadian Commitment*. Hull, Quebec, March.

– National Round Table on the Environment and the Economy. 1992. *Forest Round Table on Sustainable Development*. Edited by Steve Thompson and Allison Webb. Ottawa, April.

– Statistics Canada. *Canadian Forestry Statistics*. Annual.

Canada-British Columbia Forest Resource Development Agreement. 1986. *Renewal* 1(1). Vancouver: Communications Working Group.

Cardellichio, Peter A., Clark S. Binkley, and Vadim K. Zausaev. 1990. "Saw log Exports from the Soviet Far East." *Journal of Forestry* 88(6):12–17, 36.

Carron, L.T. 1985. *A History of Forestry in Australia*. Canberra: Australian National University Press.

– 1990. "A History of Plantation Policy in Australia." In J. Dargavel and N. Semple (eds.), *Prospects for Australian Forest Plantations*, 11–23. Canberra: Centre for Resource and Environmental Studies, Australian National University.

Caufield, Catherine. 1985. *In the Rainforest*. London: Pan Books.

– 1989. "The Cultural Annihilation of Forest Peoples." In A. Schneider (ed.), *Deforestation and "Development" in Canada and the Tropics*, 55–6. Sydney, NS: Centre for International Studies.

– 1990. "A Reporter at Large: The Ancient Forest." *New Yorker*, 14 May, 46–84.

Cavieres, Aaron C. n.d. "Conservación y utilización del bosque nativo chileno." Unpublished paper distributed by CODEFF, Santiago.

– and A. Lara. 1983. "La destrucción del bosque nativo para ser reemplazado por plantaciónes de pino insigne: Evalución y proposiciónes. I. Estudio de caso en las Provincia de Biobio." CODEFF Informe Técnico No. 1. Santiago.

– G. Martner, G. Molina, V.Paeile. 1986. "Especialización productiva, medio ambiente y migraciónes: El caso del sector forestal chileno." *Agricultura y Sociedad* 4:31–95.

Cayford, J.H. 1990. "A Brief Overview of Canadian Forestry." *Unasylva* 162(41)1:44–8.

Cenibra. n.d. *A Cenibra e o meio ambiente*.

– n.d. *Cenibra: Celulose Nipo-Brasileira S.A.*

- n.d. *O Futuro presente em cada acao.*
- 1987–88. Financial statements.
- 1987–91. *Annual Report.*
- 1991. *Fibra,* ano 7, no. 83 (novembro). Edicao especial.
- 1992. *Fibra,* ano 8, no. 89 (junho).

Chiengmai University, Thailand. Faculty of Social Sciences. 1978. "Case Study of Forest Village System in Northern Thailand 1974." Thailand Annex to the Report of the FAO/ILO/SIDA Consultation on Employment in Forest. In *Forestry for Local Community Development,* Forest Paper no. 7. Rome: FAO.

Chile. CONAF (National Forestry Corporation). 1992. "Native Forests in Chile: What's Happening Today." Unpublished paper prepared by Herand Verscheure. Santiago: CONAF, 4 Jan.

- INFOR-CORFO (Instituto Forestal–Corporación de Fomento de la Producción). 1989. "Estadísticas forestales 1988." *Boletín estadistico,* No 11. Santiago.
- INFOR-CORFO. 1990. "Estadísticas Forestales 1990." *Boletín Estadístico,* No 21. Santiago.
- INFOR-CORFO. 1991. "La pequeña empresa maderera de bosque nativo: su importancia, perspectivas, una propuesta para su desarorollo." *Informe Técnico,* No 128. Santiago.
- INFOR-CORFO. 1992. Unpublished special printout on planned mega-projects in the forestry sector, 1992–96, showing investment source, sector, and regional breakdowns. July.
- Ministerio de Agricultura. 1974. *Decreto Ley 701 de fomento forestal.* Santiago.
- 1979. *Decreto Ley 2565. Nuevo Decreto Ley 701 de fomento forestal.* Santiago.
- 1980. *Decreto Supremo 259. Reglamento del Decreto Ley 701 sobre fomento forestal.* Santiago.
- Ministerio de Secretaria General de la Presidencia. 1992. *Mensaje no. 403–323: Mensaje de S.E. el Presidente de la Republica con el que inicia un proyecto de recuperación del bosque nativo y fomento y forestal,* Santiago, 10 Abril.

Chile Forestal. 1991a. "Un plan de acción forestal para Chile." 183:8

Chile Forestal. 1991b. "Temporada de incendios 90–91." 184:25–6.

Chilean Forestry News. Various issues.

Ciriacy-Wantrop, S.V., and R.C. Bishop. 1975. "Common Property as a Concept in Natural Resources Policy." *Natural Resources Journal* 15:713–27.

Clark, David A. 1988. "European Pulp and Paper: The Outlook for Demand and Supply." In Gerard F. Schreuder (ed.), *Global Issues and Outlook in Pulp and Paper,* 46–58. Seattle: University of Washington Press.

Clark-Jones, M. 1987. *A Staple State: Canadian Industrial Resources in Cold War.* Toronto: University of Toronto Press.

Clarkson, Judith, and Jurgen Schmandt. 1993. "The Role of Air Pollution in Forest Decline." In Peter N. Nemetz (ed.), *Emerging Issues in Forest Policy,* 48–71. Vancouver: University of British Columbia Press.

Clauss, Wolfgang. 1982. *Economic and Social Change among the Simalungun Batak of North Sumatra.* Bielefelder Studien zur Entwicklungssoziologie, vol. 15. Saarbruken: Verlag Breitenbach.

Clawson, Marion. 1975. *Forests for Whom and for What?* Baltimore: Johns Hopkins Press.

– 1979. "Forests in the Long Sweep of American History." *Science* 203:1168–74.

– 1983. "Forest Depletion and Growth." In Richard C. Davis (ed.), *Encyclopedia of American Forest and Conservation History.* 1: 196–200. New York: Macmillan and Free Press.

Clayoquot Biosphere Project, British Columbia. "Data Sheet." News bulletin, various dates, 1993–94.

CMPC (Compania Manufacturera de Papeles y Cartones SA). 1988. *Annual Report* 1988, 1989.

– n.d. *Laja Pulp Mill.* Company brochure.

CODEFF (Comité National pro Defensa de la Fauna y Flora). n.d. "La conservacion del bosque nativo: compromise de todos." Brochure distributed by La Red de Monitoraeo Forestal.

– 1992. "Nota de Prensa." Rio de Janeiro, 13 junio. Photocopy.

– and international affiliated wilderness groups. 1992. "Who Is Doing What, and What you Can Do for Chile's Native Forests." Unpublished paper prepared by John Moriarty and Hernan Verscheure, Santiago, 7 May.

COFI (Council of Forest Industries). *British Columbia Forest Industry Statistical Tables.* Compiled from Statistics Canada, Ministry of Labour, and other sources. Various years.

– 1988. *Native Indian Land Claims in British Columbia: A Background Paper.* Rev. ed. Vancouver.

Colchester, Marcus. 1986a. "Banking on Disaster: International Support for Transmigration." *Ecologist.* 16(2/3):61–71.

– 1986b. "Unity and Diversity: Indonesia's Policy towards Tribal Peoples." *Ecologist* 16(2/3):89–98.

– 1986c. "The Struggle for Land: Tribal Peoples in the Face of the Transmigration Programme." *Ecologist* 16(2/3):99–110.

– 1987a. "Transmigration Update. Program Slashed in Response to Transmigration Campaign." *Ecologist* 17(1):35–41.

– 1987b. "Native Protest in Malaysian Crackdown." *Ecologist* 17(4/5):189.

Collins, Mark (ed.). 1990. *The Last Rainforests: A World Conservation Atlas.* New York: Oxford University Press.

Contreras, A. 1987. "Transnational Corporations in the Forest-Based Sector of Developing Countries." *Unasylva* 39(3/4):38–52.

Conway, Gordon R., Ibrahim Manwan, and Davis S. McCauley. 1983. "The Development of Marginal Land in the Tropics." *Nature* 304 (4 Aug.): 392.

Correa, Salvador. 1988. "Chile as a World Market Competitor in Forest Products." In G.F. Schroeder (ed.), *Global Issues and Outlook in Pulp and Paper*, 151–61. Seattle: University of Washington Press.

Corry, S. 1984. "Cycles of Dispossession: Amazon Indians and Government in Peru." *Survival International Review* 43:45–70.

Cowell, Adrian. 1990. *The Decade of Destruction: The Crusade to Save the Amazon Rainforest*. New York: H. Holt.

Cox, Thomas R. 1983. "Trade, Development, and Environmental Change: The Utilization of North America's Pacific Coast Forests to 1914 and Its Consequences." In R.P.Tucker and J.F. Richards (eds.), *Global Deforestation and the Nineteenth-Century World Economy*, 14–29. Durham, NC: Duke University Press.

– 1988. "The North American-Japanese Timber Trade." In R.P. Tucker and J.F. Richards (eds.), *World Deforestation in the Twentieth Century*, 164–88. Durham, NC: Duke University Press.

Crane, W.J.B., and R.J. Raison. 1980. "Removal of Phosphorous in Logs when Harvesting *Eucalyptus delegatensis* and *Pinus radiata* Forests on Short and Long Rotations." *Australian Forestry* 43(4):253–60.

Current State of Japanese Forestry. Contribution to IUFRO Division 4. Tokyo: Japanese Forest Economic Society. Occasional.

Dargavel, John. 1982. "Employment and Production: The Declining Forestry Sector Re-examined." *Australian Forestry* 45(4):255–61.

– 1984. "Pulp and Paper Monopolies in Tasmania." In Michael Taylor (ed.), *A Geography of Australian Corporate Power*, 69–89. London: Croom Helm.

– 1985. "The Future of Australia's Hardwood Forests." Centre for Resource and Environmental Studies Workshop, Australian National University, Canberra.

– 1989a. "Conceding to Capital: Resource Regimes in the Forests of British Columbia and Tasmania." *Australian-Canadian Studies* 6(2):93–107.

– 1989b. "Notes on the Economic Implications of the Wesley Vale Mill." Personal correspondence, 16 March.

– 1991. "Incorporating National Forests into the New Pacific Economic Order: Processes and Consequences." Paper delivered to the XVII Pacific Science Congress, Honolulu, Hawaii.

– K. Dixon, and N. Semple (eds.). 1988. *Changing Tropical Forests: Historical Perspectives on Today's Challenges in Asia, Australia and Oceania*. Canberra: Centre for Resource and Environmental Studies, Australian National University.

– Mary Hobley, and Sebastiao Kengen. 1985. "Forestry of Development and Underdevelopment of Forestry." In John Dargavel and Gary Simpson (eds.), *Forestry: Success or Failure in Developing Countries?* Working Paper 1985/20:1–37. Canberra: Centre for Resource and Environmental Studies, Australian National University.

– and Sebastiao Kengen. 1991. "Promise and Performance of Industrial Plantations in Two Regions of Australia and Brazil." Paper given at Tropical Forest History Conference, Costa Rica, February.

– and N. Semple (eds.). 1990. *Prospects for Australian Forest Plantations.* Canberra: Centre for Resource and Enviornmental Studies, Australian National University.

Darling, Craig R. 1991. "In Search of Consensus – An Evaluation of the Clayoquot Sound Sustainable Development Task Force Process." Victoria: University of Victoria Center for Dispute Resolution.

Darr, D., and D. Boulter. 1991. "Timber Trends and Prospects for North America." *Unasylva* 42:51–9.

Dauvergne, Peter. 1993–94. "The Politics of Deforestation in Indonesia." *Pacific Affairs* 66(4):497–518.

Davidson, J. 1985. "Setting Aside the Idea that Eucalypts are Always Bad." Working Paper No. 10, UNDP/FAO Project BGD/79/017, Assistance to the Forestry Sector. May.

Davis, Shelton H. 1977. *Victims of the Miracle: Development and the Indians of Brazil,* Cambridge: Cambridge University Press.

Dean, Warren. 1983. "Deforestation in Southeastern Brazil." In Richard P. Tucker and J.F.Richards (eds.), *Global Deforestation and the Nineteenth-Century World Economy,* 50–67. Durham, NC: Duke University Press.

De'Ath, Colin. 1992. "A History of Timber Exports from Thailand with Emphasis on the 1870–1937 Period." *National Bulletin Siam Sociology* 40:49–66.

de Koninck, Rodolphe. 1992. "Who Really Eats up the Forests of Southeast Asia: the Swiddeners, the Loggers, the Settlers or the State?" Paper delivered at the fifth annual conference of the Northwest Regional Consortium for Southeast Asian Studies, University of British Columbia, Vancouver, 16–18 Oct.

Denslow, J.S., and Padoch, C. (eds.). 1988. *People of the Tropical Rain Forest.* Berkeley: University of California Press.

Dickinson, Robert E. 1981. "Effects of Tropical Deforestation on Climate." In *Blowing in the Wind: Deforestation and Long Range Implications: Studies in Third World Societies,* 411–42. Williamsburg, Va: Department of Anthropology, College of William and Mary.

Dogra, Bharat. 1985. "The World Bank vs. the People of Bastar." *Ecologist* 15(1/2):44–8.

Drushka, Ken. 1990. "The New Forestry: A Middle Ground in the Debate over the Region's Forests?" *New Pacific* 4 (Fall):7–24.

– Bob Nixon, and Ray Travers (eds.). 1993. *Touch Wood: BC Forests at the Crossroads.* Madeira Park: Harbour Publishing.

Ecologist (monthly). Various issues, 1982–94.

Ecologist. 1985. "Indonesian Transmigration: The World Bank's Most Irresponsible Project." 15(5/6):300–1.

– 1986. "Indonesia's Transmigration Programme: A Special Report in Collaboration with Survival International and Tapol." 16(2/3):57–117.

Economist (monthly). Various issues, 1985–94.

Economist. 1990. "Burma Nice Guys." 7 April.

Ekachai, Sanitsuda. 1990a. "Using Buddhism to Avert Ecological Disaster." *Bangkok Post Outlook*, 21 March.

– 1990b. "Ordaining Trees to Save the Woods." *Bangkok Post Outlook*, 21 March.

Electronic Data Interchange Council of Canada. 1992. *Issues for Canada's Future.* Ottawa: The Council, June.

Environmental Defense Fund. 1987. "The Failure of Social Forestry in Karnataka." *Ecologist* 17(4/5):151–4.

Eronen, Jarmo. 1988. "Lack of resources limits growth." *Pulp and Paper International*, Oct., 49–50.

Ervin, Keith. 1989. *Fragile Majesty.* Seattle: The Mountaineers.

Esmara, Hendra. 1971. "An Economic Survey of West Sumatra." *Bulletin of Indonesian Economic Studies* 7(1):32–55.

European Parliament. 1990a. *Session Documents.* English ed., 9 June 1988. Series A Document A–0092/88. Reproduced in Sahabat Alam Malaysia, *The Battle for Sarawak's Forests*, 57–64. Penang, Malaysia: SAM.

– 1990b. Joint Motion for a Resolution, 22 Nov. 1989. B3–549/RC1 and B3–582/RC1. Reproduced in Sahabat Alam Malaysia, *The Battle for Sarawak's Forests*, 112. Penang, Malaysia: SAM.

FAO (Food and Agriculture Organization of the United Nations). 1973. *FAO World Symposium on Man-made Forests and their Industrial Importance, 1967.* Canberra: ACT/FAO.

– 1976. *Development and Forest Resources in Asia and Far East.* Rome: FAO.

– 1979. *Forestry and Forest Products Development Indonesia, Prospects for the Development of Pulp and Paper Industries in Indonesia up to the Year 2000, 1979.* FO:INS78/054, Working Paper 4. Bogor.

– 1982. *Conservation and Development of Tropical Forest Resources.* Report on Expert Meeting. Rome: FAO.

– 1985. Committee on Forest Development in the Tropics. *Tropical Forestry Action Plan.* Rome: FAO.

– 1987a. "Agriculture: Towards 2000." Conference, FAO, 24th session, Rome, 7–26 Nov. Rome: FAO.

– 1987b. *Proceedings of the Expert Consultation on World Pulp and Paper Demand and Supply Outlook, 16–18 September, 1986.* Rome: FAO.

– 1989. "Current Trends and Outlook." *Review of Field Programmes, 1988–89*, 3–26. Rome: 1989.

– 1990. "The Tropical Forestry Action Plan: Regional Priorities for Asia and the Pacific." *Unasylva* 41:49–63.

– 1991a. *Myanmar Forestry Sector: An Economic Review of Selected Issues.* Prepared for government by Sebastiao Kengen, FAO consultant. National Forest

Management and Inventory Project MYA/85/003, Union of Myanmar, Ministry of Agriculture and Forests. Yangon: FAO, January.

– 1991b. "Forestry in the 1990s: An interview with FAO Director-General Edouard Saouma." *Unasylva* 42:3–8.

– 1992. *Mixed and Pure Forest Plantations in the Tropics and Subtropics*. FAO Forestry Paper 103. Rome.

– *Yearbook of Forest Products* (annual). Geneva.

– 1993. *Forestry Statistics Today for Tomorrow, 1961–1991, 2010*. Geneva.

– n.d. "The Forest Resources of the Temperate Zones: Main Findings of the UN-ECE/FAO 1990 Forest Resources Assessment." In ECE/TIM/60, Summary, table 3, p.8.

– Coordinating Unit of the Tropical Forestry Action Plan. 1991. *TFAP Update No. 21*. Rome.

Far Eastern Economic Review (monthly). Various issues, 1985–93.

Fearnside, Philip M. 1989. "Deforestation in Amazonia." *Environment* 31(4):16–20, 39–40.

– and Gabriel de lima Ferreria. 1985. "Amazonian Forest Reserves: Fact or Fiction?" *Ecologist* 15(3): 297–300.

– and J.M. Rankin. 1985. "Jari Revisited: Changes and the Outlook for sustainability in Amazonia's Largest Silvicultural Estate." *Interciencia* 10(3): 121–9.

Feeny, David. 1988. "Agricultural Expansion and Forest Depletion in Thailand, 1900–1975." In John F. Richards and Richard P. Tucker (eds.), *World Deforestation in the Twentieth Century*, 112–46. Durham, NC: Duke University Press.

Fenton, R. 1984. "New Zealand's Exports of Logs and Sawntimber to Japan." *New Zealand Journal of Forestry* 29(2):225–48.

Fenton, R.T., and F.M. Maplesden. 1986. *The Eastern USSR: Forest Resources and Forest Products Exports to Japan*. Forest Research Institute Bulletin no. 23. Rotorua, New Zealand: New Zealand Forest Service.

Fernandes, Walter, and Sharad Kulkarni (eds.). 1983. *Towards a New Forest Policy: People's Rights and Environmental Needs*. New Delhi: Indian Social Institute.

F.L.C. Reed and Associates. 1974. *Canada's Reserve Timber Supply*. Ottawa: Industry, Trade and Commerce Canada.

– 1978. *Forest Management in Canada 1–2*. Forest Management Institute. Ottawa: Environment Canada.

Fletcher, Hugh A. 1988. "The Pulp and Paper Industry: A New Zealand Perspective." In Gerard F. Schreuder (ed.), *Global Issues and Outlook in Pulp and Paper*, 59–65. Seattle: University of Washington Press.

Flora, Donald F. 1990. "Softwood Log Trade in the Pacific Basin and Regional Economic Development." In *Proceedings of the Society of American Foresters National Convention*, Sept. 24–27, 1989, 276–378. Spokane, Washington, and Bethesda, Maryland: The Society.

Fortmann, Louise, and John W. Bruce (eds.). 1988. *Whose Trees? Proprietary Dimensions of Forestry.* Rural Study Series. Boulder and London: Westview Press.

Foster, John Bellamy. 1991. "Capitalism and the Ancient Forest." *Monthly Review* 43(5):1–16.

Foweraker, Joe. 1981. *The Struggle for Land: A Political Economy of the Pioneer Frontier in Brazil from 1930 to the Present Day.* Cambridge: Cambridge University Press.

Fox, Irving K. 1993. "Canada-United States Trade in Forest Products: Issues and Uncertainties." In Peter N. Nemetz (ed.), *Emerging Issues in Forest Policy,* 254–72. Vancouver: University of British Columbia Press.

Fraser, H.R. 1980. "PICOP: a Benchmark for Tropical Forestry." *World Wood* 21(12):13–17.

– 1981. "PICOP: New Losses Enforce Change." *World Wood* 22(3):22–3.

Friday, Laurie, and Ronald Laskey (eds.). 1989. *The Fragile Environment.* Cambridge: Cambridge University Press.

Funakoshi, Shoji. 1988. "The Process and Keynote of Forest Policy in the Pre-World War II Era." In R. Handa (ed.), *Forest Policy in Japan,* 16–21. Chosaki, Japan: Nippon Ringyo.

Furer-Haimendorf, C. Von. 1982. *Tribes of India: The Struggle for Survival.* Berkeley: University of California Press.

Gall, N. 1979. "Ludwig's Amazon Empire." *Forbes,* 14 May, 127–44.

Gamboni, Jorge Morales. 1989. *El desarrollo forestal en Concepción.* Serie: "Abriendo Caminos." Concepción: GEA.

GEA (Groupo de Estudios Agro-Regionales de la Universida Academicia de Humanismo Christiano y el Consejo Latinoamericano de Ciencias Sociales). 1990. *Estudios Agrarios Bolétin 25.* Contributions by Miriam Medina, Lucia Molina Lagos, Joge Morales Gamboni. Santiago, Chile: Junio.

Geertz, Clifford. 1963. *Peddlars and Princes: Social Development and Economic Change in Two Indonesian Towns.* Chicago: University of Chicago Press.

– 1975. *The Social History of an Indonesian Town.* Greenwood, Conn.: Yale University Press.

Gentry, A.H., and Lopez-Parodi, J. 1980. "Deforestation and Increased Flooding of the Upper Amazon." *Science* 210:1354–6.

Ghosh, R.C., O.N. Kaul, and B.K. Subba Rao. 1978. "Some Aspects of Water Relations and Nutrition in *Eucalyptus* Plantations." *Indian Forester* 104(78):517–24.

Gilbert, L.E. 1977. "The Role of Insect-Plant Coevolution in the Organization of Ecosystems." In V. Labyrie (ed.), *Comportement des Insectes et Milieu Trophique,* 399–482. Paris: CNRS.

– 1980. "Food Web Organization and Conservation of Neotropical Diversity." In M.E. Soule and B.A. Wilcox (eds.), *Conservation Biology: An Evolutionary-Ecological Perspective,* 11–33. Sunderland, MS: Sinauer Associates, Inc.

Ginting, M., and R. Daroesman. 1986. "An Economic Survey of North Sumatra." *Bulletin of Indonesian Economic Studies* 22(2):80–102.

Goldsmith, Edward. 1985. "Open letter to Mr. Clausen, President of the World Bank." *Ecologist* 15(1/2):4. Also correspondence p.78.

Goldstone, A. 1974. "Indonesia: Asia's Timber Giant." *Far Eastern Economic Review* 85(26):57.

Gomez-Pompa, A., C. Vazquez-Yanes and S. Guevara. 1972. "The Tropical Rainforest: a Non-renewable Resource." *Science* 177(4051):762–5.

Goodland, Robert. 1981. "Indonesia's Environmental Progress in Economic Development." In *Where Have All the Flowers Gone?* 215–76. Williamsburg, Va: Department of Anthropology, College of William and Mary.

– 1982. *Tribal Peoples and Economic Development: Human Ecologic Conserations.* Washington, DC: World Bank, May.

– 1985. "Brazil's Environmental Progress in Amazonian Development." In J. Hemming (ed.), *Change in the Amazon Basin,* Vol. 1: *Man's Impact on Forests and Rivers,* 3–35. Manchester: Manchester University Press.

– (ed.). 1990. *Race to Save the Tropics: Ecology and Economics for a Sustainable Future.* Washington, DC: Island Press.

– and H.S. Irwin. 1975. *The Amazon Jungle: Green Hell to Red Desert?* New York: Elsevier Scientific.

Grainger, Alan. 1984. "Quantifying Changes in Forest Cover in the Humid Tropics: Overcoming Current Limitations." *Journal of World Forest Resource Management.* 1(1):1–63.

Greaves, A. 1979. "Gmelina – Large Scale Planting, Jarilandia, Amazon Basin." *Commonwealth Forestry Review* 58(4):267–9.

Greenpeace-Brasil-"Agapan." 1992. "Information on Aracruz." Photocopy distributed to participants in Rio Earth Summit Conference, May.

Greenpeace Canada. 1989. *A Blind Eye: Federal Regulation of the Pulp and Paper Industry in Canada.* Vancouver: Greenpeace Canada.

Greenpeace International. 1992. *Greenwash.* Washington, DC: Greenpeace.

Grey, Dennis. 1990. "Last of World's Great Teak Forest Falling to Greed in Burma." *Vancouver Sun,* 4 Aug., B7.

Gross, D.R., G. Eiten, N.M. Flowers, F.M. Leoi, M.L. Ritter, and D.W. Werner. 1979. "Ecology and Acculturation among Native Peoples of Central Brazil." *Science* 106:1043–50.

Grossman, Rachel, and Lenny Siegel. 1977. "Weyerhaeuser in Indoncsia." *Pacific Research* 9(1):1–12.

Gro-Thong, Yupin. 1988. "Eucalyptus – Large-Scale Planting Still in Doubt." Translated from Thai in *Bangkok Review,* May 1988: 224–30.

Guha, Ramachandra. 1983. "Forestry in British and Post-British India." Parts I and II, 1882–96; parts III and IV, 1940–47. *Economic and Political Weekly,* 5, 12, Nov.

Guppy, Nicholas. 1983a. "Proposals for an Organisation of Timber Exporting Countries (OTEC)." *Malaysian Forester* 46(2):1–19.
– 1983b. "The Case for an Organisation of Timber Exporting Countries (OTEC)." In S.L. Sutton, T.C. Whitmore, and A.C. Chadwick, (eds.), *Tropical Rain Forest: Ecology and Management.* Oxford: Blackwell Scientific Publications.
– 1984. "Tropical Deforestation: A Global View." *Foreign Affairs* 62(4–5):928–65.
Gurmit Singh, K.S. 1981. "Destroying Malaysian Forests." In *Where Have all the Flowers Gone?* 181–90. Williamsburg: Department of Anthropology, College of William and Mary.
Hagner, Stig O.A. 1980. "Forest Management under New Conditions: What Has Been Done in Scandinavia." In *The Forest Imperative: Proceedings, Canadian Forest Congress, Ontario Science Centre, Toronto, Sept. 22–23 1980,* 43–8. Montreal: Canadian Pulp and Paper Association.
Haley, David. 1981. "A Regional Comparison of Stumpage Values in British Columbia and the United States Pacific Northwest." Vancouver. Photocopy.
– 1985. "The Forest Tenure System as a Constraint on Efficient Timber Management: Problems and Solutions." *Canadian Public Policy,* 11, supplement, 315–20.
– 1994. "New Directions in Swedish Forest Policy." *B.C. Professional Forester Forum,* no. 5 (Oct./Nov.).
– and Martin K. Luckert. 1986. *The Impact of Tenure Arrangements on Forest Management and Forestry Investment in Canada.* Commissioned Report for Canadian Forestry Service. Ottawa: Supply and Services, May.
Hall, Claes. 1988. "Global Issues in Pulp and Paper: A Brazilian Perspective." In Gerard F. Schreuder (ed.), *Global Issues and Outlook in Pulp and Paper,* 66–8. Seattle: University of Washington Press.
Halvorsen, Robert. 1990. "Give Away the 'Forest' and Save the Trees." *Economics Thammasat* (Bangkok), 7 May.
Hammond, Herb. 1991. *Seeing the Forest among the Trees: The Case for Wholistic Forest Use.* Vancouver: Polestar Books.
Handa, R. 1988a. "Timber Economy and Forest Policy after the World War II." In R. Handa (ed.), *Forest Policy in Japan,* 22–35. Tokyo: Nippon Ringyo Chosakai.
– (ed.). 1988b. *Forest Policy in Japan.* Tokyo: Nippon Ringyo Chosakai.
Hansing, J., and S. Wibe. 1993. "Rationing the Supply of Timber: The Swedish Experience." In Peter N. Nemetz (ed.), *Emerging Issues in Forest Policy,* 157–70. Vancouver: University of British Columbia Press.
Hardin, Gerrett. 1968. "The Tragedy of the Commons." *Science* 162:1243–8.
Hartler, Nils. 1987. "Technological Advances in Relation to the Future Supply of Pulp Fibres." In FAO, *Proceedings,* 1987:125–33.
Hartshorn, Gary S. 1979. *Report on Activities as a Forest and Man Fellow to Institute of Current World Affairs, Sept. 1979,* Sponsored by Institute of Current World Affairs, Wheelock House, Hanover, NH.

– 1980. "Neotropical Forest Dynamics." *Biotropica* 12, supplement, 23–31.
– 1981. *Report to Institute of Current World Affairs, December, 1981, on Activities as a Forest and Man Fellow, Sponsored by That Institute.* Wheelock House, Hanover, NH.
– 1983. "Ecological Implications of Tropical Plantation Forestry: A Global Assessment." In R. Sedjo (ed.), *The Comparative Economics of Plantation Forestry: A Global Assessment,* 84–93. Washington, DC: Resources for the Future.
Hay-Roe PaperTree Letter. Vancouver: Hay-Roe. Various issues.
Hayter, Roger. 1985. "The Evolution and Structure of the Canadian Forest Product Sector: An Assessment of the Role of Foreign Ownership and Control." *Fennia* 163:439–50.
– 1986. "Export Performance and Export Potentials: Western Canadian Exports of Manufactured End Products." *Canadian Geographer* 30:26–39.
– 1992. "International Trade Relations and Regional Industrial Adjustment: The Implications of the 1982–86 Canadian–US Softwood Lumber Dispute for British Columbia." *Environment and Planning* 24:153–70.
– 1993. "International Trade and the Canadian Forest Industries: The Paradox of the North American Free Trade Agreements." Paper prepared for meetings of the Canadian Studies Association for German Speaking Peoples, Grainan, Germany, February.
– and T.J. Barnes. 1990. "Innis' Staple Theory, Exports and Recession: British Columbia, 1981–86." *Economic Geography* 66:156–73.
– and T.J. Barnes. 1992. "Labour Market Segmentation, Flexibility, and Recession: A British Columbian Case Study." *Environment and Planning: Government and Policy* 10:333–53.
– and John Holmes. 1993. "Booms and Busts in the Canadian Paper Industry: The Case of the Powell River Paper Mill." Discussion paper 27, Simon Fraser University, November. Photocopy.
– and John Holmes. 1994. "Recession and Restructuring at Powell River, 1980–94: Employment and Employment Relations in Transition." Discussion paper 28, Simon Fraser University, May. Photocopy.
Hecht, S.B. 1982. "Cattle Ranching Development in the Eastern Amazon." PHD diss., University of California at Berkeley.
Hecht, Susanna, and Alexander Cockburn. 1989. *The Fate of the Forest: Developers, Destroyers and Defenders of the Amazon.* New York: Verso.
Henderson-Sellers, A. 1981. "The Effect of Land Clearance and Agricultural Practices Upon Climate." In *Blowing in the Wind: Deforestation and Long-Range Implications,* 443–85. Williamsburg, Va: Department of Anthropology, College of William and Mary.
Herrera, R., C. F. Jordan, E. Medina, and H. Klinge. 1991. "How Human Activities Disturb the Nutrient Cycles of a Tropical Rainforest in Amazonia." *Ambio* 10:109–14.
Higgs, Richard. 1992a. "Watch It Grow." *Paper,* Sept., 36–54.

- 1992b. "Chile Taps In." *Paper*, Sept., 21–9.

Hill, Hal. 1988. *Forest Investment and Industrialization in Indonesia*. New York: Oxford University Press.

- (ed.). 1989. *Unity and Diversity: Regional Economic Development in Indonesia Since 1970*. New York: Oxford University Press.

Hiraoka, Mario. 1992. "*Caboclo* and *Ribereno* Management in Amazonia: A Review." In Kent H. Redford and Christine Padoch (eds.), *Conservation of Neotropical Forests: Working from Traditional Resource Use*, 134–57. New York: Columbia University Press.

Hochfarber, Eugenio Bobenrieth. 1990. "El sector forestal en la VIII region: balance de una decada." *Informa Economico Regional* 6 (agosto):53–82.

Hohol, Roman. 1991. "The Impact of Societal Change on North American Groundwood Papers Consumption." Address to 5th Groundwood Papers Conference, Chicago, 5 Nov.

Horne, Garry, and Charlotte Penner. 1992. "British Columbia Local Area Economic Dependencies." Paper presented at 26th Annual Pacific Northwest Regional Economic Conference, 30 April–2 May 1992, Victoria, BC.

Humphrey, Craig R. 1988. "Timber-Dependent Communities." In A.E. Luloff and Louis E. Swanson (eds.), *American Rural Communities*, 34–60. Boulder: Westview Press.

Hunter, Lachlin. 1984. "Tropical Forest Plantations and Natural Stand Management: A National Lesson from East Kalimantan?" *Bulletin of Indonesian Economic Studies* 20(1):98–116.

Hurst, Philip. 1987. "Forest Destruction in South-East Asia." *Ecologist* 17(4–5):170–4.

- 1990. *Rainforest Politics: Ecological Destruction in South-East Asia*. London: Zed Books.

Hurwitz, Roger. 1989. "Patterns of Media Use in Developed and Developing Countries." In FAO, *Proceedings*, 1989: 89–101.

Indonesia. [1988?]. "Selected Statistics from Central Bureau of Statistics." In *Agriculture, Fisheries, Forestry and Environment.*

- Department of Forestry. 1966. *Forest Resources of Indonesia*. Jakarta.

- Department of Industry. 1979. *Investment Policies in the Industrial Sector During the Third Five Year Development Plan/Pelita (1979/1980–1983/1984): Brief Explanations on Some Main Projects*. Jakarta.

Indonesian Forestry Community. 1990. "Indonesia: Tropical Forests Forever." *Journal of Forestry* 88(9):26–31.

Instituto de Technologia, Universidade Federal do Espírito Santo, Brazil. 1987. *Analise do relatorio de impacto ambiental. Interessado: Aracruz Celulose A/ A. e Arucruz Florestal S/A*. Vitória.

International Institute for Environment and Development. 1985. *Forest Policies in Indonesia: The Sustainable Development of Forest Lands*. 4 vols. Jakarta, Indonesia.

Irwan, Alexander. 1989. "Business Patronage, Class Struggle, and the Manufacturing Sector in South Korea, Indonesia, and Thailand." *Journal of Contemporary Asia* 19(4):398–434.

Ishii, Yutaka. 1985. "Development Process of Forestry in Hokkaido." *Current State of Japanese Forestry*, 4:30–8. Tokyo: Japanese Forest Economic Society.

– 1990. "Development of the Pulp and Paper Industry in Japan and the Importation of Pulpwood." Photocopy.

– and Arai, Hotoshi. 1986. "A Study on the Present State of Plantation Forestry in Hokkaido." Paper presented to XVIII IUFRO, Division IV.

IUCN (International Union for the Conservation of Nature and Natural Resources). 1980. *The World Conservation Strategy*. Gland, Switzerland: United Nations Environmental Program, World Wildlife Fund.

Iwai, Y. 1989. "The Movement of the Lumbering Industry in U.S.A. and Its Influence on Japanese Forest Industry." *Current State of Japanese Forestry*, 6:12–22. Tokyo: Japanese Forest Economic Society.

– 1978. "Forests Are for People – Once?" *Tiger Paper* 5:4.

– 1979. "A Plea for S.E. Asia's Forests." *Habitat* 7:4.

Jakarta Post. 1990. "Two Myanmarese, One Thai Die in Border Town Bomb Blast." 7 May, 5.

Japan Economic Journal. 1991. "Japanese Firms to Build Pulp Plant in Canada." 11 May.

JATAN (Japan Tropical Forest Action Network). 1993. "Report on Eucalyptus Plantation Schemes. Investment Activities by Japan's Paper Industry in Brazil and Chile. Period of visit. 1992." Tokyo.

Japan Lumber Journal. Tokyo. Various issues.

Japan Pulp and Paper. Tokyo: Tec Times Co. Ltd. Various issues.

Jasria, Asriel. 1990. Unpublished notes for brief on government policy for companies in Indonesia, submitted by Putra group.

Jelvez, A., K.A. Blatner, and R.L. Govett. 1990. "Forest management and Production in Chile." *Journal of Forestry* 88:30–4.

Jesson, Bruce. 1980. *The Fletcher Challenge: Wealth and Power in New Zealand*. Pokeno, NZ: Jesson.

JETRO (Japanese External Trade Organizaton). 1988. *JETRO Magazine*. Sept. Tokyo.

JFTA (Japan Forest Technical Association). 1983. *Wood Demand and Supply in Japan*. Tokyo: Nippon Mukuzai Bichiku Kiko.

– *1985. Forestry and wood industry in Japan.* Tokyo: Nippon Mokuzai Bichiku Kiko.

Johnson, Jay A., and W. Ramsay Smith (eds.). 1988. *Forest Product Trade: Market Trends and Technical Developments*, Seattle: University of Washington Press.

Johnson, Robert. 1989. "Supply Will Outstrip Demand to 1992." *Pulp and Paper International* 32(2):47–51.

Johnston, R.J. 1989. *Environmental Problems: Nature, Economy and State.* London and New York: Belhaven Press.

Jomo, K.S. 1992a. "The Continuing Pillage of Sarawak's Forests." In Malyasia, Institute of Social Analysis, *Logging against the Natives of Sarawak*, i–ix. 2nd ed. Selangor, Malysia: The Institute.

– 1982b. "Logging, Politics, Business and the Indigenous People of Sarawak." Paper delivered to the fifth annual conference of the Northwest Regional Consortium for Southeast Asian Studies, University of British Columbia, Vancouver, 16–18 Oct.

Jordan, Carl F. 1987a. "Shifting Cultivation. Case Study No. 1: Slash and Burn Agriculture Near San Carlos de Rio Negro, Venezuela." In Jordan (ed.), *Amazonian Rain Forests*, 9–23. New York: Springer-Verlag.

– (ed.). 1987b. *Amazonian Rain Forests: Ecosystem Disturbance and Recovery,* New York: Springer-Verlag.

– Jiragorn Gajaseni, and Hiroyuki Watanabe. 1992. *Taungya: Forest Plantations with Agriculture in Southeast Asia.* Oxford: C.A.B. International.

– and R. Herrera. 1981. "Tropical Rain Forests: Are Nutrients Really Critical?" *American Naturalist* 117:167–80.

JPA (Japan Paper Association). 1991. "Pulp and Paper Statistics, 1991." Tokyo.

– 1994. "Pulp and Paper Statistics." Tokyo.

Julian, D. 1988. "The Paper Industry." Tokyo: Kleinwort Grieveson Securities Japanese Research, May.

Kallio, Markku, Dennis P. Dykstra, and Clark S. Binkley (eds.). 1987. *The Global Forest Sector: An Analytical Perspective.* New York: Wiley.

Kardell, Lars, Eliel Steen, and Antonio Fabiao. 1986. "Eucalyptus in Portugal: a Threat or a Promise." *Ambio* 15(1):643–50.

Kartasubrata, Junus. 1987. "Social Forestry Research at IPB." In *Planning and Implementation of Social Forestry Programmes in Indonesia.* Workshop organized by the Faculty of Forestry GMU in collaboration with FONC and FAO/WEDP, 1–3 Dec., Yogyakarta, Indonesia.

– 1990. "Research Support to Community Forestry Projects on Forestland in Java Indonesia." Paper presented at Seminar on Research Policy for Community Forestry, organized by RECOFTC. 8–11 Jan., Bangkok.

Kartawinata, Kuswata. 1981. "The Enviirional Consequences of Tree Removal from the Forest in Indonesia." In *Where Have All the Flowers Gone?* 191–214. Williamsburg, Va: Department of Anthropology, College of William and Mary.

Kato, Takashi. 1981. "A Regional Comparison of Forest Productivity, Stumpage Prices, Logging and Regeneration Costs Among Japan, Canada, and the United States." *Current State of Japanese Forestry* 1:19–40.

– 1982. "Comparison of Softwood Lumber Manufacturing and Selling Costs between the Pacific Coast of North America and Japan." *Current State of Japanese Forestry* 2:30–9.

– 1985. "Outlook for Japan Imports of Wood Chips and Pulp and Paper." In *Pacific Rim Markets for Forest Products, 1985–2000*. Vancouver: PaperTree Economics Ltd.

Katsuhisa, Hikojiro. 1984. "Forest Products Trade of Japan." *Current State of Japanese Forestry* 3:10–16, 19–22.

Kawake, Jiro. 1988. "Characteristics and Tasks of Japan's Pulp and Paper Industry." In Gerard F. Schreuder (ed.), *Global Issues and Outlook in Pulp and Paper*, 20–8. Seattle: University of Washington Press.

Keihakutown, Japan. 1988. Information/statistics pamphlet distributed by the town administration in 1988.

Kellison, R.C., and B.J. Zobel. 1987. "Technological Advances to Improve the Wood Supply for the Pulp and Paper Industry in Developing Countries." In FAO, *Proceedings*, 1987:136–44.

Kemf, Elizabeth. 1990. "Month of Pure Light: The Regreening of Viet Nam." World Wide Fund for Nature.

Kengen, Sebastiao. 1985. "Industrial Forestry and Brazilian Development: A Social, Economic and Political Analysis with Special Emphasis on the Fiscal Incentives Scheme and the Jequitinhonha Valley in Minas Gerais." PHD thesis, Australian National University.

– 1991a. "Forest Management in Brazil: A Historical Perspective." Paper given at Forest History Conference, Tropical Forest History, San Jose, Costa Rica, February.

– 1991b. "National Forest Management and Inventory Project, Myanmar Forestry Sector." Report prepared for government of the the Union of Myanmar. Yangoon: FAO.

Kent, Francis B. 1968a. "Brazilians Indignant at Indian Genocide Report." *Los Angeles Times*, 22 March.

– 1968b. "Brazil Gets Inquiries on Alleged Indian Slayings." *Los Angeles Times*, 29 March.

Kimmins, Hamish. 1992. *Balancing Act: Environmental Issues in Forestry*. Vancouver: University of British Columbia Press.

Kinkead, G. 1981. "Trouble in D.K. Ludwig's Jungle." *Fortune*, 21 April, 102–17.

Klabin (Industrias Kalbin de Papel e Celulose SA). 1990. *Annual Report*.

– n.d. "Demonstracoes Financeiras as em 32 de Dezembro de 1990 e. de 1989."

– n.d. *Industrias Klabin Papel e Celulose, Paraná Division*.

– n.d. *Klabin, In Partnership with Nature*.

Knight, Patrick. 1988. "Caima: Set to Diversify and Optimize." *Pulp and Paper International* 30(8):40–1.

– 1992. "Simao Sows Seeds of Market Advantage." *Pulp and Paper International* 34(5):74.

Komarov, Boris. 1980. *The Destruction of Nature in the Soviet Union*. White Plains, NY: M.E. Sharpe.

Komkris, Thiem. "Forestry Aspects of Land Use in Areas of Swidden Cultivation." In Peter Kunstadter et al. (eds.), *Farmers of the Forest*, 61–70. Honolulu: University of Hawaii Press.

Korten, D.C., and R., Klauss. 1984. *People-Centered Development*. West Harford, Conn.: Kumarian Press.

Kucinski, Bernardo. 1982. "Brazil's Amazonian Dream Ends Up as a Nightmare." *Guardian* (London), 13 Jan.

Kumar, Raj. 1986. *The Forest Resources of Malaysia*. Oxford: Oxford University Press.

Kumazaki, Minoru. "Japanese Economic Development and Forestry." In R. Handa (ed.), *Forest Policy in Japan*, 1–35. Tokyo: Nippon Ringyo Chosakai.

Kunstadter, Peter. 1988. "Hill People of Northern Thailand." In Julie Sloan Denslow and Christine Padoch (eds.), *People of the Tropical Rain Forest*, 93–110. Berkeley: University of California Press, in association with the Smithsonian Institution Traveling Exhibition Service, Washington, DC.

– E.C. Chapman, and Sanga Sabhasri (eds.). 1978. *Farmers in the Forest: Economic Development and Marginal Agriculture in Northern Thailand*. Honolulu: University of Hawaii Press.

Kuntijoro-Jakti, Hero Utomo. 1988. "External and Domestic Coalitions of the Bureaucratic Authoritarian State in Indonesia." PHD diss., University of Washington.

Kyoto (Japan) Prefectural Government. 1987. "This is Kyoto Prefecture." Population data, section 14:61–2.

Laarman, Jan. G. 1984. "Timber Exports from Southeast Asia: Away from Logs towards Processed Wood." *Columbia Journal of World Business* 19:77–82.

– 1988. "Export of Tropical Hardwoods in the Twentieth Century." In John F. Richards and Richard P. Tucker (eds.), *World Deforestation in the Twentieth Century*, 147–63. Durham, NC: Duke University Press.

– Gerard F. Schreuder, and Erik T. Anderson. 1988. "An Overview of Forest Products Trade in Latin America and the Caribbean Basin." In Jay A. Johnson and W. Ramsay Smith (eds.), *Forest Products Trade: Market Trends and Technical Developments*, 3–22. Seattle: University of Washington Press.

– and Roger A. Sedjo. 1992. *Global Forests: Issues for Six Billion People*. New York: McGraw-Hill.

Lanly, Jean-Paul. 1982. "Tropical Forest Resources." FAO Forestry Paper 30. Rome: FAO.

– and Clement, J. 1979. "Present and Future National Forest and Plantation Areas in the Tropics." *Unasylva* 31(123):FAO 12–20.

Lara, Antonio. 1985. "Los ecosistemas forestales en el desarrollo de Chile." *Ambiente y desarrollo* 1(3):81–99.

– 1992. "Temperate Forests in Chile: Conservation Problems and Challenges." Unpublished photocopy.

– L. Araya, J. Capella, and M. Fierro. 1987. *Evaluation of the Destruction of Native Forests in Chile*. Final Report Project #3181 WWF–International/CODEFF. Santiago.

– P. Donoso, and M. Cortes. 1991. *Development of Conservation and Management Alternatives for Native Forests in South-Central Chile*. Final Report Project #3181 WWF–US/CODEFF. Santiago.

– and Thomas T. Veblen. 1993. "Forest Plantations in Chile: a Successful Model?" in Alexander Mather (ed.), *Afforestation: Policies, Planning and Progress*, 118–39. London: Belhaven Press.

Lewis, Norman. 1969. "Genocide – From Fire and Sword to Arsenic and Bullets, Civilization Has Sent Six Million Indians to Extinction." *Sunday Times* (London), 23 Feb.

Lindblad, J.T. 1985. "Economic Change in Southeast Kalimantan, 1880–1940." *Bulletin of Indonesian Economic Studies* 21(3):69–103.

Lindsay, Holly. 1989. "The Indonesian Log Export Ban: An Estimation of Foregone Export Earnings." *Bulletin of Indonesian Economic Studies* 25(2):111–23.

Lintu, L. 1991. "The World Market Situation and New Trends in Newsprint." *Unasylva* 42(167):43–50.

Lönnstedt, Lars. 1984. "Stability of Forestry and Stability of Regions: Contradictory goals? The Swedish Case." *Canadian Journal of Forest Resources* 14:707–11.

Lovelock, James. 1985. "Are We Destabilising World Climate? The Lessons of Geophysiology." *Ecologist* 15(1/2):52–5.

Lowe, R.G. 1977. "Experience with the Tropical Shelterwood System of Regeneration in Natural Forest in Nigeria." *Forest Economics and Management* 1:193–212.

Luckert, Martin K. 1988. "Summary of a Study on the Role of Canadian Forest Tenures in Affecting Social Welfare, the Distribution of Rents, the Allocation of Resources, and Investments in Silviculture." Based on PHD thesis in progress. Photocopy, January.

Ludwig, Donald, Ray Hilborn, and Carl Walters. 1993. "Uncertainty, Resource Exploitation, and Conservation: Lessons from History." *Science* 260 (2 April): 17, 36.

Lugo, A.E., and S. Brown. 1982. "Conversion of Tropical Moist Forests: A Critique." *Interciencia* 7:89–93.

Luhde, Friedrich. 1988. "An Investor Views the Opportunities." *Pulp and Paper International* 30(7):39–44.

Lundgren, B., and J.B. Raintree. 1983. "Agroforestry." In B. Nestel (ed.), *Agricultural Research for Development, Potential and Challenges in Asia*, 37–49. The Hague, Netherlands: ISNAR.

Lutzenberger, Jose. 1982. "The Systematic Destruction of the Tropical Rainforest in the Tropical Rain Forest in the Amazon." *Ecologist* 12(6):248–52.

- 1987a. "Who is Destroying the Amazon Rainforest?" *Ecologist* 17(4–5):155–60.
- 1987b. "Brazil's Amazonian Alliance." *Ecologist* 17(4–5):190–1.
Lyons, Bob. 1989. *Dire Straits: Pollution in the Strait of Georgia, British Columbia, Canada.* Vancouver: Greenpeace Canada.
MAART (Monsoon Asia Agroforestry Joint Research Team). 1986. *Comparative Studies on the Utilization and Conservation of the Natural Environment by Agroforestry Systems.* Report on joint research among Indonesia, Thailand, and Japan. Nishinomiya, Japan: Asahi.
McClintock, Wayne, and Nick Taylor. 1983. "Pines, Pulp and People: A Case Study of New Zealand Forestry Towns." Centre for Resource Management, University of Canterbury and Lincoln College, New Zealand.
MacDonald, Cecil. 1983. "Tropics Could Be Major Fiber Source." *Pulp and Paper International* 25(7):26–8.
McDonald, H. 1976. "Indonesia: A New Hope for Loggers." *Far Eastern Economic Review* 92(21) (21 May), 66–7.
- 1990. "Partners in Plunder." *Far Eastern Economic Review* 147(8) (22 Feb.): 16–18.
McGaughey, Stephen E., and Hans M. Gregersen (eds.). 1983. *Forest-Based Development in Latin America: An Analysis of Investment Opportunities and Financing Needs.* Washington, DC: Inter-American Development Bank, March.
M'Gonigle, Michael. 1986. "From the Ground Up: Lessons from the Stein River Valley." In Warren Magnusson et al. (eds.), *After Bennett: A New Politics for British Columbia,* 169–91. Vancouver: New Star Books.
- 1987. "Local Economies Solve Global Problems." *New Catalyst,* Spring.
- 1988–90. "Developing Sustainability: A Native/Environmentalist Prescription for Third-Level Government." *B.C. Studies,* no. 84:65–99.
- 1990. "Borealis Nix: The Political Economy of the Pulp Mill Bonanza in Canada's Northern Forests." Paper delivered to the Canadian Political Science/ Canadian Anthropology and Sociology Association joint meetings, Learned Societies of Canada, 27 May.
McInnis, John. 1994. "Japanese Investment in Alberta's Taiga Forests." Photocopied notes, 25 Feb.
Mackie, Cynthia. 1986. "Disturbances and Succession Resulting from Shifting Cultivation on Upland Rainforest in Indonesian Borneo." PHD diss., Rutgers University.
McLennan, Marshall Seaton. 1972. "Peasant and Hacendero in Nueva Ecija: The Socio-Economic Origins of a Philippine Commercial Rice-Growing Region." PHD diss., University of California.
McNamara, Robert S. 1982. "The Year 2000 Committee: A New Initiative." *Focus* 4(3):1.
McNeill, John R. 1988. "Deforestation in the Araucaria Zone of Southern Brazil, 1900–1983." In John F. Richards and Richard P. Tucker (eds.), *World*

Deforestation in the Twentieth Century, 15–32. Durham, NC: Duke University Press.

Malaysia. INSAN (Institute of Social Analysis). 1990. *Logging against the Natives of Sarawak.* 2d ed. Selangor, Malaysia.

Manning, Chris. 1971. "The Timber Boom." *Bulletin of Indonesian Economic Studies* 7(3):30–60.

Marchak, M. Patricia. 1983. *Green Gold: The Forest Industry in British Columbia.* Vancouver: University of British Columbia Press.

– 1986. "The Rise and Fall of the Peripheral State: The Case of British Columbia." In Robert J. Brym (ed.), *Regionalism in Canada*, 124–59. Toronto: Irwin.

– 1988–89. "What Happens When Common Property Becomes Uncommon?" *B.C. Studies* 80:3–23.

– 1989. "British Columbia: 'New Right' Politics and a New Geography." In M.S. Whittington and Glen Williams (eds.), *Canadian Politics in the 1990s*, 45–59. Toronto: Nelson.

– 1990. "Brief to the B.C. Forest Resources Commission re: Corporate Concentration and Community Stability."

– 1991. "For Whom the Tree Falls: Restructuring of the Global Forest Industry." *B.C. Studies* 90:3–24.

– 1991–92. "Global Markets in Forest Products: Sociological Impacts on Kyoto Prefecture and British Columbia Interior Forest Regions." *Journal of Business Administration* 20 (1/2):339–69.

Margolick, Michael, and Russell S. Uhler. 1991–92. "The Economic Impact on British Columbia of Removing Log Export Restrictions." In *Journal of Business Administration* 20(1/2):273–96.

Maser, Chris. 1990. *The Redesigned Forest.* Toronto: Stoddart.

Mather, A.S. 1990. *Global Forest Resources.* London: Belhaven.

May, P. 1986. *A Modern Tragedy of the Non-Commons: Agro-Industrial Change and Equity in Brazil's Babacu Palm Zone.* Latin American Program Dissertation Series. Ithaca, NY: Cornell University Press.

May, Peter H. "Common Property Resources in the Neotropics: Theory, Management Progress, and an Action Agenda." In K.H. Redford and Christine Padoch (eds.), *Conservation of Neotropical Forests: Working from Traditional Resource Use*, 359–78. New York: Columbia University Press.

Meggars, B.J. 1971. *Amazonia: Man and Culture in a Counterfeit Paradise.* Chicago: Aldine-Atherton.

Meyer, Philip. 1990. "Outline of Some Key Socio-economic Issues Affecting Development of an Old Growth Policy for British Columbia." Submission to Old Growth Values Team, Province of British Columbia, July.

Meyer Resources. 1990. "A Preliminary Framework for Identification of Socio-economic Values Relevant to Old Growth Policy Development in British Columbia." A report to the Old Growth Values Team, Province of British Columbia, July.

Mitchell, Andrew. 1994. "Indah Kiat Lays Foundations for Future Growth." *Pulp and Paper International,* Oct., 24–31.

Miracle, M.P. 1973. "The Congo Basin as a Habitat for Man." In B.J. Meggers, E.S. Ayensu, and W.D. Duckworth (eds.), *Tropical Forest Ecosystems in Africa and South America: A Comparative Review,* 335–44. Washington, DC: Smithsonian Institution Press.

Mitsuda, Hisayoshi, and Charles C. Geisler. 1988. "Environmentalism as if People Mattered: A Case Study of the Japanese National Trust Movement." Paper presented at the XII World Congress of Rural Sociology, Bologna, Italy, 26 June–1 July.

Miyamacho, Japan. 1988. Information/statistics pamphlet distributed by the town administration in 1988.

Mohr, E.C.J., F.A. van Baren, and J. van Schuylenborgh. 1972. *Tropical Soils: A Comprehensive Study of Their Genesis.* The Hague, Paris, Djakarta: Mouton-Ichtiar Baru-Van Hoeve.

Molleda, Julio. 1987. "The Implications for Raw Material Supply." FAO *Proceedings,* 1987:159–62.

– 1988. "The Role of the Iberian Pulp Producers in the World Markets." In Gerard F. Schreuder (ed.), *Global Issues and Outlook in Pulp and Paper,* 76–88. Seattle: University of Washington Press.

Moore, Barrington, Sr. 1910. "Forest Problems in the Philippines." *American Forestry,* February:75–81; March:149–50.

Moraga, Z., Mara Eugenia, and Juan Rodriguez U. 1989. "Diagnostico socio-economico de la region del Bío-Bío." Section on forestry, unpublished working papers, 32–52. Concepción: University of Concepción, agosto.

Moran, E. 1981. *Developing the Amazon.* Bloomington: Indiana University Press.

Morris, Brian. 1986. "Deforestation in India and the Fate of Forest Tribes." *Ecologist* 16(6):253–7.

Murphy, Rhoads. "Deforestation in Modern China." In Richard P. Tucker and J.F. Richards (eds.), *Global Deforestation and the Nineteeth-Century World Economy,* 111–28. Durham, NC: Duke University Press.

Muthoo, M.K. 1990. "Economic Considerations and Environmental Policy Implications in the Management of Renewable Natural Resources." *Unasylva* 163(41):50–6.

– 1991. "An Overview of the FAO Forestry Field Programme." *Unasylva* 166(42):30–9.

Myers, Norman. 1980. *Conversion of Moist Tropical Forests.* Washington, DC: National Academy of Sciences.

– 1981. "Deforestation in the Tropics: Who Gains, Who Loses?" In *Where Have All the Flowers Gone? Deforestation in the Third World,* 1–25. Williamsburg, Va: Department of Anthropology, College of William and Mary.

– 1984. *The Primary Source: Tropical Forests and Our Future.* New York: Norton.

Nahuz, Marcio A.R. 1988. "The Latin American Southern Cone's Role as a New Wood Supplying Region." In Jay A. Johnson and W. Ramsay Smith (eds.), *Forest Products Trade: Market Trends and Technical Developments*, 23–37. Seattle: University of Washington Press.

Nan (Province of, Thailand). n.d. "Green Nan Project." Project and Planning Office, Nan. Photocopy.

Nation (Bangkok). 1990a. "Foresters Question Eucalyptus." 7 April.

– 1990b. "US Senate Approves Bill to Ban Burmese Goods." 27 April.

– 1990c. "Log Bill: Bt 2bn Export at Stake." 27 April.

Nations, James D. 1992. "Xateros, Chicleros, and Pimenteros: Harvesting Renewable Tropical Forest Resources in the Guatemalan Peten." In Kent H. Redford and Christine Padoch (eds.), *Conservation of Neotropical Forests: Working from Traditional Resource Use*, 208–19. New York: Columbia University Press.

– and Daniel I. Komer. 1987. "Rainforests and the Hamburger Society." *Ecologist* 17(4–5):161–7.

Nectoux, Francois, and Yoichi Kuroda. 1989. "Timber from the South Seas: An Analysis of Japan's Tropical Timber Trade and Its Environmental Impact." Gland, Switzerland: World Wildlife Federation International.

Nemetz, Peter N. (ed.). 1993. *Emerging Issues in Forest Policy*. Vancouver: University of British Columbia Press, 1993. Previously published as special edition of *Journal of Business Administration* 20, no. 1–2 (1991–92).

Neves, A.R. 1979. "Avaliacao socio-economica de um programa de reflorestamento na Regiao de Carbonita, Vale de Jequitinhonha, MG." MSC thesis, Universidade Federal de Vicosa.

New Zealand. Development Finance Corporation. 1980. *Forest Industry Study*. Wellington, March 1980.

– Ministry of Works and Development. 1983. *Central North Island Planning Study Findings*. Study Project Team, Wellington, April.

New Zealand Forestry Council. 1980. "Policy Guidelines for Private Indigenous Forests." Wellington, December.

– 1981. "Report of the Working Party on Industry Strategy, 1981 New Zealand Forestry Conference. Exotic Forests: Their Importance to New Zealand's Future." Wellington.

– 1982. "Final Report on the Second Session of the 1981 New Zealand Forestry Conference." Wellington, July.

New Zealand Forest Service. 1977. "Management Policy For New Zealand's Indigenous State Forests." Wellington.

– 1981. *Policy on Exotic Special Purpose Species*. Wellington.

New Zealand Public Service Association. 1984. "Controlling Interest." Special issue of *Economic Information*. Wellington: Research Division of the Association. 8(1).

Ngau, Harrison, Thomas Jalong Apoi, and Chee Yoke Ling. 1987. "Malaysian Timber: Exploitation for Whom?" *Ecologist* 17(4–5):175–84.

Nigh, R.B., and J. Nations. 1980. "Tropical Rain Forests." *Bulletin of the Atomic Scientists* 12–19 March.

Nikiforuk, Andrew, and Ed Struzik. 1989. "The Great Forest Sell-Off." *Globe and Mail Report on Business Magazine*, Nov.:57–68.

Nilsson, Sten. 1985. *An Analysis of the British Columbia Forest Sector around the Year 2000.* Vancouver: Forest Economics and Policy Analysis Project, University of British Columbia, April.

– and Peter Duinker. 1987. "A Synthesis of Survey Results: The Extent of Forest Decline in Europe." *Environment* 29(9):4–31.

Noble, Kimberley. 1986. "Lessons from Our Neighbors of the North." *Globe and Mail Report on Business Magazine*, Nov.:50–61.

Nogueira, U.B. 1980. "Charcoal Production from Eucalyptus in Southern Bahjai from Iron and Steel Manufacture in Minas Gerais, Brazil." PHD thesis, Michigan State University.

Nordin, C.F., and Meade, R.H. 1992. "Deforestation and Increased Flooding of the Upper Amazon." *Science* 215:426–7.

Norman, Herbert. 1975. *Origins of the Modern Japanese State: Selected Writings of E.H. Norman.* Ed. John W. Dower. New York: Pantheon.

Norse, E.A., K.L. Rosenbaum, D.S. Wilcove, B.A. Wilcox, W.H. Romme, D.W. Johnston, and M.L. Stout. 1986. *Conserving Biological Diversity in Our National Forests.* Washington, DC: Wilderness Society.

North Cariboo Community Futures Group. 1987. *Report.*

O'Brian, Hugh. 1988a. "The Non-Wood Leader Takes the Lead." *Pulp and Paper International* 30(10):84.

– 1988b. "Japan's Market Faces Changing Times." *Pulp and Paper International* 30(11):63–72.

Oldfield, Margery L. 1981. "Tropical Deforestation and Genetic Resources Conservation." In *Blowing in the Wind: Deforestation and Long Range-Implications*, Studies in Third World Societies, 277–345. Williamsburg, Va: Department of Anthropology, College of William and Mary.

Oliveira, Pedro. 1987. "On Western Europe's Last Frontier." *Pulp and Paper International* 29(11):83–5.

Olsson, Anders. 1993. "Southeast Asia: Land of Plenty." Speech to delegates at *PPI*'s first Asian Paper conference, 1992; as published in *Pulp and Paper International* 35(2):5.

Orgill, Brian. 1988. "Pacific Rim Report." *Tappi Journal*, Dec.:23.

Osako, Masako M. 1983. "Forest Preservation in Tokugawa Japan." In Richard P. Tucker and J.F. Richards (eds.), *Global Deforestation and the Nineteenth-Century World Economy*, 129–45. Durham, NC: Duke University Press.

Oshima, H. 1986. *Indonesia's Investment Climate – Opportunities for Japanese Investors.* Jetro Jakarta Centre, August.

Otero, Luis A. 1984. "Caracterización laboral, estudio de las condiciónes de trabajo yu analysis ocupaciónal de los trabajadores forestales en la Octava Region del país." Thesis, Universidad de Chile, Santiago.

– 1989. "La silvicultura como factor del desazrrollo social en la region del Bío-Bío." *Ambiente y Desarrollo* 5(1):55–65.

– 1990. "Impacto de la actividad forestal en comunidades locales en la VIII Region." *Ambiente y Desarrollo* 6(2):61–9.

– 1992. "La transformación de los bosques y el asentamiento de población rural: una representación grafica." *Renarres* 8(31):18–22.

Otten, Mariel. 1986. "'Transmigrasi': From Poverty to Bare Subsistence." *Ecologist* 16(2–3):71–6.

Padoch, Christine, and Wil de Jong. 1992. "Diversity, Variation, and Change in Ribereno Agriculture." In Kent H. Redford and Christine Padoch (eds.), *Conservation of Neotropical Forests: Working from Traditional Resource Use*, 158–74. New York: Columbia University Press.

Palmer, Ingrid. 1978. *The Indonesian Economy Since 1965.* London: Frank Cass.

Palmer, Randall. 1990. "Efforts on to Revive Vietnam's Wasteland." *Nation* (Bangkok), 27 April.

Panayotou, Theodore. 1989. "The Economics of Man-Made Natural Disasters: a Framework for Rehabitation and Prevention." In *Safeguarding the Future, Restoration and Sustainable Development in the South of Thailand.* Report by team of the National Operations Center, National Economic and Social Development Board, U.S. Agency for International Development. Bangkok.

– and Peter S. Ashton. 1992. *Not By Timber Alone: Economics and Ecology for Sustaining Tropical Forests.* Washington, DC: Island Press.

Paoliello, Jose Luiz. 1988. "The Brazilian Pulp Industry: Background, Present Structure, and Future Outlook." In G.F. Schreuder (ed.), *Global Issues and Outlook in Pulp and Paper*, 69–75. Seattle: University of Washington Press.

Paper (bi-weekly). Various issues, 1990–94.

Papermaker. Oct. 1994.

PaperTree (monthly). Various issues, 1989–94.

Parsons, S.A. 1982. "Developments in the Forest-Based Industries of Indonesia, Malaysia and the Philippines, Implications for Australia." Project 1675, Bureau of Agricultural Economics, Canberra, Occasional Paper No. 67. Canberra: Australian Government Publishing Service.

Patterson, David. 1989. "Asia: Consumption Growth Fuelled Mainly by New Domestic Production." *Pulp and Paper International* 31(2):55–7.

Pauletti, Hamlet. 1981. "The Forest Closes in on Millionaire Ludwig." *WorldPaper*, Jan.:3.

Pearse, Peter. 1980. Address to annual meeting of the B.C. Truck Loggers' Association, Vancouver.

– 1984. "Obstacles and Incentives for Silviculture in Canadian Forest Policies." Vancouver: Forest Economics and Policy Analysis Project, University of British Columbia, 9 Aug.

Penna, Ian. 1987. "Australia's Timber Industry: Promises and Performance." Melbourne: Australian Conservation Foundation, November.

– 1992. "Japan's Paper Industry: An Overview of its Structure and Market Trends." Tokyo: Chikyu no Tomo/Friends of the Earth Japan, Asia-Pacific Forest Conservation Project, May.

Percy, M.B., and C. Yoder. 1987. "The Softwood Lumber Dispute and Canada–
US Trade in Natural Resources." Halifax: The Institute for Research on
Public Policy.

Perez-Sainz, J.P. 1979. *Transmigration and Accumulation in Indonesia*. Geneva:
International Labor Office.

Petmak, P. 1983. "Validity of Agroforestry System in Northeast Thailand."
Agroforestry, no. 1. Silvicultural Research Sub-division, Royal Forest Department, Thailand.

Phantumvanit, Dhira. 1986. "Clean Technologies for the Pulp and Paper
Industry, the Textile Industry and Metal Coating and Finishing in Thailand." TDRI (unpublished papers).

– and Khunying Suthawan Sathirathai. 1988. "Thailand: Degradation and
Development in a Resource-Rich Land." *Environment* 30(1):10–15, 30–36.

– and Suthawan Sathirathai. 1986. "Profiles of the Pulp and Paper, Textile
and Metal Coating Industries in Thailand – a Country Report for the
International Symposium on Clean Technologies." Presented at meeting on
Treatments of Hazardous Chemicals and Wastes, Wiesbaden/Hessen, Federal Republic of Germany, April.

– and Suthawan Sathirathai. 1987. "Pressure on Forests – the Thai Experience." Paper presented at Regional Meeting to Plan Linkages between
Forestry and Other Sectors of the Economy, FAO Regional Office for Asia
and the Pacific, Bangkok, 15–18 Dec.

Pinchot, Gifford. 1947. *Breaking New Ground*. New York: Harcourt, Brace and
World.

Plotkin, Mark, and Lisa Famolare (eds.). 1992. *Sustainable Harvest and Marketing of Rain Forest Products*. Washington, DC: Island Press.

Plumwood, Val, and Richard Routley. 1982. "World Rainforest Destruction –
The Social Factors." *Ecologist* 12(1):4–22.

Poore, Duncan, Peter Burgess, John Palmer, Simon Rietberger, and Timothy
Synnots. 1989. *No Timber without Trees: Sustainability in the Tropical Forest*. A
study for ITTO. London: Earthscan.

Poore, M.E.D., and C. Fries. 1985. *The Ecological Effects of Eucalyptus*. FAO Paper
59. Rome.

Posey, C.E. 1980. Statement of Dr. Clayton E. Posey on Jari Florestal, in Hearings
before the Subcommittee on Foreign Affairs, House of Representatives, 96th
Congress, Second Session, May 7, June 19, and September 18, 1980: Tropical
Deforestation, 428–33. Washington, DC: U.S. Government Printing Office.

Posey, Darrell A. 1981. "Ethnoentomology of the Kayapó Indians of Central
Brazil." *Journal of Ethnobiology* 1(1):165–74.

– 1983a. "Indigenous Knowledge and Development: An Ideological Bridge to
the Future." *Ciencia e Cultura* 35(7):877–94.

– 1983b. "Indigenous Ecological Knowledge and Development of the Amazon." In E. Moran (ed.), *The Dilemma of Amazonian Development*, 225–57.
Boulder: Westview Press.

– 1984. "A Preliminary Report on Diversified Management of Tropical Forest by the Kayapó Indians of the Brazilian Amazon." In G.T. Prance and J.A. Kallunki (eds.), *Ethnobotany in the Neotropics,* 112–26. Bronx: New York Botanical Garden.

– 1985a. "Indigenous Management of Tropical Forest Ecosystems: the case of the Kayapó Indians of the Brazilian Amazon." *Agroforestry Systems* 3(2):139–58.

– 1985b. "Native and Indigenous Guidelines for New Amazonian Development Strategies: Understanding Biological Diversity through Ethnoecology." In J. Hemming (ed.), *Change in the Amazon Basin,* 1:156–81. Manchester: Manchester University Press.

– and William Balée (eds.). 1989. *Resource Management in Amazonia: Indigenous and Folk Strategies. Advances in Economic Botany.* Vol. 7. New York: New York Botanical Garden.

Postel, Sandra, and John C. Ryan. 1991. "Reforming Forestry." In *State of the World, 1991: A Worldwatch Institute Report,* 74–92. New York: Norton.

PPI (*Pulp and Paper International,* monthly). Various issues, 1985–94.

Prance, G.T., and T.S. Elias (eds.). 1977. *Extinction Is Forever.* New York: New York Botanical Garden.

Pratt, Larry, and Ian Urquhart. 1993. "The Last Great Forest: Political Economy and Environment in Alberta." Paper presented to annual meetings of the Canadian Political Science Associaton, Ottawa, 6–8 June.

– and Ian Urquhart. 1994. *The Last Great Forest: Japanese Multinationals & Alberta's Northern Forests.* Edmonton: NeWest Press.

Price, David. 1985. "The World Bank vs Native Peoples – A Consultant's View." *Ecologist* 15(1/2):73–7.

Price-Waterhouse Consultants Ltd. 1989. *The Forest Industry in British Columbia 1988.* Vancouver: Price Waterhouse.

– 1991. *Forest Products Survey 1990.* Vancouver and Toronto.

Prinz, Bernard. 1987. "Major Hypotheses and Factors: Causes of Forest Damage in Europe." *Environment* 29(9):11–15, 32–7.

Proctor, John. 1983. "Mineral Nutrients in Tropical Forests." *Progress in Physical Geography* 7(3):422–31.

Pryde, P.R. 1972. *Conservation in the Soviet Union.* Cambridge: The University Press.

– 1983. "The Decade of the Environment in the U.S.S.R." *Science* 220:274–9.

Pulp and Paper Canada (monthly). Various issues, 1985–94.

Pulp and Paper Week (weekly). Various issues, 1985–94.

Puntasen, Apichai, Somboon Siriprachai, and Chaiyuith Punyasavatsut. 1990. "Political Economy of Eucalyptus: A Case Study of Business Cooperation by the Thai Government and its Bureaucracy." Paper presented to conference on the Political Economy of the Environment in Asia, Simon Fraser University, Burnaby, 11–13 Oct.

Pura, Raphael. 1990. "In Sarawak, a Clash Over Land and Power." *Asian Wall Street Journal,* 7 Feb.

Raison, R.J., P.K. Khanna, and W.J.B. Crane. 1982. "Effects on Intensified Harvesting on Rates of Nitrogen and Phosphorous Removal from *Pinus radiata* and *Eucalyptus* forests in Australia and New Zealand." *New Zealand Journal of Forestry Science* 12(2):394–401.

Randers, J., and L. Lonnstedt. 1979. "Transition Strategies for the Scandinavian Forestry Sector." In Lonnstedt and Randers (eds.), *Wood Resource Dynamics in the Scandinavian Forestry Sector,* Studies Forestry Sueccica, 152:19–37.

Rankin, J.M. 1985. "Forestry in the Brazilian Amazon." In G.T. Prance and T.E. Lovejoy (eds.), *Key environments: Amazonia,* 369–92. Oxford: Pergamon Press.

Rao, Y.S., and C. Chandrasekharan. 1982. "The State of Forestry in Asia and the Pacific." *Unasylva* 35(140):11–21.

Raumolin, Jussi. 1982. "The Relationship of Forest Sector to Rural Development: Some Reflections on the Theory and Practice of Forest-Based Development." University of Oulu, The Research Institute of Northern Finland, no. 24 (May). Photocopy.

– 1983–84. "The World Economy of Forest Products and the Comparative Study of the Development Impact of the Forest Sector: An Exploratory Study." *Yearbook of the Finnish Society for Economic Research,* 188–211. Helsinki: The Society.

– 1984a. "The Formation of the Sustained-Yield Forestry System in Finland." In H.K. Steen (ed.), *History of Sustained Yield Forestry: A Symposium, Western Forestry Center, Portland, Oregon, October, 1983,* 115–69. Santa Cruz, Calif.: Forest History Society.

– 1984b. "The Impact of Forest Sector on Economic and Social Development in Finland." Research Institute of Northern Finland, paper C51.

– 1985. "The Impact of Forest Sector on Economic Development in Finland and Eastern Canada." *Fennia* 163(2):395–437.

– 1986a. "Introduction to Comparative Studies betweeen Finland and Canada." In J. Ramoulin (ed.), *Natural Resource Exploitation and Problems of Staples-Based Industrialization in Finland and Canada,* special ed. of *Fennia* 163(2):387–94.

– 1986b. "Recent Trends in the Development of the Forest Sector in Finland and Eastern Canada." *Zeitschrift der Gesellschaft fur Kanada-Studien* 11:89–114.

– (ed). 1987. *Suomen Antropologi,* 12(4). Special issue on swidden cultivation. Helsinki: Finnish Anthropological Society.

– 1988. "Restructuring and Internationalization of the Forest, Mining and Related Engineering Industries in Finland." Etla (Helsinki) unpublished paper, no. 267.

– 1990. "The Transfer and Creation of Technology in the World Economy with Special Reference to the Mining and Forest Sectors." Etla (Helsinki) unpublished paper, no. 313.

Redford, Kent H., and Christine Padoch (eds.). 1992. *Conservation of Neotropical Forests: Working from Traditional Resource Use.* New York: Columbia University Press.

Reed, F.L.C. 1987. "Forest Policy and Community Stability in British Columbia." Paper prepared for National Conference on Community Stability in Forest-Based Economies, Portland, Oregon, 16–18 Nov.

Reid, Collins and Associates Ltd. 1987. *Mission Tree Farm Review.* Vancouver, BC. Photocopy.

Remrod, Jan. 1987. "Sweden's Forests, a Growing Resource." *WorldWood.* 31(6):S4–S5.

Repetto, Robert. 1988. *The Forest for the Trees? Government Policies and the Misuse of Forest Resources.* Washington, DC: World Resources Institute.

Revkin, Andrew. 1990. *The Burning Season: The Murder of Chico Mendes and the Fight for the Amazon Rain Forest.* Boston: Houghton Mifflin.

Ribeiro, Darcy. 1967. "Culturas e linguas indigenas do Brasil." In *Educacao e Ciencias Sociais,* 1–102. Rio de Janeiro. Translated and reprinted in Janice H. Hopper (ed.), *Indians of Brazil in the Twentieth Century,* 79–165. Washington, DC.

Rich, Bruce M. 1985. "Multi-Lateral Development Banks: Their Role in Destroying the Global Environment." *Ecologist* 15(1/2):56–68.

Richards, E.G. 1987. *Forestry and the Forest Industries: Past and Future.* Dordrecht: Martinus Nijhoff for the United Nations.

Richards, J.F., and R.P. Tucker (eds.). 1988. *World Deforestation in the Twentieth Century.* Durham, NC: Duke University Press.

Richards, Paul. 1973. "The Tropical Rain Forest." *Scientific American* 229:6.

– 1981. *The Tropical Rain Forest.* Cambridge: Cambridge University Press.

Richardson, Mary, Joan Sherman, and Michael Gismondi. 1993. *Winning Back the Words: Confronting Experts in an Environmental Public Hearing.* Toronto: Garamond Press.

Riocell. *Annual Reports,* 1991, 1992.

– n.d. *Riocell: An Open Book*

– 1991. *Relatorio de Administracao, 1991*

– 1992. *Riocell's Figures*

Ripasa SA Celulose e Papel. *Annual Reports,* 1991, 1992.

– In-house promotional literature.

Rivera, R., and Cruz, M.E. 1983. "La realidad forestal chilena." In *GIA Resultados de Investigaciones no 15.* Santiago: GIA.

Roberts, Ralph W., Stanley L. Pringle, and George S. Nagle. 1991. "Leadership in World Forestry: Discussion Paper." Hull, Que.: Canadian International Development Agency, September.

Robbins, William G. 1988. *Hard Times in Paradise: Coos Bay, Oregon, 1850–1986.* Seattle: University of Washington Press.

Robinson, W. 1985. "Imperialism, Dependency and Peripheral Industrialisation: The Case of Japan in Indonesia." In R. Robison, K. Hewison, and R. Higgot (eds.), *Southeast Asia: Essays in the Political Economy of Structural Change*. London: Routledge.

Robison, Richard. 1986. *Indonesia: The Rise of Capital*. Canberra: Asian Studies Association of Australia.

– Kevin Hewison, and Richard Higgot (eds.). 1987. *Southeast Asia in the 1980s: The Politics of Economic Crisis*. North Sydney: Allen & Unwin.

Roche, Michael M. 1991. "Privatising the Exotic Forest Estate: The New Zealand Experience." Paper presented to the History of the Forest Economy of the Pacific Basin Symposium, XVII Pacific Science Congress, Honolulu, Hawaii.

Rolo, Luis Bernardo. 1987. "Viewpoint: Portugal Must Protect Its Wood." *Pulp and Paper International* 29(7):27.

Romm, Jeff. 1986. "Forest Policy and Development Policy." *Journal of World Forest Resource Management* 2:85–103.

Ross, M.S., and D.G. Donovan. 1986. "The World Tropical Forestry Action Plan: Can It Save the Tropical Forests?" *Journal of World Forest Resource Management* 10:119–35.

Roth, Dennis M. 1983. "Philippine Forests and Forestry: 1565–1920." In Richard P. Tucker and J.F. Richards (eds.), *Global Deforestation and the Nineteenth-Century World Economy*, 30–49. Durham, NC: Duke University Press.

Rowley, Anthony. 1977. "Forests: Save or Squander." *Far Eastern Economic Review* 2:48.

Rubinoff, Ira. 1982. "Tropical Forests: Can We Afford Not to Give Them a Future?" *Ecologist* 12(6):253–8.

Russell, Charles E. 1987. "Plantation Forestry. Case Study No. 9: The Jari Project, Pará, Brazil." In C.F. Jordan (ed.), *Amazonian Rain Forests*, 76–89. New York: Springer-Verlag.

Sagarra, Alberto Fernandez. 1988. "Latin American Notes." *Tappi Journal*, Dec., 25–6.

Sahabat Alam Malaysia (World Rainforest Movement). 1987. *Forest Resources Crisis in the Third World*. Proceedings, conference, 6–8 Sept. 1986. Penang, Malaysia: SAM.

– 1990a. *The Battle for Sarawak's Forests*. Penang, Malaysia: SAM.

– 1990b. *Solving Sarawak's Forest and Native Problem*. Penang, Malaysia: SAM.

Sahunalu, Pongsak, and Wuthipol Hoamuangkaew. 1986a. "Agroforestry in Thailand: Present Condition and Problems." In Monsoon Asia Agroforestry Joint Researach Team, *Comparative Studies on the Utilization and Conservation of the Natural Environment by Agroforestry Systems: Report on Joint Research among Indonesia, Thailand, and Japan*, 66–92. Kyoto University.

– 1986b. "Thailand." In Monsoon Asia Agroforestry Joint Researach Team, *Comparative Studies on the Utilization and Conservation of the Natural Environment by Agroforestry Systems: Report on Joint Research among Indonesia, Thailand, and Japan*, 223–48. Kyoto University.

Saldarriaga, Juan G. 1987. "Recovery Following Shifting Cultivation. Case Study No. 2: A Century of Succession in the Upper Rio Negro." In C.F. Jordan (ed.), *Amazonian Rain Forests*, 24–33. New York: Springer-Verlag.

Salonen, Heikki J.W., and Pekka Niku. 1988. "The Future of the Forest Products Industry: A Worldwide Perspective." In Gerard F. Schreuder (ed.), *Global Issues and Outlook in Pulp and Paper*, 285–300. Seattle: Washington Press.

Sanchez, Pedro A. 1981. "Soils of the Humid Tropics." In *Blowing in the Wind*, 347–410. Williamsburg, Va: Department of Anthropology, College of William and Mary.

– and D.E. Bandy. 1983. "Soil Fertility Dynamics after Clearing a Tropical Rainforest in Peru." *Soil Science Society of America Journal* 47:1171–8.

– D.E. Bandy, J.H. Villachica, and J.J. Nicholaides. 1982. "Amazon Basin Soils: Management for Continuous Crop Production." *Science* 216:821–7.

Sandwell Management Consultants. 1978. *Forest Feasibility Study*. Reports x4040/1 to x4040/6. Yangon, Myanmar.

Sarawak Study Group. 1992. "Logging in Sarawak: The Belaga Experience." In Institute of Social Analysis, *Logging against the Natives of Sarawak*, 1–30. Selangor, Malaysia: The Institute.

Sathirathai, Surakiart. 1987. "Laws and Regulations Concerning Natural Resources, Financial Institutions and Export: Their Effects on Economic and Social Development." Paper presented at the Management of Economic and Social Development Workshop I, 13–15 March. Pattaya, Bangkok: TDRI.

– 1989a. "Potential of Commercial Fast-growing Tree Plantations in Thailand," Paper presented at seminar on "Economic Forest: Myth or Reality!" Feb. 1989. Bangkok: TDRI.

– 1989b. "Employment Effects of Reforestation Programs." Paper presented at seminar on "Economic Forest: Myth or Reality!" Feb. 1989. Bangkok: TDRI.

Savage and Associates. 1993. "Cariboo Forest Sector Employment Trends Project: A Tool for Community Review of Employment and Population Trends." Report prepared for the BC Ministry of Economic Development, Small Business and Trade, April.

Schmidt, Harald, and Antonio Lara. 1985. "Descripcion y potencialidad de los bosques nativos de Chile." *Ambiente y Desarrollo*, 1(2):91–108.

Schneider, Aaron (ed). 1989. *Deforestation and "Development" in Canada and the Tropics: The Impact on People and the Environment*. Sydney, NS: University College of Cape Breton.

Schodde, R. 1973. "General Problems of Fauna Conservation in Relation to the Conservation of Vegetation in New Guinea." In A.B. Costin and R.H.

Groves (eds.), *Nature Conservation in the Pacific*, 123–44. Canberra: Australian National University Press.

Schreuder, Gerard F. (ed). 1988. *Global Issues and Outlook in Pulp and Paper.* Seattle: University of Washington Press.

– and Erik T. Anderson. 1988. "International Wood Chip Trade: Past Developments and Future Trends, with Emphasis on Japan." In G.F. Schreuder (ed.), *World Trade in Forest Products*, 162–84. Seattle: University of Washington Press.

Schwindt, Richard. 1979. "The Pearse Commission and the Industrial Organization of the British Columbia Forest Industry." *B.C. Studies* 41:3–35.

– 1985. *An Analysis of Vertical Integration and Diversification Strategies in the Canadian Forest Sector.* Vancouver: Forest Economics and Policy Analysis Project, University of British Columbia, June.

Science Council of Canada. 1983. *Canada's Threatened Forests.* Ottawa, The Council.

Scott, Geoffrey. 1987. "Shifting Cultivation Where Land Is Limited." In C.F. Jordan (ed.), *Amazonian Rain Forests: Ecosystem Disturbance and Recovery*, 34–45. New York: Springer-Verlag.

Searle, Graham. 1975. *Rush to Destruction: An Appraisal of the New Zealand Beech Forest Controversy.* Wellington: A.H. & A.W. Reed.

Sears, John, and Katherine Bragg. 1987. "Bio-Bio: A River Under Threat." *Ecologist* 17(1):15–20.

Sebire, R.A. 1980a. "Forests and Forest Industries of the Philippines." *Australian Forest Industries Journal* 46(10):10–19.

– 1980b. "Bright Future Forecast for Philippines." *Australian Forest Industries Journal* 46(11):10–19.

Secrett, Charles. 1986. "The Environmental Impact of Transmigration." *Ecologist* 16(2/3):77–88.

Sedjo, Roger A. 1983. *The Comparative Economics of Plantation Forestry: A Global Assessment*, Washington, DC: Resources for the Future.

– 1988. "Native Forests, Secondary Species, Plantation Forests and the Sustainability of Indonesia's Forest Industry." Report prepared for "Assistance to Forestry Development Planning," FAO/UNDP/INS/83/019. Jakarta, Indonesia.

– 1989. "The Expanding Role of Plantation Forestry in the Pacific Basin." Paper presented to a national conference on "Prospects for Australian Plantations," Canberra, Australia, 21–25 Aug.

– and Marion Clawson. 1983. "How Serious Is Tropical Deforestation?" *Journal of Forestry* 81:792–4.

– and Kenneth Lyon. 1989. *The Long-Term Adequacy of World Timber Supply.* Washington, DC: Resources for the Future.

Segarra, Alberto Fernandez. 1980. "Latin American Notes." *Tappi Journal*, Dec., 25.

Sesser, Stan. 1991. "A Reporter at Large: Logging The Rain Forest." *New Yorker,* 27 May, 42–67.

Sharma, B.D. 1978. *Industrial Complexes and Their Tribal Hinterlands: Forests, Tribal Economy, and Regional Development.* Occasional Papers on Tribal Development, nos. 16, 26. New Delhi: Ministry of Home Affairs.

Sherman, George. 1981. "The Culture-Bound Notion of 'Soil Fertility': On Interpreting Non-Western Criteria of Selecting Land for Cultivation." In *Blowing in the Wind,* 487–514. Williamsburg, Va: Department of Anthropology, College of William and Mary.

Shimotori, Shigeru, and Yukio Akibayashi. 1989. "The Structure of Forestry Employment in Mountain Villages in a Period of Slow Economic Growth and Revitalization of the Village." *Current State of Japanese Forestry* 6:76–87. Tokyo: The Japanese Forest Economic Society.

Shiva, Vandana. 1987. "Forestry Myths and the World Bank." *Ecologist* 17(4/5):142–9.

– and J. Bandyopadhyay. 1983. "Eucalyptus – a Disastrous Tree for India." *Ecologist* 13(5):184–7.

– H.C. Sharatchandra, and J. Bandyopadhyay. 1982. "Social Forestry – No Solution within the Market." *Ecologist* 12(4):158–68.

Sidaway, S. "The Availability and Use of Eucalyptus Pulps." *Tappi Journal,* Dec., 47–51.

Simmons, Ian G. 1989. *Changing the Face of the Earth: Culture, Environment, History.* New York: Verso.

Simon Fraser University. Natural Resources Management Program. 1990. *Wilderness and Forestry: Assessing the Cost of Comprehensive Wilderness Protection in British Columbia.* Burnaby, BC: SFU, January.

Sinclair, William. 1988. *Federal Regulation of the Pulp and Paper Industry in Canada,* Ottawa: Environment Canada.

– 1991. "Controlling Effluent Discharges from Canadian Pulp and Paper Manufacturers." *Canadian Public Policy* 17(1):86–105.

Singh, Gurmit K.S. 1981. "Destroying Malaysian Forests." In *Where Have All the Flowers Gone? Deforestation in the Third World,* Studies in Third World Societies, 181–90. Williamsburg, Va: Department of Anthropology, College of William and Mary.

Sioli, Harald. 1973. "Recent Human Activities in the Brazilian Amazon Region and their Ecological Effects." In B.J. Meggers, E.S. Ayensu, and W.D. Duckworth (eds.), *Tropical Forest Ecosystems in Africa and South America: A Comparative Review,* 321–44. Washington, DC: The Smithsonian Institution.

– 1987. "The Effects of Deforestation in Amazonia." *Ecologist* 1(4/5):134–8.

Sirin, Lockman M. 1990. "Malaysia." In Asian Productivity Organization, *Forestry Resources Management,* 247–68. Tokyo: APO.

Skogsstyrelsen. The National Board of Forestry, Sweden. 1994a. *The Forestry Act. Valid from January 1, 1994.*

- 1994b. *Sweden's New Forest Policy.*
Slocan Valley Community Forest Management Project. 1975. *Final Report.* Winslow, BC. Photocopy.
Slocum, Ken. 1986. "Cutting Criticism, Forest Service's Sales of Timber Below Cost Stir Increasing Debate." *Wall Street Journal,* 18 April, 1, 14.
Smith, Nigel/J.H. 1981. "Colonization Lessons from a Tropical Forest." *Science* 214:755–61.
- 1982. *Rainforest Corridors: The Transamazon Colonization Scheme.* Berkeley: University of California Press.
- 1990. Review of *The Fate of the Forest.* In *Economic Geography* 66:174–6.
Smucker, Philip. 1990. "Trees 'Ordained' in Forest Fight." *Globe and Mail,* 13 July.
Soetikno, Abubakar, and Peter Sutton. 1988. "Time to Talk in Terms of Millions." *Pulp and Paper International* 30(1):40–4.
Sommer, A. 1976. "Attempts at an Assessment of the World's Tropical Moist Forests." *Unasylva* 28:112–13.
Sonnenfeld, David. 1990. "Origins of Indonesia's Log Export Ban: A Struggle for Local Control of Natural Resources?" Unpublished manuscript.
Sopow, Eli. 1985. *Seeing the Forest: A Survey of Recent Research on Forestry Management in British Columbia.* Working paper prepared for the Western Resources Program of the Institute for Research on Public Policy, October.
- 1987. "How to Lobby the Government." *BC Business,* July, 104–19.
Spears, John S. 1979. "Can the Wet Tropical Forests Survive?" *Commonwealth Forestry Review* 57(3):1–16.
- 1980. "The World Bank's Forestry Lending Program." Geneva: UN Conference on Trade and Development.
Statistics Canada. "Canadian Forestry Statistics." Cat 25–202. Annual, 1979–86.
Steen, Harold K. (ed.). 1984. *History of Sustained-Yield Forestry: A Symposium.* Santa Cruz, Calif.: Forest History Society.
Sterling Wood Group Inc. 1984. *Status of the British Columbia Coast Forest Industry.* Victoria: Queen's Printer, September.
Stevenson, Susan. 1988. "Fifty Years for Canfor: An Intriguing History and an Exciting Future." *Pulp and Paper Canada* 89(10):13–22.
Stolar, A.L. 1985. *Capitalism and Confrontation in Sumatra's Plantation Belt, 1870–1979.* New Haven: Yale University Press.
Stone, R.D. 1985. *Dreams of Amazonia.* New York: Viking.
Strong, D.R., Jr. 1974. "Rapid Asymptotic Species Accumulation in Phytophagus Insect Communities: The Pests of Cacao." *Science* 185:1064–6.
Strueck, Wendy. 1989. "Indians to Get $2.75 Million." *Vancouver Sun,* 5 Aug., A1.
Suryo, Sudjono. 1986. "Forestry Development: Indonesia." In *Five Perspectives on Forestry for Rural Development in the Asia-Pacific Region,* Bangkok: Regional Office for Asia and the Pacific, FAO.

Sutton, Peter. 1984. "PPI Interview: Portugal's 4th Pulpmaker Starts Up." *Pulp and Paper International* 26(7):32–3.

– 1988a. "Indah Kiat Aims for a Million." *Pulp and Paper International* 30(2):45–6.

– 1988b. "Indonesian Innovation at Inti Indorayon." *Pulp and Paper International* 30(2):48–50.

Sutton, W.R.J. 1975. "The Forest Resources of the USSR: Their Exploitation and Their Potential." *Commonwealth Forestry Review* 54(2):110–38.

Sveriges Officiella Statistik. Skogsstyrelsen Jonkoping. 1993. *Skogsstatistik Arsbok, 1993.* Stockholm, Sweden.

Taiga Rescue Network. 1992. *The Boreal Forests of the World: Ecology, Biodiversity and Sustainable Use.* International Scientific Meeting organized by the Swedish Society for Nature Conservation, Abstracts and Proceedings. Jokkmokk, Sweden.

Takahashi, Akira, et al. 1980. "Recent Trends in the Wood Industry of Japan." Parts 1–2. *Forest Products Journal* 30(5):28–34; 30(6):21–26.

Takeuchi, Kenji. 1974. *Tropical Hardwood Trade in the Asia-Pacific Region.* Occasional Paper no. 17. Baltimore: Johns Hopkins University Press for the International Bank for Reconstruction and Development.

Tappi Journal. Various issues.

Tappi Journal. 1988. "Bleaching Said to be Key to CTMP from Eucalyptus." Dec., 11, 14.

Taylor, Duncan, and Jeremy Wilson. 1992. "Environmental Health – Democratic Health: An Examination of Proposals for Decentralization of Forest Management in British Columbia." Paper presented to annual meetings of the Canadian Political Science Association, University of Prince Edward Island, 31 May.

TDRI (Thailand Development Research Institute). 1989. "Potential of Commercial Fast-Growing Tree Plantations in Thailand." Bangkok.

– 1990. "Eucalyptus: For Whom and For What?" Bangkok.

Temple, S.A. 1977. "Plant-Animal Mutualism: Coevolution with Dodo Leads to Near Extinction of Plant." *Science* 197:885–6.

Thailand. FIO/Ministry of Agriculture and Cooperatives. 1981. *FIO Progress Report of Reforestation Programme for Forest Village System.* Bangkok.

Thaitawat, Nusara. 1990. "A Tiny Nomadic Forest Tribe That Faces Extinction." *Bangkok Post*, 16 April.

Thalib, Dahlan. 1967. "Timber Development." *Bulletin of Indonesian Economic Studies* 7:91–5.

Thongtham. 1987. "From Mangroves to Prawn Ponds: Is It Worth the Price We Pay?" *Asia Magazine*, 14 June.

– ("Normita"). 1989. "Battling Soil Erosion in the South." *Bangkok Post*, 3 Sept.

– ("Normita"). 1990. "A Taste of Southern Hospitality." *Bangkok Post*, 18 March.

Tillman, David A. 1985. *Forest Products. Advanced Technologies and Economic Analyses.* Orlando: Acadaemic Press.

Time. 1976. "Ludwig's Wild Amazon Kingdom." 15 Nov., 59–59A.

– 1979. "Billionaire Ludwig's Brazilian gamble." 10 Sept., 76–8.

– 1982. "End of a Billion-Dollar Dream." 25 Jan., 59.

Tin Wis Coalition Forestry Working Group. 1991. *Community Control, Developing Sustainability, Social Solidarity.* Vancouver: Tin Wis Coalition.

Toronto Stock Exchange. 1989. "Fletcher Challenge Limited." *Review,* March.

Totman, Conrad. 1984. "From Exploitation to Plantation Forestry in Early Modern Japan." In Harold K. Steen (ed.), *History of Sustained-Yield Forestry: A Symposium,* 270–9. Santa Cruz, Calif.: Forest History Society.

– 1985. *The Origins of Japan's Modern Forests: The Case of Akita.* Asian Studies at Hawaii, No. 31. Honolulu: Center for Asian and Pacific Studies, University of Hawaii, University of Hawaii Press.

– 1986. "Tokugawa Peasants: Win, Lose, or Draw?" *Monumenta Nipponica* 41(4):458–76.

Tracey, Jacqueline L. 1990. "Deforestation and Cattle Ranching in the Amazon." Graduating essay for MA degree, Faculty of Forestry, University of British Columbia.

Treece, Dave. "Brazil's Greater Carajas Programme." *Ecologist* 17(12–13):75.

Tsutsumi, T., K. Yoda, P. Sahunalo, P. Dhanmanonda, and B. Prachaiyo. 1983. "Forest: Felling, Burning and Regeneration." In *Shifting Cultivation, an Experiment at Namphrom, Northeast Thailand and Its Implications for Upland Farming in the Tropics,* 13–62. Kyoto University.

Tucker, Richard P. 1983. "The British Colonial System and the Forests of the Western Himalayas, 1815–1914." In Richard P. Tucker and J.F. Richards (eds.), *Global Deforestation and the Nineteenth-Century World Economy,* 146–66. Durham, NC: Duke University Press.

– 1988. "The British Empire and India's Forest Resources: the Timberlands of Assam and Kumaon, 1914–1950." In John F. Richards and Richard P. Tucker (eds.), *World Deforestation in the Twentieth Century,* 91–111. Durham, NC: Duke University Press.

– and J.F. Richards (eds.). 1983. *Global Deforestation and the Nineteenth-Century World Economy.* Durham, NC: Duke University Press.

Turner, John, and Marcia J. Lambert. 1986. "Effects of Forest Harvesting Nutrient Removals on Soil Nutrient Reserves." *Oecologia* (Berlin) 70:140–8.

Udarbe, Marcelo Pangan. 1990. "Sabah." In Asian Productivity Organization, *Forestry Resources Management,* 217–25. Tokyo: APO.

Uhler, Russell S. 1991. "Why Not Log Markets?" University of British Columbia, Forest Economics and Policy Analysis Research Unit, *FEPA Newsletter* 7(1):2–3.

Umprai, Thach-chai (chief forester, Chachoensao Provincial Forest Office, Thailand). 1990. "How to Increase the National Forest Resource for the Rural Development of Chachoengsao Province." Photocopy.

UNCED (United Nations Conference on Environment and Development). 1991. "Protection of Land Resources: Deforestation." Progress Report by the Secretary-General of the Conference, Second Session, Geneva, 18 March–5 April, Working Group I. Item 2(a) of the provisional agenda. Geneva: UNCED.

UNCTAD (United Nations Conference on Trade and Development). 1976–83. "Preparatory Meetings on Tropical Timber." TD/Timber Series. Geneva: United Nations.

– 1983. "International Tropical Timber Agreement, 1983." TD/Timber/11. Geneva: United Nations.

UNESCO. 1972. "Convention Concerning the Protection of the World Cultural and Natural Heritage." General Conference, 17th Session, Paris.

UNESCO/UNEP/FAO. 1978. *Tropical Forest Ecosystems: A State-of-Knowledge Report.* Paris: UNESCO.

United States. 1952. *Resources for Freedom: The Report of the President's Materials Policy Commission.* (The Paley Report). 5 vols. Washington: Office of the President.

– Department of Agriculture. Forest Service. 1988. *The South's Fourth Forest: Alternatives for the Future.* Forest Resource Report no. 24. Washington, DC: Forest Service.

– Department of Agriculture, Forest Service, and Department of the Interior. Bureau of Land Management. 1993. *Forest Ecosystem Management: An Ecological, Economic, and Social Assessment. Report of the Forest Ecosystem Management Assessment Team.* Also *Final Supplemental Environmental Impact Statement*, vols. I and II. Washington, July.

– Department of State. Council on Global Environmental Quality. 1981. *Global Future: Time to Act, Report to the President on Global Resources, Environmental and Population.* Washington, DC: U.S. Government Printing Office.

United States Interagency Task Force on Tropical Forests. 1980. *The World's Tropical Forests: A Policy Strategy and Program for the United States, Report to the President.* Department of State Publication 9117. Washington, DC.

United States International Trade Commission. 1982. *Conditions Relating to the Importation of Softwood Lumber into the United States.* Report to the Senate Committee on Finance on Investigation no. 332–134, under Section 3.32 of the Tariff Act of 1930. Washington, DC, April.

"Urgencia de una politica para el manejo y la conservacion del bosque nativo." 1991. Statement issued by forty-one faculty members at Chilean universities, 1 Aug.

Usher, Ann Danaiya. 1989a. "Wonder Tree or Ecological Manace?" *Nation* (Bangkok), 8 Feb.

– 1989b. "Speaking of Forests." *Nation* (Bangkok), 24 March.

– 1989c. "The Dilemma of the 1990's: Eucalyptus vs Land Rights." *Nation* (Bangkok), 5 May.

- 1989d. "The Gap Between Policy and Reality." *Nation* (Bangkok), 29 Dec.
- 1990a. "Regulating the Eucalyptus Boom." *Nation* (Bangkok), 31 Jan.
- 1990b. "The Most (Mis)quoted Forester." *Nation* (Bangkok), 14 Feb.
- 1990c. "The Notorious Camaldulensis." *Nation* (Bangkok), 16 Feb.
- 1990d. "What Price the Eucalyptus Tree?" *Nation*, 22 Feb.
- 1990e. "Thailand's Battle of the Eucalyptus Trees." *Globe and Mail*, 2 April, A8.
- 1990f. "Clapping with Only One Hand." *Nation* (Bangkok), 11 April.
- 1990g. "After the Forest ..." Series of five articles in the *Nation* (Bangkok), June: "After the Forest ...," "Eucalyptus – Widening the Gap," "The Tree Farm that Never Was," "Pulp Links from Oz to Siam," and "The Shaping of a Master Plan."
- 1990h. "Merits of New Forestry Bill Questioned." *Nation* (Bangkok), 18 June.
- 1990i. "Call for Radical Forestry Reforms." *Nation* (Bangkok), 6 Dec.
- 1991a. "A Finn-ancial Harvest." *Nation* (Bangkok), 10 Feb.
- 1991b. "Conservationists Call for Rethink of Forestry Plan." *Nation* (Bangkok), 11 Feb.
- 1991c. "Thai Forest Plan at a Crossroads." *Nation* (Bangkok), 15 Feb.
- 1991d. "The Changing Face of Security." *Nation* (Bangkok), 21 April.
Ushiomi, Toshitaka. 1964. *Forestry and Mountain Village Communities in Japan: A Study in Human Relations*. Tokyo: Kokusai Bunka Shinkokai.
Veblen, T.T. 1983. "Degradation of Native Forest Resources in Southern Chile." In H.K. Steen (ed.), *History of Sustained-Yield Forestry: A Symposium*, 344–52. Durham, NC: Forest History Society.
Vernon, C.G., and Christy, E.J. 1981. "The UNESCO Program on Man and the Biosphere." In E.J. Kormondy and J.F. McCormick (eds.), *Handbook of Contemporary Development in World Ecology*, 701–20. Westport, Conn.: Greenwood Press.
Village of Hazelton. 1991. *Framework for Watershed Management (formerly the Forest Industry Charter of Rights)*. Hazelton, BC: The Corporation of the Village of Hazelton.
Vincent, Jeffrey R. 1992. "The Tropical Timber Trade and Sustainable Development." *Science* 256:1651–5.
Vohra, B.B. 1985. "Why India's Forests Have Been Cut Down." *Ecologist* 15(1/2):50–1.
von Furer-Haimendorf, C. 1982. *The Tribes of India: The Struggle for Survival*. Berkeley, Los Angeles, and London: University of California Press.
Wacharakitti, Sathit. 1988. "Forest Development in Thailand." In National Institute for Research Advancement, The Japan Institute of Systems Research, Proceedings of International Symposium, *The Development of Regional Agriculture and Environmental Conservation in Southeast Asian Countries, Kyoto, March 15–17, 1988*, ed. by Teitaro Kitamura, 122–36. Kyoto University.

Wagner, William Leroy. 1987. "Privateering in the Public Forest? A Study of the Expanding Role of the Forest Industry in the Management of Public Forest Land in British Columbia." MA thesis, University of Victoria.

– 1988. "An Emerging Corporate Nobility? Industrial Concentration of Economic Power on Public Timber Tenures." *Forest Planning Canada* 4(2):14–19.

Wahana Lingkungan Hidup Indonesia (Indonesian Forum for the Environment). 1992. *Sustainability and Economic Rent in Indonesian Forestry Sector.* Jakarta.

– and Yayasan Lembaga Bantuan Hukum Indonesia (Indonesian Legal Aid Institute). 1992. *Mistaking Plantations for the Forest: Indonesia's Pulp and Paper Industry, Communities, and Environment.* Jakarta, July.

Washington State University. Agricultural Research Center. College of Agriculture and Home Economics, Pullman. 1986. *Trends in Japanese Imports of Selected Forest Products, by Supplier.* Research Bulletin 1986. Based on computer tapes obtained from the Organization of Economic Cooperation and Development.

Watkins, M.H. 1963. "A Staple Theory of Economic Growth." *Canadian Journal of Economics and Political Science* 29:141–58.

Watson, Ian. 1990. *Fighting Over the Forests.* Sydney: Allen & Unwin.

Webb, Kernaghan. 1988. *Pollution Control in Canada: The Regulatory Approach in the 1980's.* Ottawa: Law Reform Commission of Canada.

Wertheim, W.F. (ed.) 1986. *Indonesian Economics: The Concept of Dualism in Theory and Policy.* The Hague: W. van Hoeve.

West Kalimantan (Province, Indonesia). Regional Investment Coordinating Board. 1989. *Map of Regional Investment, West Kalimantan Province.*

Westoby, Jack C. 1983. "Keynote Address." Presented to the Institute of Foresters of Australia, 10th Triennial Conference, University of Melbourne.

– 1987. "Forest Industries in the Attack on Underdevelopment." *Unasylva* 16(4):168–201. Reprinted in *The Purpose of Forests: Follies of Development* (Oxford: Blackwell, 1987).

– 1989. *Introduction to World Forestry: People and Their Trees.* Oxford: Basil Blackwell.

Where Have All the Flowers Gone? Deforestation in the Third World. 1981. Studies in Third World Societies. Williamsburg, Va: Department of Anthropology, College of William and Mary.

White, W., B. Netzel, S. Carr, and G.A. Fraser. 1986. "Forest Sector Dependence in Rural British Columbia 1971–81." Information Report BC–X–278, Pacific Forestry Centre. Ottawa: Canadian Forestry Service.

Widman Management Ltd. 1984. *British Columbia Log Exports: An Analysis of the Real Economic Issues.* November.

Wilbert, J. 1972. *Survivors of Eldorado,* New York: Praeger.

Williams, Douglas. 1986. "The Economic Stock of Timber in the Coastal Region of British Columbia." Forest Economics and Policy Analysis Project, University of British Columbia, Report 86–11, vol. I and II.

Williams, G. 1983. *Not for Export: Towards a Political Economy of Canada's Arrested Industrialization*. Toronto: McClelland and Stewart.

Williams, Michael. 1988. "The Death and Rebirth of the American Forest: Clearing and Reversion in the United States, 1900–1980." In John F. Richards and Richard P. Tucker (eds.), *World Deforestation in the Twentieth Century*, 211–9. Durham, NC: Duke University Press.

– 1989. *Americans and Their Forests: A Historical Geography*. Cambridge: Cambridge University Press.

Williams, Ward D. 1988. "International Scene." *Tappi Journal*, Dec.: 24–6.

Williston, Ed. 1989. "The Industy in the 1990s: Where Technology Is Headed." *World Wood* 30(1):20–1.

Wilson, Donna (ed.). 1985. "Participants' Report: The Economy." In *Democratic Socialism: The Challenge of the Eighties and Beyond*, 128–42. Vancouver: New Star Books.

Wilson, Jeremy. 1987. "Resolution of Wilderness vs. Logging Conflicts in British Columbia: A Comparison of Piecemeal and Comprehensive Approaches." Paper presented to the Canadian Political Science Association annual meetings, Hamilton.

– 1987–88. "Forest Conservation in British Columbia, 1935–1985: Reflections on a Barren Political Debate." *B.C. Studies* 76:3–32.

– 1990. "Wilderness Politics in B.C.: The Business Dominated State and the Containment of Environmentalism." In William Coleman and Grace Skogstad (eds.), *Policy Communities and Public Policy in Canada: A Structural Approach*, 141–69. Mississaugua: Copp Clark Pitman.

Wilson, Robert A. 1993. "Eucalyptus: Paradigm or Protagonist." Highlights from speech at 1993 market pulp conference in Vancouver. Printed in *Papertree*, Oct. (no pagination).

Wise, P.K., and M.G. Pitman. 1981. "Nutrient Removal and Replacement Associated with Short Rotation Eucalypt Planations." *Australian Forestry* 44(3):142–52.

Wood, Paul M. 1989. "The Management of Non-Timber Values within Tree Farm Licences in B.C.: An Overview Analysis from a Social Justice Perspective." Unpublished manuscript, April.

– 1990. "A Summary of Old Growth Forest Values." Submission to Old Growth Working Group, BC Ministry of Forests, May.

Woodbridge, Reed and Associates. 1984. *British Columbia's Forest Products Industry, Constraints to Growth*. Ottawa: Minister of State for Economic and Regional Development, May.

– 1986. "World Market Pulp Demand with Special Reference to Eucalyptus." Edmonton, Alta: Canadian Forestry Service and Alberta Forest Service.

– 1988. *Canada's Forest Industry: The Next 20 Years, Prospects and Priorities*. Prepared for the Canadian Forestry Service. Ottawa.

Woodland Resource Services Ltd. 1987. "A Report on the Prince George Timber Supply Area Public Inquiry for the Ministry of Forests and Lands." (The Ewing Report). Victoria, BC, February.

World Bank. *World Development Report* (annual). New York: Oxford University Press.

– 1982. *Tribal Peoples and Economic Development: Human Ecological Considerations.* Washington, DC: World Bank.

– 1989. *Papua New Guinea – The Forestry Sector; a Tropical Forest Action Plan Review.* Washington, DC: World Bank, October.

World Commission on Environment and Development (chaired by Gro Harlem Brundtland). 1987. *Our Common Future.* Oxford: Oxford University Press.

World Paper, June 1994.

World Rainforest Movement. *See* Sahabat Alam Malaysia

Worldwatch Institute. *State of the World: A Worldwatch Institute Report on Progress toward a Sustainable Society* (annual). New York: Norton.

World Wood (monthly). Various issues, 1986–94.

Worster, Donald. 1990. "Toward an Agroecological Perspective in History." *Journal of American History* 76:1087–106.

WRI (World Resources Institute). 1985. *Tropical Forests: A Call for Action.* Part III. Country Investment Profiles. World Bank, International Task Force, United Nations Development Programme. New York: WRI, October.

– 1991. *World Resources 1990–91.* New York: Oxford University Press.

Wright, A.W. 1983. "Soviet Natural Resource Exports and the World Market." In R.G. Jensen, T. Shabad, and A.W. Wright (eds.), *Soviet Natural Resources in the World Economy,* 617–22. Chicago: University of Chicago Press.

Wright, Roger. 1992. "What's happening Out There?" Notes for Canadian Broadcasting Corporation current affairs broadcast, 15 June.

Yoda, K., T. Tsutsumi, P. Sahunalo, P. Dhanmanonda, and B. Piachaiyo. 1983. "Vegetation Management on the Abandoned Land After Shifting Cultivation." In K. Kyuama and C. Panintara (eds.), *Shifting Cultivation and Experiment at Nam Phrom, Northeast Thailand and its Implications for Upland Farming in the Tropics,* 205–19. Kyoto University.

Young, W. 1980. "Increasing Demands on the Forest Resource – the Challenge of the '80's." In *The Forest Imperative, Proceedings, Canadian Forest Congress, Ontario Science Centre, Toronto, Sept 22–23, 1980,* 22–5. Montreal: Canadian Pulp and Paper Association.

Youngblood, Ruth. 1990. "Lumber Blades Threaten S.E. Asian Forests." *Nation* (Bangkok), 22 April, 7.

Yung Whee Rhee and Therese Belot. 1990. "Export Catalysts in Low-Income Countries: A Review of Eleven Success Stories." World Bank Discussion Papers, no. 72. Washington, DC: World Bank.

Zhang, Daowei. 1992. "Community Forestry in Canada: Is It Economically Feasible?" Faculty of Forestry, University of British Columbia, April. Photocopy.

– Jeannette Leitch, and Peter H. Pearse. 1992. "Trends in Foreign Investment in Canada's Forest Industry." Unpublished paper, Faculty of Forestry, University of British Columbia, January.

Zinke, Paul. 1989. "Forest Influences on the Floods of 2531 and Flood Hazard Mitigation." In *Safeguarding the Future, Restoration and Sustainable Development in the South of Thailand.* Report by a team of the National Operations Center, National Economic and Social Development Board, U.S. Agency for International Development. Bangkok.

Zobel, Bruce. 1988. "Eucalyptus in the Forest Industry." *Tappi Journal,* Dec., 42–6.

Index